STRUCTURAL GEOLOGY

An Introduction to Geometrical Techniques

Third Edition

DONAL M. RAGAN

Department of Geology
Arizona State University

JOHN WILEY & SONS, INC.

New York   Chichester   Brisbane   Toronto   Singapore

Library of Congress Cataloging in Publication Data:

Ragan, Donal M.
Structural geology.

Includes index.
1. Geology, Structural. I. Title.

QE601.R23   1984   551.8   84-19658
ISBN 0-471-08043-8

Printed in the United States of America

10 9 8 7 6 5

Printed and bound by Malloy Lithographing, Inc..

Preface

The first steps in the study of geologic structures are largely geometrical. This was true in the historical development of our knowledge of such structures. It is also true in the initial stages of any field investigation, and in the education of a structural geologist. This concern for geometry includes the methods of describing and illustrating the form and orientation of geologic structures, and the solution of various dimensional aspects of these structures. Even at these early stages, though, it is important to realized that geometry is not the end. The final goal, however elusive, is a complete understanding of the products and processes of rock deformation.

This book attempts to fill a need for an introduction to the geometrical techniques used in structural geology. I have sought an approach which is at once basic and modern. The topics covered include well-established techniques, newer approaches which hold promise, and an introduction to certain fundamental mechanical concepts and methods. Students who go no further in structural geology should have a working knowledge of the basic geometrical techniques and at least some appreciation of where structural geology is headed. At the same time, those who do go on, either in advanced courses or on their own, should have the necessary foundation.

The first few chapters apply the methods of orthographic projection to the solution of simple structural problems. For an introduction or review an outline is given in Appendix A. Applications to geologic and topographic maps are included and extensive use is made of Mackin's powerful method of visualization-- the down-structure view of geologic maps.

Stereographic projection and the stereonet, and the methods of plotting and solving angular problems, are introduced fairly early, and many of the same elementary problems as well as some more advanced ones are solved with their use.

Next, faults are described and classified, and problems of displacement are solved by combining orthographic and stereographic methods. The geometry of states of stress in two dimensions is then described. With this as background, the Coulomb criterion of shear failure is applied to the interpretation of shear and extensional fractures in rocks.

Concepts of deformations are treated, and card-deck models are utilized to illustrate many important aspects of general homogeneous and heterogeneous states of strain, as well as the flow which leads to such states. The difficulties of describing states of inhomogeneous strain leads to the geometrical description and classification of folds and related structures. Methods of relating folds to topography are treated, and the down-structure view of geologic maps is reemphasized. The geometric properties of parallel and similar folds are explored in some detail, again with the aid of card-deck models. The methods of structural analysis are then applied to folds and to tectonites. The basic of the equal-area projection is included, as is the method of constructing and contouring point diagrams.

Finally, methods of presenting and analyzing geologic data, including maps, cross sections and block diagrams, are developed. Throughout, emphasis is given to the geologic maps, the single most important tool in structural geology.

About the same time the second edition of this book was published, there appeared on the scene an amazing device--the hand-held calculator. Over the past decade there has been a remarkable increase in its power and an equally remarkable decrease in its price. As a result, many things will never be the same, including the way structural geology is done. The calculator has, for example, made obsolete a number of techniques, including the use of alignement diagrams and nomograms. Wherever possible, I have included the detailed derivation of a number of mathematical expressions which allows important problems to be solved quickly and efficiently. An exceedingly attractive blend of the old and the new involves the use of the stereonet, spherical triginometry, and the calculator (see Appendix B).

An important by product of this more analytical approach is that students are introduced to a wider variety of techniques than is usually the case; perhaps it will also increase their mathematical appreciation and motivation. The personal computer is on the same path, and without question, it too will have a profound effect on the way many things are done. In recognition of this trend, I will have available a collection of programs which will manipulate data or solve problems efficiently and quickly.

Donal M. Ragan

Acknowledgements

A book such as this is, by its very nature, eclectic, and I acknowledge my debt to a large number of authors whose material I have adapted. If this book has a single recurring theme it is the down-structure view of geologic maps; and I continue to owe the late Hoover Mackin my gratitude for a thorough grounding in the method. Over the years, a large number of readers and reviewers have made many important comments, suggestions, and corrections. I cannot list them all, but this does not diminish my thanks to them. Not least I continue to learn from my students.

Ed Stump has shared responsiblility for the introductory class in structural geology here at A.S.U. for the past several years, and was always available for advise and counsel as the third edition evolved. Dick Threet, in an incredible act of generosity, allowed me free access to his unpublished lab manual, and it strongly influenced my approach, especially in the first few chapters. Sharon Moser reviewed the entire manuscript and made a number of suggestions which lead to significant improvements. I continue to be influenced by Neville Price and John Ramsay, as will be evident on many pages.

I also thank the periodicals and publishers who permitted me to reproduce copyrighted material:

Figs. 1.2, 1.15, and 2.2   R.R. Compton, 1962, *Manual of Field Geology,* John Wiley & Sons, Inc.

Fig. 6.5 F.A. Donath, 1962, *Bulletin of the Geological Society of America* and the author.

Fig. 6.13 and 6.14 B.E. Hobbs, W.D. Means and P.F. Williams, 1976, *An Outline of Structural Geology,* John Wiley & Sons, Inc.

Figs. 11.13 and 11.14 G.J. Borrodaile, 1976, *Proceedings,* Koninklijke Nederlandse Akademie van Wetenschappen, Ser. B, v. 79(5), p. 332, 334.

Fig. 11.16 and 20.9 F.J. Turner and L.E. Weiss, 1963, *Structural Analysis of Metamorphic Tectonites,* McGraw-Hill Book Co.

Fig. 12.12 S.W. Carey, 1962, *Journal of the Alberta Society of Petroleum Geologists.*

Fig. 12.13 C.D.A. Dahlstrom, 1969, *Canadian Journal of earth Science* (National Research Council of Canada).

Fig. 12.14 J. Goguel, 1952, *Traite de Tectonique,* Masson et Cie.

Fig. 12.16 C.D.A. Dahlstrom, 1969, *Bulletin of Canadian Petroleum Geology.*

Fig. 12.17 L.U. De Sitter, 1964, *Structural Geology,* McGraw-Hill Book Co.

Fig. 12.18 C.W. Barnes and R.W. Houston, 1969, *Contribution to Geology.*

Fig. 13.4 and 13.13 J.G. Ramsay, 1967, *Folding and fracturingof Rocks,* McGraw-Hill Book Co.

Fig. 14.7 and 14.8 J.H. Mackin, 1950, *Journal of Geology.*

Fig. 15.4 F. Kalsbeek, 1963, *Neues Jahrbuch für Mineralogie Monatshefte.*

Fig. 16.1 G. Oertel, 1962, *Bulletin of the Geological Society of America* and the author.

Fig. 16.3 K.E. Lowe, 1946, *American Mineralogist.*

Fig. 16.5 E.Den Tex, 1954, *Journal of the Geological Society of Australia.*

Fig. 18.4 J.W. Low, 1957, *Geologic Field Methods, Harper & Row.*

Fig. 19.3 R. Balk, 1937, *Bulletin of the Geological Society of America.*

Fig. 20.2 F.E. Wright, 1911, *Carnegie Institute of Washington.*

Fig. 20.10 Argand, 1911, *Beiträge zur Geologischen Karte der Schweiz.*

A Note To The Student

The accompanying exercise and map problems generally can be solved with a minimum of equipment and material. The equipment needed should include a drafting quality compass, a semicircular protractor, an accurate metric scale, and a straight edge (a large triangle will do). Those contemplating the purchase of these items or those using drafting equipment for the first time are advised to read the section in Appendix A on accuracy in mechanical drawing. As with all things, a better job can be done faster with more elaborate equipment. A drafting machine and a light table are especially desirable additions.

To those contemplating the purchase of a calculator, I strongly recommend that serious consideration be given to the Hewlett-Packard HP-41C. While more expensive than most, it is a first-class machine and well worth the additional cost.

Printed plotting nets of several sorts, together with some exercise material, are grouped at the back of the book on perforate pages. You will find it very convenient to permanently mount the two 15-cm stereonets. One successful method is to glue the printed nets, being careful not to alter their dimensions, to each side of a 20-cm square of particle board which has been well coated with shellac. The surface of each net can then be protected with a sheet of self-adhesive clear plastic. A small hole is made exactly at the center to accept a map pin.

One important skill that should be cultivated along with an increasing understanding of structural geometry is the ability to produce an effective diagram. The requirements are, in varying degrees, both technical and artistic. For many purposes a technically competent diagram is adequate; the literature is full of examples. With care and a little experience, the necessary atttributes of clarity and accuracy can be developed. However, an artistic touch invariables raises the quality of an illustration above the merely adequate. This skill can be acquired, and you are urged to practice by making quick sketches of three-dimensional structural features as they are encountered. The importance of developing this habit cannot be overemphasized. By visualizing the form of a structure and then making that visualization concrete by drawing, you will sytrengthen your ability.

In a number of exercises you will be asked to solve a problem graphically, and then to check yourself using the appropriate mathematical formula. This is done so that you will gain experience in both methods, and because the feedback you receive from both approaches will increase your understanding. Unless you are a rare exception, you will have some difficulties as you encounter these unfamiliar graphical techniques and the visualization they require. You then may be tempted to get a quick answer using a cookbook approach by plugging number into equations. Be warned that this is a high-risk endeavor, and in the long run may be the worst thing that you can do. You must and can develop the ability to visualize in three dimensions; it is the only way you can be assured that you are solving the right problem. Many hints and tricks have been included to h elp you along this road. Use them. In the final analysis the responsibility is yours—keep at it and you will succeed!

Contents

1
Attitude of Planes

1.1 INTRODUCTION

Especially in the early stages of an investigation of the geologic structure of an area, much attention is paid to determining and recording of the orientation of various structural elements. Structural planes are the most common of these elements, and they are also a useful starting point in the introduction to the geometrical methods of structural geology.

1.2 DEFINITIONS

Attitude: the general term for the orientation of a plane or line in space, usually related to geographic coordinates and the horizontal. Both trend and inclination are components of attitude.

Trend: the direction of a line in a horizontal plane, commonly specified by its bearing or azimuth.

Bearing: a horizontal angle measured east or west of true north or south.

Azimuth: a horizontal angle measured clockwise from true north.

Strike: the trend of a horizontal line on an inclined plane.

Inclination: the general term for the vertical angle, measured downward, from the horizontal to a plane or line.

Dip: the inclination of the line of greatest slope on an inclined plane. It is measured perpendicular to the strike.

Apparent dip: the inclination of a line on an inclined plane measured in a direction oblique to the strike direction. It is always less than true dip.

1.3 DIP AND STRIKE

The terms dip and strike apply to any planar structure, and together their values constitute a statement of the attitude of the plane. The planar feature most frequently encountered in many areas is the bedding plane; other structural planes include cleavage, schistosity, foliation, joints and faults.

Each type of plane has a special map symbol, called a dip and strike symbol, which in general has three parts:

1. a <u>strike line</u> plotted long enough so that its trend can be accurately measured from the map,

2. a <u>dip mark</u> at the midpoint of one side of the strike line to indicate the direction of downward inclination of the plane,

3. a <u>dip angle</u> written close to the dip mark and on the same side of the strike line.

The only exceptions are the special cases of horizontal and vertical planes. The most common symbols are shown in Fig. 1.1, and their usage is fairly well established by convention. However, it is sometimes necessary to use these or other symbols in special circumstances, so that the exact meaning of all symbols must be explained in the map legend.

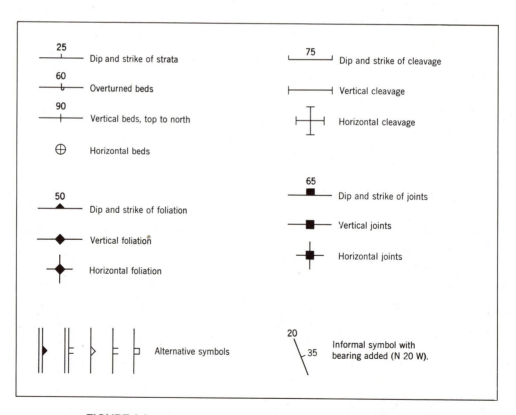

FIGURE 1.1 Map symbols for structural planes.

Dip and strike angles are also often referred to in text, although the usage is considerably less standard. Each of the following forms refers to exactly the same plane:

1. N65°W, 25°S: the bearing of the strike direction is 65° west of north, and the dip is 25° in a southerly direction. For any given strike, there are only two possible dip directions, one on each side of the strike line, hence it is necessary only to identify which side by a single or at most two letters.

2. 295°, 25°S: the azimuth of the strike direction is 295° and the dip is 25° to the south. Conventionally, the trend of the northernmost end of the strike line is given, but the azimuth of the opposite end of the line may also be used, as in 115°, 25°S.

3. 25°, S25°W: the dip is 25° and the trend of the dip direction has a bearing of 25° west of south.

4. 25°/205°: the dip is 25°, and the trend of the dip direction has an azimuth of 205°.

The first two forms, called the strike notation, are the most common, with the difference usually depending on whether the compass used to make the measurement is divided into quadrants or a full 360°, and on personal preference. The advantage of the quadrant method of presentation is that most people find it easier to grasp a mental image of a trend more quickly with it.

The third and fourth forms, called the dip notation, are more generally reserved for the inclination and trend of lines rather than planes, although when the line is the dip line, it may apply to both. The fourth method gives the attitude of the plane unambiguously without the need for letters, and is, therefore, particularly useful for the mathematical treatment of attitude data.

It is essential to learn to read the attitudes expressed in all these shorthand forms with confidence, and to this end we will use them in examples and problems. However, they are not the best way of recording attitude data in the field. It is a very common mistake to read or record the wrong cardinal direction, for example, to write E when W was intended for the dip or strike direction, and thus to introduce an error which may be as large as 180°. One way which helps avoid such errors is to use the right hand rule (Barnes, 1981, p. 59): record the strike of your index finger when the thumb points down dip. Alternatively, sketching the actual dip and strike symbol in the field notebook and adding the bearing or azimuth of the strike direction (see the informal symbol in Fig. 1.1) permits a visual check on the correctness of the recorded attitude. Stand facing north and simply see that the structural plane and its symbolic representation are parallel. When the attitude is recorded in this way it is also easier to transfer the symbol to a base map.

There are a number of ways to determine the attitude of a structural plane, and all are based on field measurements of one kind or another. The most direct method is to hold the compass directly against an exposed plane surface at the outcrop. For strike, one edge of the open compass case is placed against plane and the compass rotated until it is horizontal (Fig. 1.2a). The trend, given by either its bearing or azimuth, in this position gives the strike direction. Dip is determined by placing one side of the compass box and lid directly against the exposed plane perpendicular to the previously measured strike. The clinometer bubble is leveled and the dip angle read (Fig. 1.2b). Indirect methods are also available, and are the subject of this and several later chapters. All the techniques dealt with here are concerned with determining the relationship between the components of the attitude of planes--the true and apparent dips, and the strike.

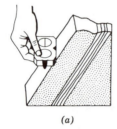 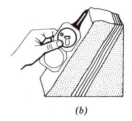

(a)   (b)

FIGURE 1.2 Attitude of a plane with a Brunton Compass (from Compton, 1962): (a) measurement of strike; (b) measurement of dip.

1.4 GRAPHIC METHODS

Of several different possible approaches to solving problems involving strike, true and apparent dip, at the start we choose an entirely graphical technique--the method of orthographic projection (see Appendix A). There are two reasons for this choice. First, with it we may readily and simply obtain solutions to a wide variety of problems, and second, it will allow the various components of the problems to be visualized in three dimensions. This visualization is of crucial importance in developing the ability to solve all geometric problems in structural geology.

By way of introduction, consider a simple geological situation. An inclined structural plane of interest strikes due north, and dip 35° to the east (N0°, 35°E). First, suppose that the earth's surface is perfectly flat and horizontal. On this surface, the trace of the inclined plane will then be a line of strike. Next, we imagine that a rectangular glass box is constructed with its top coinciding with the earth's surface, and its vertical sides parallel and perpendicular to this outcrop trace (see Fig. 1.3).

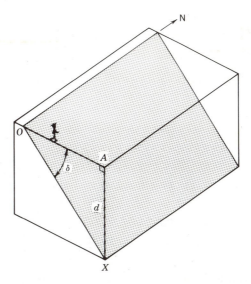

FIGURE 1.3 Block diagram showing the angle of true dip.

To do this imagine standing at a point O on the surface trace of the plane and then walking some distance, say 100 m, due east, that is, in the direction of the dip, to another surface point A. As we make this traverse, the vertical distance to the inclined plane steadily increases from zero at point O to some depth d beneath point A. Since we know the traverse length w we can easily make a scaled drawing of the top surface of the box showing its proper dimensions (Fig. 1.4a). We then imagine that we are actually looking vertically

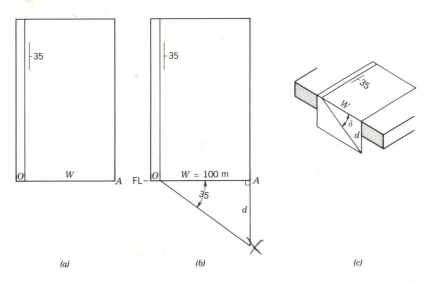

FIGURE 1.4 Section showing the true dip angle: (a) map view with strike line and traverse OA in the dip direction; (b) required section by folding about the line OA; (c) visualization by folding the draw-ing over the edge of a table.

downward at the top of the box; from this view point, we cannot see the vertical sides of the box. In order to view the side perpendicular to the strike, we imagine turning it upward as if it were hinged along the edge OA; this hinge is called a folding line, abbreviated FL. We can now easily construct a view of this side.

CONSTRUCTION (Fig. 1.4b)

1. Knowing the angle of true dip $\delta = 35°$, draw a line on this now upturned side making this angle with the horizontal line OA.

2. From point A, at a scale distance w = 100 m from point O in a direction normal to the strike, draw in a vertical line perpendicular to OA, to intersect the inclined line at point X. The distance AX is then the depth d to the inclined plane at point A, which we then may measure using the same scale.

A very useful way of visualizing what we have accomplished by this construction is to actually bend the drawing along the folding line over the edge of a table top (Fig. 1.4c), and then to view the result from several directions, including horizontally.

Once this visualization can be made with confidence, we can, of course, relate the three elements of this problem with the simple trigonometric equation,

(1.1) $\tan \delta = d/w$ TOA

Then knowing any two elements, the third can always be determined.

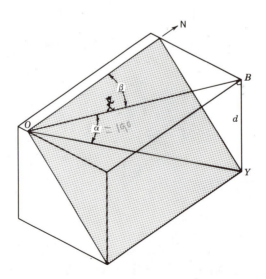

FIGURE 1.5 Block diagram showing an angle of apparent dip.

A closely related situation involves constructing a section in an oblique direction to show an angle of apparent dip α (alpha). For example, using the same inclined plane, we now go from the base point 0 toward N30°E to corner B (Fig. 1.5). As we make this traverse, the inclined plane falls away from our path according to the angle of apparent dip α = 19°. We now wish to reconstruct a view of the vertical plane containing this angle of inclination. In such cases it is more convenient to define the traverse direction (and the trend of the apparent dip) by the horizontal angle β it makes with the strike direction. This gives the direction of the line of interest relative to the strike direction, rather than to true north; for this reason we may refer to β (beta) as the structural bearing of the line. It should be noted that by the choice of strike in this example, the true bearing and the structural bearing are coincidently equal; generally this will not be so.

The length ℓ of the traverse required to arrive at corner B of our imaginary box is given by the relationship,

(1.2) $$\sin \beta = w/\ell$$

In this example, where w = 100 m, we find ℓ = 200 m. Knowing the apparent dip along this line we may once again construct a view of the required vertical plane with the aid of a folding line along OB.

CONSTRUCTION

1. From the data given, draw the top of the imaginary box, showing the line of the oblique tranverse making an angle of β = 30° with the line of strike (Fig. 1.6a).

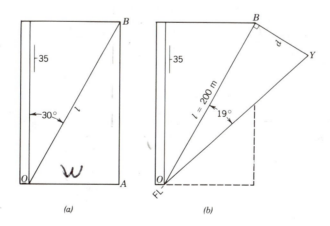

(a)   (b)

FIGURE 1.6   Construction of an apparent dip angle: (a) map view with strike line and traverse oblique to the dip direction; (b) required section by folding about line OB.

2. With this traverse line as FL1, plot an inclined line making an angle
   of $\alpha = 19°$ with OB (Fig. 1.6b). A vertical line at B then intersects
   this inclined line at point Y, and the distance BY is the depth d.

As before, bending the drawing over the edge of a table top along the folding line
allows complete visualization of the construction. With this picture clearly in
mind, we may then relate the three elements of the problem by the equation,

(1.3) $\tan \alpha = d/\ell$

Because the oblique traverse length ℓ is always greater than the strike-normal traverse length w , a comparison of this equation with Eq. 1.1 clearly shows that the
angle of apparent dip is always less than the angle of true dip.

1.5 FINDING APPARENT DIP

In the previous problem, the angle of apparent dip was given. However, this
angle can not always be measured in the field, and if inclined planes are to be depicted in a vertical section not perpendicular to the strike direction we need a
method for finding it.

PROBLEM

In the same structural situation as before, and knowing only that the true
dip $\delta = 35°$, find the apparent dip in a direction having a structural bearing of $\beta = 30°$.

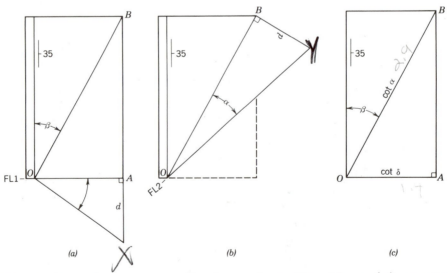

FIGURE 1.7   Finding the apparent dip in a given direction: (a) construction
of section showing the true dip and the depth to the inclined
plane at point A; (b) a vertical section in the required direction with apparent dip determined from the known depth d; (c)
apparent dip by the cotangent method.

CONSTRUCTION

1. Repeat the construction of Fig. 1.4b to find the depth d at point A
   (Fig. 1.7a). The length w of the strike-normal traverse may be chosen arbitrarily, but its scaled length should be conveniently large,
   say greater than about 10 cm, but greater yet for small angles of true
   dip.

2. Draw a line of strike passing through the surface point A.

3. Draw a second line from point O in a direction making an angle of β
   with the strike direction to intersect this second line of strike at B.

4. An auxiliary view of the vertical plane along OB = FL2 is then constructed to show the point Y at this same depth d below the surface point
   B (Fig. 1.7b). In this vertical plane, the line OY is inclined to the
   horizontal at the angle of apparent dip.

ANSWER

The angle BOY is measured and found to be α = 19°.

Several additional points should be noted. We need not ever know the actual
depth d; its scaled length may be transferred directly from AX to BY with a compass
or a divider. Also, in practice, the entire construction is usually accomplished on
one drawing; two are used here to avoid the confusion of too many crossing construction lines. Finally, the direction of upward folding is immaterial, though it is
usually best to chose it in the direction of the greatest open space on the drawing
in order to avoid interfering lines.

A useful short-cut method for determining an apparent dip combines a geometrical construction with trigonometric data and may be illustrated with the same
problem.

CONSTRUCTION (Fig. 1.7c)

1. Draw a map view showing a line of strike, the true dip direction perpendicular to the strike, and the apparent dip direction making an angle β
   with the strike direction.

2. Using a convenient scale, measure a distance OA = cot δ = 1.43 along
   the line in the true dip direction; remember that cot x = 1/tan x.

3. At point A construct a second line of strike perpendicular to OA and
   intersecting the apparent dip line at point B. The length of OB, measured using the same scale, is then equal to cot α.

ANSWER

In terms of the fully graphical technique, this use of cotangents is equivalent to choosing the depth d = 1. This method gives a solution more quickly, while still retaining the visual advantage of the completely graphical approach. It is especially useful when dealing with small dip angles which are difficult to construct accurately at any reasonable scale.

If the strike and an apparent dip are known, it is a simple matter to reverse this construction to find the angle of true dip. A closely related problem involves finding the structural bearing of a line whose apparent dip is specified.

PROBLEM

A plane with dip δ = 35° contains a line whose apparent dip is α = 19°. What is the trend of this line?

APPROACH

To determine the structural bearing of a line we must construct the horizontal right triangle OAB, as in Fig. 1.8a. We may easily find the length of side OB from the angle of apparent dip; the problem is then reduced to discovering its orientation. To show how this may be done most simply, it is advantageous to use the cotangent method.

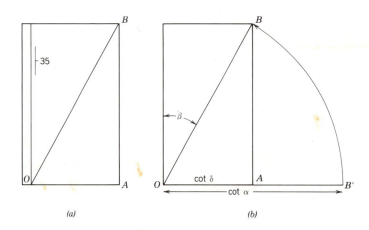

FIGURE 1.8 Finding the bearing of a line of apparent dip: (a) map of the problem; (b) solution by the cotangent method.

CONSTRUCTION (Fig. 1.8b)

1. Draw a map view showing the strike and direction of true dip.

2. In this dip direction measure off a line whose scaled length is equal to cot δ, locating the surface point A. A line from point A perpendicular to the dip line represents the second side of the required triangle.

3. Along this same dip line measure a length equal to cot α, thus determining the length OB'. Then with the base point O as center and OB' as radius, swing an arc to intersect the strike line at point B. The orientation of the hypotenuse OB of the resulting triangle gives the trend of the line of apparent dip.

ANSWER

The angle between the strike direction and the oblique line has a structural bearing of β = 30°.

These problems may also be solved with the aid of a trigonometric equation. From Eqs. 1.1, 1.2 and 1.3,

(1.4) $$\tan \alpha = \tan \delta \sin \beta$$

As defined, the angle of apparent dip is unambiguous, and with reasonable care no difficulties should be encountered. There are, however, some situation where an apparent "apparent dip" angle may be observed (see Exercise 1.6), and this may be confusing. For this reason that some prefer to speak of a dip component rather than an apparent dip.

1.6 FINDING TRUE DIP AND STRIKE

In some field situations it is not possible to measure the true dip and strike directly. However, if apparent dips in two directions are known, the complete attitude of the inclined plane can be found.

PROBLEM

From two apparent dips (10°, N72°W and 25°, N35°E) determine the true dip and strike.

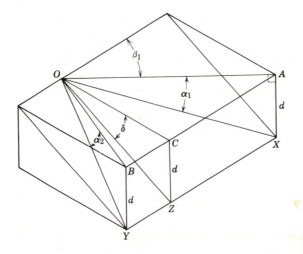

FIGURE 1.9 Block diagram showing the angles involved in the problem of determining the true dip and strike from two apparent dips.

APPROACH (Fig. 1.9)

The two lines whose inclinations are given by the apparent dip angles intersect on the inclined plane at a point. Three points determine a plane, so two additional points must be found. A second point is located from a vertical triangle containing one of the apparent dips. A third point, associated with the second apparent dip, could be found in like manner. However, it is advantageous to locate this third point at the same elevation as the second. Then a line joining these two point of equal elevation is, by definition, a line of strike. The true dip is measured perpendicular to this strike line.

CONSTRUCTION (Fig. 1.10)

1. Plot the trends of the two apparent dip directions in map view, intersecting at a point 0.

2. Using one of these as folding line FL1 construct the vertical triangle AOX with the associated apparent dip angle, and thus determining the

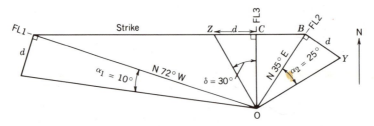

FIGURE 1.10 Graphical construction of dip and strike from known apparent dips.

depth d at the surface point A. The length of the traverse along
this line is chosen arbitrarily, but it should be long enough so that
the angles can be accurately measured.

3. Using the other apparent dip direction as a second folding line FL2,
   together with its associated apparent dip angle, construct a second
   vertical triangle BOY. This time the traverse length is determined by
   using the same depth found in Step 2, thus locating the position of
   surface point B.

4. Because points X and Y are both located at identical vertical distances
   below the base point O, they have equal elevations. A line through the
   two corresponding surface points A and B is then a line of strike.

5. A line from point O perpendicular to this strike line and intersecting
   it at point C establishes the direction of the true dip. At a depth d
   below C, point Z lies on the horizontal line XY. With this true dip direction as folding line FL3 locate Z at the same depth d. The line OZ
   is then inclined at the true dip angle.

ANSWER

The attitude of the inclined plane represent by the two apparent dips is
N90°E, 30°N.

This same type of problem may be solved even more quickly by an extension of
the semigraphical cotangent method of Fig. 1.7c. Because it avoids the plotting of
dip angles, this is particularly useful way of dealing with small inclinations.

CONSTRUCTION (Fig. 1.11)

1. In map view, plot rays from a single point O in the two apparent dip
   directions.

2. Using a convenient scale, plot distances equal to the values of the cotangents of the apparent dip angles along their respective rays. The
   line joining the two points so located is a line of strike.

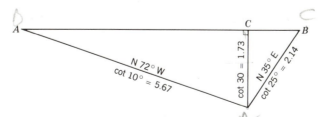

FIGURE 1.11 Graphic-trigonometric method for determining true dip and strike
from two apparent dips (after Kitson, 1929).

3. The perpendicular distance, measured at the same scale, from O to this
   strike line is equal to the cotangent of the true dip angle.

A solution to the two apparent dip problem may also be obtained trigonometrically. This is useful when routinely solving a number of closely related problems, and it can also be an aid when dealing with small dip angles. The following derivation is based on the geometry of Fig. 1.12. From the vertical right triangles BOY and AOX, each containing an apparent dip angle, we find the lengths of the two corresponding oblique traverses to be,

(1.5) $$\ell_1 = d/\tan \alpha_1 \quad \text{and} \quad \ell_2 = d/\tan \alpha_2$$

where the subscripts identify each apparent dip direction. From the horizontal right triangle COA, and using the identity $\sin x = \cos(90° - x)$, the length of the strike normal traverse OC is

(1.6) $$w = \ell_1 \cos[\phi - (90° - \beta_1)]$$

β is Structural bearing of line

Also, from the horizontal right triangle COB,

(1.7) $$w = \ell_2 \sin \beta_2$$

where ϕ is the horizontal angle between the two apparent dip directions. With the identity for the cosine of the difference of two angles, Eq. 1.6 may be rewritten as

(1.8) $$w = \ell_1(\cos \phi \sin \beta_1 + \sin \phi \cos \beta_1)$$

Equating Eqs. 1.7 and 1.8, and eliminating the two traverse lengths by using Eqs. 1.5, we obtain,

$$\frac{\sin \beta_1}{\tan \alpha_1} = \frac{\cos \phi \sin \beta_1 + \sin \phi \cos \beta_1}{\tan \alpha_2}$$

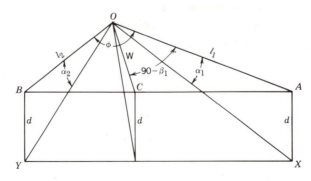

FIGURE 1.12   Angles and lengths involved in the derivation of Eq. 1.8.

and this simplifies to

$$(1.9) \qquad \cot \beta_1 = \frac{\tan \alpha_2}{\tan \alpha_1 \sin \phi} - \cot \phi$$

An ambiguity may arise from the use of this equation when the true dip does not lie between the two apparent dip directions. This difficulty may be avoided by labeling the smaller apparent dip angle α_1.

With the structural bearing β_1 of the α_1 direction, the true dip angle is then obtained from Eq. 1.4. $\qquad \tan \alpha = \tan \delta \sin \beta$

1.7 ACCURACY OF DIP AND STRIKE MEASUREMENTS

With reasonble care, horizontal angles may be read on the dial of a compass to the nearest degree, especially if the magnetic needle is equipped with induction damping. Vertical angles may also be read on the clinometer scale to the nearest degree, or better using the venier. Such results, however, should not be taken as an indication that the dip and strike angles are accurately known, because all such measurements are subjects to errors from several sources.

One possible source of error is an improperly adjusted compass; if the magnetic declination is incorrectly set, all trend angles will be systematically in error. Another error is introduced if there are magnetic materials in the vicinity, either due to the presence of magnetite in the rocks, or a piece of iron, such as a hammer, near the compass.

Still another source of error lies with the operator. It is never possible to align the compass in exactly the correct direction when making these angular measurements. In most cases, such departures are small, probably only a degree or two. There are, however, certain situations where small errors may become magnified, and these cases should be treated with special care. One of these is associated with measurement of the strike direction. By definition, the strike is the trend of a horizontal line on the inclined plane. If the compass is not exactly horizontal, then a direction other than the true strike will be measured. The geometry of this situation is shown in Fig. 1.13. An operator error ε_o, which is an angular departure from the horizontal, goes uncorrected.

The result is that an apparent strike s', rather than the true strike s is recorded. The angle between these two trends is the strike error ε_s, and its magnitude as a function of the dip angle may be evaluated. From Fig. 1.13 three relationships are:

$$w = d/\tan \delta, \qquad \ell = d/\tan \varepsilon_o \qquad \text{and} \qquad w/\ell = \sin \varepsilon_s$$

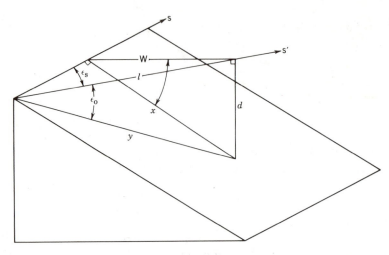

FIGURE 1.13 The angles and lengths used in the derivation of Eq. 1.9.

Combining the first two with the third, we have

$$(1.10) \qquad \sin \varepsilon_s = \frac{\tan \varepsilon_o}{\tan \delta}$$

This result, first obtained by Müller (1933, p. 232; see also Woodcock, 1976), has been solved for values of the maximum operator error for angles of 1°–5°, and the results shown in Fig. 1.14. From this graph it can be seen that for small dip angles the strike error may be large.

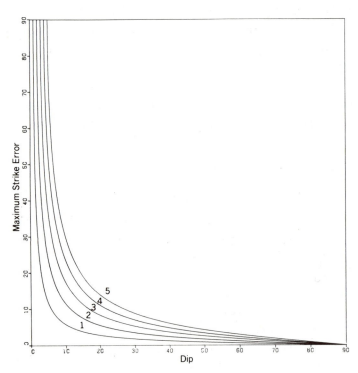

FIGURE 1.14 Graph showing the maximum strike error arising from operator errors $\varepsilon_o = 1°-5°$ as a function of the angle of dip.

One approach to controlling this source of error is to measure two apparent dips using long base lines, and calculate the true dip and strike with the methods of §1.5. If the dip angle is very small but still needs to be accurately determined, surveying methods may be necessary.

The graph of Fig. 1.14 shows that the strike error predicted for a vertical plane is zero. There are, however, still other sources of error. One of these is particularly important. Irregularities of the bedding, or other plane surfaces which effect the placement of the compass make it difficult to be sure that it is properly aligned. This and related matters are treated in further detail in Chapter 5. A useful approach which helps eliminate the effects of small-scaled irregularities to stand back from the outcrop several metres and determine the trend of a horizontal line of sight parallel to the bedding, and the inclination of the bedding perpendicular to this line (Fig. 1.15). Although it takes practice to become proficient, this is probably the most accurate method of determining dip and strike at the scale of a single outcrop.

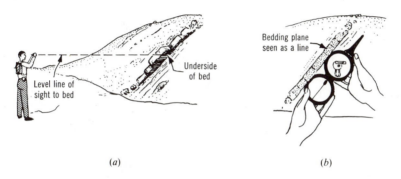

(a)   (b)

FIGURE 1.15   Measuring dip and strike to avoid minor irregularities on the
               bedding planes (from Compton, 1962): (a) sighting a level line
               in the plane of the bedding surface; (b) measuring the dip of
               bedding.

EXERCISES

1. Using the following data determine the unknown component graphically, and check
   your results trigonometrically. Each graphical results should be within 1° of
   the calculated answer; if it is not then repeat your construction using greater
   care, or making it larger, or both.

   a. If the attitude of a plane is N75°W, 22°N, what is the apparent dip in
      the direction N50°E?

   b. An apparent dip is 33°, N47°E, and the true strike is N90°E. What is
      the true dip?

c. The true dip is 40° due north.   In what direction will an apparent dip
   of 30° be found?

2. A certain bed dips 40° due north.   In what direction will the apparent dip be exactly half as great?   Will this same relationship hold if the bed dip 10°, 20°,
   50°, 80°?   If not, why not? 3/5/2

3. Three points A, B and C on an inclined plane have elevations of 150 m, 75 m, and
   100 m, respectively.   The map distance from A to B is 110 m in a direction of
   N10°W, and from A to C it is 1560 m in a direction N40°E.   What is the dip and
   strike of the plane?   (Hint: use Eq.1.3, to determine two apparent dips).

4. The most important need for the apparent dip arise during the construction of
   structure sections.   Fig. X1.1 is a simple geologic map of an inclined sequence
   of sedimentary strata intruded by a basalt dike, and the whole cut by a fault.
   Construct a vertical section along the line XX' showing the three structural
   planes with the correct inclination and proper position.

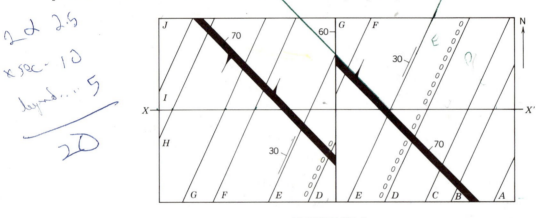

FIGURE X1.1

5. What is the maximum potential error in determining the strike direction if the
   maximum operator error is 2°, and the dip is 5°.

6. A smooth, vertical, east-west trending cliff exposes the traces of strata incline
   uniformly at 45°.   There is no hint of the third dimension.   What can be said
   about the true attitude of the beds with no further information?   What is the appearance of these inclined traces viewed from different positions on the ground.
   (Unless you have an unusual ability to visualize you will not be able to answer
   this question by just thinking about it.   You may find it useful to pin a sketch
   of such a cliff face on a wall at eye level; then by viewing it as you walk past,
   you can actually see the required angular relationships).

2
Thickness and Depth

2.1 DEFINITIONS

<u>Thickness</u>: the perpendicular distance between the two parallel planes bounding a tabular body of rock.

<u>Depth</u>: the vertical distance from a specified level (commonly the earth's surface) downward to a point, line or plane.

2.2 THICKNESS BY DIRECT MEASUREMENT

Although the geologist may determine the thickness of any stratiform body of rock, most often the concern is with the <u>stratigraphic</u> <u>thickness</u> of sedimentary and volcanic rocks. In this context "measuring a section" generally refers to a litho-logic description of the rock units as well as a determination of their thicknesses (Kottlowski, 1965). In structural geology, the concern is with thickness alone.

The thickness of a layer of rock may be determined in a number of ways. In the most favorable situations it may be possible to measure thickness directly; other-wise it must be determined from indirect measurements.

Several examples will illustrate how thickness may be measured directly. In the simplest case, the thickness of a horizontal layer exposed on a vertical cliff

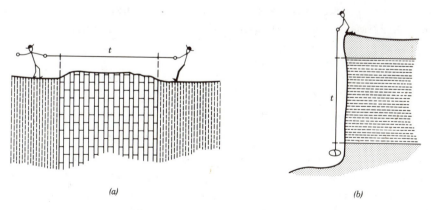

(a) (b)

FIGURE 2.1 Direct measurement of thickness: (a) vertical layer; (b) horizon-tal layer.

face may be obtained by hanging a measuring tape over the edge of the cliff (Fig. 2.1a). Alternatively, if the elevation of the top and bottom of the horizontal layer can be determined accurately, the thickness is simply the difference of the elevation figures regardless of the slope angle. Another special case involves the exposure of a vertical layer on a horizontal surface; a tape extended perpendicular to the strike allows the thickness to be obtained directly (Fig. 2.1b).

More generally, thickness may be measured directly regardless of the relationship between the slope and dip with a Jacob's staff (a light pole with gradations and clinometer attached at the top; see Robinson, 1959; Hansen, 1960). The staff is tilted toward the dip direction through the dip angle and a point on the ground is then sighted in. The thickness of the layer or portion of the layer between the base of the staff and the sighted point is equal to the length of the staff (Fig. 2.2). For layers less than staff height the gradations are used, and by occupying successive positions units of any thickness may be measured.

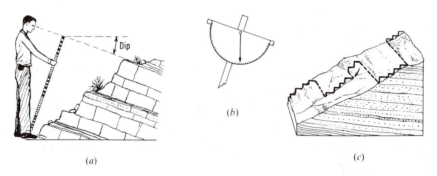

FIGURE 2.2   Thickness of an inclined strata using a Jacob's staff (from Compton, 1962): (a) sighting down the dip; (b) simple clinometer; (c) stepwise course of measurements.

The principle common to each of these approaches is that if a line of sight can be obtained parallel to the dip direction, the layer appears in edge view, and the thickness can be obtained by measuring across this view perpendicular to the two parallel bounding planes.

2.3   THICKNESS FROM INDIRECT MEASUREMENTS

When direct measurement of thickness is not possible, there are several alternatives. Which of these is adopted depends on the field situation, on the equipment at hand, on the accuracy required, and finally, on personal preference. Given a choice, always make the most nearly direct measurements possible.

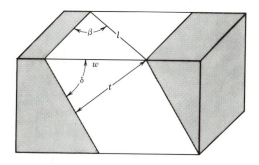

FIGURE 2.3 Block diagram showing the outcrop width on a horizontal surface.

The simplest of the indirect approaches is to measure the width of the exposed layer perpendicular to the strike direction on a horizontal plane. From this outcrop width w and the angle of dip δ, the thickness t can be determined by constructing a scaled triangle of the vertical plane containing the line of true dip (Fig. 2.3), or it may be calculated from

$$(2.1) \qquad t = w \sin \delta$$

Because of obstructions or lack of outcrops, it is not always possible to measure outcrop width in the strike-normal direction. For an oblique horizontal traverse another correction is required (Fig. 2.3). In effect, the traverse length ℓ is too long and must be reduced to the equivalent outcrop width. As before, this adjustment can be made with a scaled drawing, or with

$$(2.2) \qquad w = \ell \sin \beta$$

where β is the structural bearing of the traverse. Substituting this result into Eq. 2.1 yields a direct expression for thickness.

$$(2.3) \qquad t = \ell \sin \beta \sin \delta$$

When dealing with the more complicated problem of measuring thickness on sloping ground, there are two alternatives. With the first slope distance and slope angle along the line of the traverse are used. The second involves determining the vertical and horizontal distances between the two ends of the traverse. Both methods have advantages and disadvantages. The first method yields somewhat simpler relationships. The second is convenient when highly irregular slopes are involved, and it can also be used to obtain thickness from measurements made on geologic maps.

When the outcrop width is measured directly, the approach is closely related to the result of Eq. 2.1, except that thickness is a function of both dip angle δ and slope angle σ. Several cases are shown in Fig. 2.4, together with the corresponding trigonometric equations. The general formula is

$$(2.4) \qquad t = w \sin(\delta \pm \sigma)$$

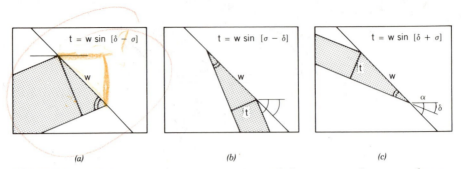

FIGURE 2.4 Thickness from outcrop width measured on a slope.

In this equation the sum is used if the slope and dip are inclined in the opposite directions as shown in Fig. 2.4a and the difference used if they are inclined in the same directions as shown in Fig. 2.4b,c. This same convention will be used in a number of equations derived in the chapter. In the second case it should be noted that if the slope angle is greater than the dip angle, this equation will give a thickness which is negative. It is therefore understood that the difference $(\delta - \sigma)$ is always to be taken as a positive quantity.

The second approach involves measuring the horizontal h and vertical v distances between two points in a strike-normal direction. Three subcases are illustrated in Fig. 2.5, and the general equation is

$$(2.5) \qquad t = h \sin \delta \pm v \cos \delta$$

where, again, the sum is taken if the slope and dip are inclined in opposite directions, and the difference if they are in the same direction.

If, in addition, the traverse is made oblique to the strike, an additional correction involving the structural bearing similar to that made in Eq. 2.3 can be introduced, giving

$$(2.6) \qquad t = h \sin \beta \sin \delta \pm v \cos \delta$$

In this same case, if the slope length and angle are measured, it would seem a simple matter to introduce a similar correction. Unfortunately, because they are

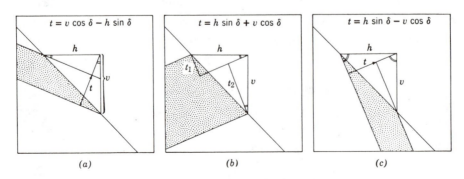

FIGURE 2.5 Thickness from vertical and horizontal components.

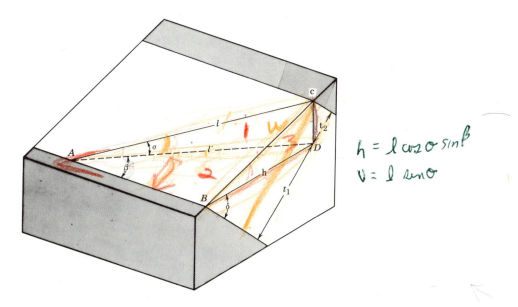

FIGURE 2.6 Thickness in general case of oblique traverse made on a slope.

not in a horizontal plane, there is no easy way of measuring the appropriate angles in the field (BAC or BCA of Fig. 2.6). It is therefore necessary to take a different approach. In the notation of Fig. 2.6,

$$h = \ell' \sin \beta \qquad \text{and} \qquad \ell' = \ell \cos \sigma$$

where ℓ' is the horizontal projection of the inclined traverse length and σ is the slope angle in the direction of this oblique traverse. Combining these two expressions then gives

$$h = \ell \cos \sigma \sin \beta$$

Also from Fig. 2.7,

$$v = \ell \sin \sigma$$

Using these expressions for h and v in Eq. 2.5 then yields the equation, first derived by Mertie (1922, p. 41),

(2.7) $$t = \ell(\cos \sigma \sin \beta \sin \delta \pm \sin \sigma \cos \delta)$$

2.4 THICKNESS BETWEEN NON-PARALLEL PLANES

Previously the measured layer was assumed to be strictly homoclinal, that is, had the same attitude at the two bounding planes. However, often the attitudes are different at the upper and lower ends of a traverse. Besides the possibility of measurement error which we will treat later, there are two basic reasons for such divergencies: the bounding planes may not in fact be parallel because the rock body is slightly wedge-shaped, or the layer may be folded.

If the departure from parallelism is small, thickness may be approximated by using the mean of the two dip angles

$$\delta = (\delta_1 + \delta_2)/2$$

and the mean of the two structural bearings

$$\beta = (\beta_1 + \beta_2)/2$$

in Eqs. 2.3 or 2.7. If the deviation from parallelism is greater, shorter intervals with more nearly parallel boundaries can be treated separately, and the results summed to give an estimate of the total thickness.

If the beds are folded, then the boundaries are curved surfaces rather than planes, and the matter is considerably more complicated. If it can be assumed that the bounding surfaces are still parallel, that is, the distance between the two surfaces measured perpendicular to the surfaces is constant, then the thickness can be estimated by a simple construction (Hewett, 1920).

PROBLEM

A strike-normal traverse is made on a slope; the measured strike directions at the upper and lower ends of the traverse are the same, but the dip angles are not. Estimate the thickness of the folded bed between the two stations.

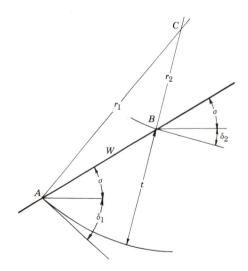

FIGURE 2.7 Construction of the thickness of a folded layer.

CONSTRUCTION (Fig. 2.7)

1. Draw a scaled cross-section showing the slope angle σ along the traverse line, the two measured dip angles at stations A and B, where the measured slope distance w = AB.

2. At each of the two stations construct perpendiculars to the dip lines to intersect at a point C.

3. With C as center, draw arcs with radii of AC and BC. The thickness is then the distance between the two concentric arcs as measured along either radius.

Thickness in this case may also be obtained trigonometrically. Labeling the dip angles so that $\delta_1 > \delta_2$ and the corresponding radii are r_1 and r_2, by the Law of Sines,

$$r_1/\sin A = r_2/\sin B = w/\sin C$$

The lengths of the two radii are then

$$r_1 = (w \sin A)/\sin C \qquad \text{and} \qquad r_2 = (w \sin B)/\sin C$$

Because $t = (r_1 - r_2)$,

$$t = w(\sin B - \sin A)/\sin C$$

With the angular relationships

$$\sin A = \sin[90° - (\delta_1 + \sigma)] = \cos(\delta_1 + \sigma)$$

$$\sin B = \sin[90° + (\delta_2 + \sigma)] = \cos(\delta_2 + \sigma)$$

$$\sin C = \sin[180° - (A+B)] = \sin(\delta_1 + \delta_2)$$

this expression for the thickness becomes

(2.8)
$$t = \frac{w[\cos(\delta_2 \pm \sigma) - \cos(\delta_1 \pm \sigma)]}{\sin(\delta_1 - \delta_2)}$$

There several important limitations to this method. First, it assumes that the traces of the folded beds on a vertical plane have the form of circular arcs, which is unlikely. Mertie (1940) has described the use of parallel curves of a more general nature which takes into account additional dip measurements. This gives a better representation of the thickness of the layer, but constructing these curves is somewhat involved, and it is apparently little used.

Errors arising from the assumption of circular arcs can be minimized by confining attention to short traverse lengths, for the smaller the difference between the two dip angles, the closer the arcs will be to the true curves.

A second limitation is imposed if the two strike directions differ, a situation which suggests that the folds are not horizontal. True thickness then can no longer be represented in a vertical section. This and other matters related to the geometry of folded strata are considered in greater detail in Chapters 11 and 14.

2.5 DEPTH TO A PLANE

Once the relationships involved in the determination of thickness can be visualized, and the methods used in its calculation are understood, the solution of depth problems should present little additional difficulty.

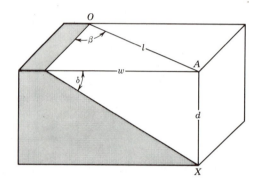

FIGURE 2.8 Depth to an inclined plane from a traverse perpendicular or oblique to the strike direction.

As with thickness, the simplest case is the depth to an inclined plane from a horizontal surface at a distance w measured from a point on the outcrop trace of the plane in a strike-normal direction. The depth may be found by constructing the scaled triangle, as in Fig. 2.8, or by using the formula

$$(2.9) \qquad d = w \tan \delta$$

If the distance ℓ is measured oblique to the strike, the apparent dip in the traverse direction can be used in Eq. 2.9, or, by substituting

$$w = \ell \sin \beta,$$

in Eq. 2.9 we have an expression for the depth directly in terms of the true dip,

$$(2.10) \qquad d = \ell \sin \beta \tan \delta$$

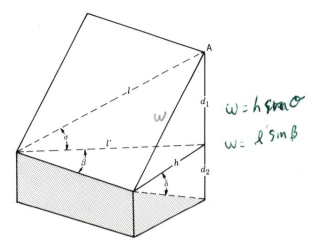

Handwritten notes in figure: $w = h \, \mathrm{sen} \, \sigma$; $w = \ell' \sin \beta$

FIGURE 2.9 Depth from a measured oblique traverse made on a slope.

The general case involves the depth to a plane of known attitude, given the length of an oblique traverse made on a slope. From Fig. 2.9, the two partial depths are

$$d_1 = \ell \sin \sigma \qquad \text{and} \qquad d_2 = h \tan \delta$$

Also from Fig. 2.9

$$h = \ell' \sin \beta \qquad \text{and} \qquad \ell' = \ell \cos \sigma$$

Using these last two results in the expression for the second depth then gives

$$d_2 = \ell \cos \sigma \sin \beta \tan \delta$$

Because the total depth is the sum of the partial depths, we obtain the equation (after Mertie, 1922, p. 48),

$$(2.11) \qquad d = \ell(\tan \delta \cos \sigma \sin \beta \pm \sin \sigma)$$

2.6 DISTANCE TO A PLANE

A closely related measure is the distance D to a plane in a direction other than vertical, as, for example, along an inclined drill hole. This problem may be solved graphically by constructing the scaled drawing, or it may be solved analytically. From the map of Fig. 2.10b,

$$w = \ell \sin \beta$$

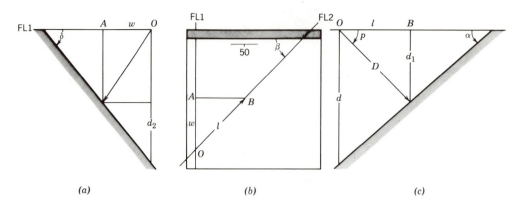

FIGURE 2.10 Distance to an inclined plane.

where β is the bearing of the drill hole. From the section along the trend of the drill hole of Fig. 2.10c,

$$\ell = D \sin p$$

where p is the plunge of the drill hold. From the strike-normal section of Fig. 2.10a, and using these two results, the second partial depth is

$$d_2 = w \tan \delta = D \sin p \sin \beta \tan \delta$$

From Fig. 2.10c, the first partial depth is

$$d_1 = \ell \tan p = D \sin p \tan p$$

Again, the total is the sum of these two partial depths. Adding the two expressions and solving the result for the distance D, yields (after Mertie, 1922, p. 48),

$$(2.12) \qquad D = \frac{d}{\cos p(\sin \beta \tan \delta \pm \tan p)}$$

As before, the sum is taken when the slope and plunge are in the opposite directions, and the difference when they are in the same directions. By combining this result with Eq. 2.11, the distance to any plane along any inclined line can be calculated from easily measured quantities.

2.7 MEASUREMENT ERRORS

Because none of the measurements made in the field can be absolutely accurate, all determinations of thickness and depth will also be approximate. To get some feeling for the role of such errors, consider the following situations (Mertie, 1947, p. 799). The thickness of a homoclinal layer is required. Measurements of h, v, β, and δ are made, and t is calculated using Eq. 2.6. By using a plane table or a transit, assume that the measurements of h and v can be made so accurately that their errors are inconsequential. We then wish to determine how measurement errors of dip and strike effect the calculated thickness. To do this, the difference between the true thickness and the calculated thickness dt is required, taking h and v as constants, and δ and β as variables. Then

$$(2.13) \qquad dt = \frac{\partial t}{\partial \beta} d\beta + \frac{\partial t}{\partial \delta} d\delta$$

From Eq. 2.6, we have

$$\partial t / \partial \beta = h \cos \beta \sin \delta$$

$$\partial t / \partial \delta = h \sin \beta \cos \delta \mp v \sin \delta$$

Therefore

$$(2.14) \qquad dt = (h \sin \beta \cos \delta)d\beta + (h \sin \beta \cos \delta \mp v \sin \delta)d\delta$$

As an example, suppose that

$$\beta = 57°, \quad \delta = 30°, \quad h = 147.08 \text{ m, and } v = 12.86 \text{ m}$$

and let the dip be in a direction opposite to that of the slope. Then from Eq. 2.6,

$$t = 61.68 + 11.14 = 72.81 \text{ m}$$

Assume that the errors of observation of strike and dip are as small as 1°. Substituting this 1° error, expressed in radians (1° = .01745 rad), and using increments, rather than differentials of β and δ in Eq. 2.14, gives

$$d\beta = d\delta = \Delta\beta = \Delta\delta = 1° = .01745$$

$$dt = .01745[40.05 + (106.83 - 6.43)] = 2.45 \text{ m}$$

The possible percentage error in thickness is then

$$dt/t = 2.45/72.81 = .034 = 3.4\%$$

Thus even with perfect geographic measurements, the thickness of this simple homoclinal sequence may have any value between 70.4 and 75.3 m. If we are confident that the errors in β and δ are no larger than 1°, the thickness might then be written as t = 72.8 ± 2.5 m.

The relative importance of strike and dip measurements may also be determined by assuming that the total error in the calculated thickness is equally divided between two terms of Eq. 2.13, or

$$\frac{\Delta t}{2} = \frac{\partial t}{\partial \beta} \Delta \beta = \frac{\partial t}{\partial \delta} \Delta \delta$$

Therefore

$$\Delta \beta = \frac{\Delta t}{2(\partial t/\partial \beta)} \quad \text{and} \quad \Delta \delta = \frac{\Delta t}{2(\partial t/\partial \delta)}$$

If we are willing to tolerate an error in the thickness of no more than 5%, that is,

$$\Delta t = (72.81)(0.05) = 3.64 \text{ m}$$

Then

$$\Delta \beta = 3.64/80.10 = .04544 = 2.6°$$

and

$$\Delta \delta = 3.64/200.80 = .01813 = 1.0°$$

In other words, the dip has to be correct to 1° in order for the stratigraphic thickness to be correct to 5%, and the dip measurement is more than twice as critical as the measurement of the strike. Of course these figures apply only to the sample problem, but they do show how measurement errors effect the determination of thickness. Similar results will be obtained for other equations for thickness, and also for the equations giving depth.

EXERCISES

1. The attitude of a sandstone unit is N65°E, 35°S. A horizontal traverse with a bearing of S10°E made from the bottom to the top of the unit measured 125 m. Determine the thickness graphically and check your result using Eq. 2.3.

2. The following information is from a geologic map. The attitude of a basalt sill is N5°W, 38°W. An eastern point on the lower contact has an elevation of 900 m, and a western point on the upper contact has an elevation of 1025 m. The line connecting these two points has a bearing of N85°W. Determine the thickness of the sill graphically and check your result using Eq. 2.6.

3. The geologic sketch map of Fig. X2.1 shows a thick shale formation between two limestone units exposed on a south-facing slope. A trail angles up brushy slope in a N30°E direction at a nearly constant 20° angle; the traverse length in crossing the entire shale unit is 366 m. The strata have a consistent attitude of N80°E, 35°N. If the shale-limestone contact lines are approximately horizontal how steep is the shale slope? What is the difference in elevation between the

beginning and ending of the traverse? What is the thickness of the shale.

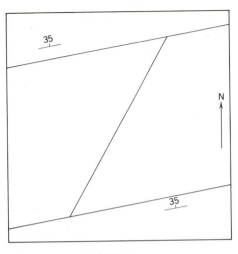

FIGURE X2.1

4. A south-to-north, strike-normal traverse made across a series of badland beds dipping 50° due north yielded the following data (the setting is shown in Fig. X2.2):

Lithologic unit	slope distance	slope angle
(5) upper sandstone	6.7 m	0°
(4) upper purple mudstone	17.7 m	17°
(3) lower sandstone	8.8 m	−13°
(2) pink claystone	8.0 m	15°
(1) lower purple mudstone	8.2 m	12°

Determine the total thickness.

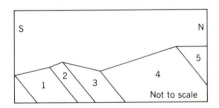

FIGURE X2.2

5. A mineralized vein with an attitude of N37°W, 50°SW is exposed on a ridge crest. How far down a 22° slope in a N82°W direction would it be necessary to go to find a point at which the vein lies at a depth of 100 m? At that point, what is the minimum length and inclination of a shaft to reach the vein?

6. A 125 m long strike-normal traverse made up a 15° slope between the bottom and top of a limestone strata gave the following information: the dip at the bottom of the unit is 55° and at the top it is 65°, both in a downslope direction. The strike directions at both points are the same. Using the method of tangent arcs of Fig. 2.7 estimate the thickness of the unit; check your result with Eq. 2.8. Compare this with the result obtained from Eq. 2.7 using the mean of the two dip angles.

3
Structural Planes and Topography

3.1 EXPOSURES ON HORIZONTAL SURFACES

In the previous chapter, the simplest examples of the determination of thickness assumed that the earth's surface was a horizontal, geometrically perfect plane. The intersection of inclined structural planes and layers with is surface result in an <u>outcrop pattern</u>, and the trace of this pattern represented in map view is a simple <u>geologic map</u>. The width of the outcrop bands depends on two factors: the actual thickness of the layers, and the angle of dip of each layer. The effect of each of these factors is shown in Fig. 3.1. In essence, these same relationships apply to less-than-perfect real horizontal topographic surfaces.

(a)   (b)

FIGURE 3.1 Relationship between outcrop width and thickness and dip: (a) with dip constant, outcrop width varies with thickness; (b) with thickness constant, outcrop width varies with dip angle.

In the special case of a vertical layer, the outcrop width in map view is equal to the thickness of the layer. This unique relationship results from the fact that the map shows such layers in edge view, that is, a line of sight in viewing the map coincides with a line which is parallel to the layers. In estimating thickness, one intinctively seeks just such a view of tabular object.

Of the infinite number of such lines of sight, one is always readily identifiable on a geologic map; it is in the dip direction. For layers inclined at angles of less than 90°, an auxiliary view perpendicular to this line could easily be constructed that would show the layers in edge view, and therefore in true thickness (Fig. 3.2a). However, it is unnecessary to make this construction because the same information can be obtained directly from the map itself. Simply rotate the geologic map so that the dip direction is "north", and then view the map pattern along a line of sight inclined to the plane of the map at the dip angle. In this view, the outcrop width, which is greater than the thickness, is foreshortened by just the right amount to appear as true thickness (Fig. 3.2b). In adopting this oblique, down-dip view of the map, it may help to reduce depth perception by closing one eye. Clearly, this method of viewing the outcrop bands of inclined layers is limited to cases involving significant dip angles, for it is physically impossible to view horizontal strata in edge view along an oblique line of sight of a map.

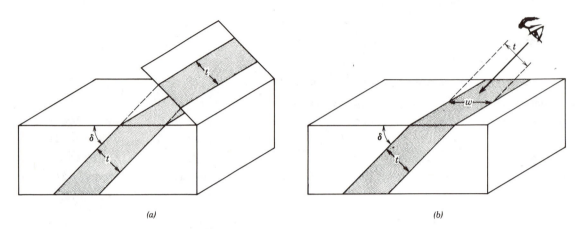

(a)   (b)

FIGURE 3.2   An auxiliary view showing true thickness: (a) constructed perpendicular to the line of true dip; (b) obtained by down-dip view of the map.

This principle is used in reverse for traffic signs painted on streets. By purposely distorting the letters as viewed vertically (map view) the foreshortening which accompanies the driver's oblique view of the road surface exactly compensates for the distortion and the warnings appear in normal proportions and are perfectly readable (see Fig. 3.3).

FIGURE 3.3   View from right along a line of sight inclined about 20° to the plane of the page.

In effect, twisting such simple geologic maps so that the inclined strata are viewed in the down-dip direction restores the sedimentary beds to their original horizontal attitude. Contacts on the map then cease to be just lines separating stratigraphic units on the earth's surface; they come to life as the depositional and erosional surfaces which they once were. A geologic map viewed down-dip represents a kind of cross-section, such as might be seen in the walls of the Grand Canyon. As such, the important dimension of sequence of deposition in time is added to the map. Unconformities become buried landscapes, and this view facilitates comparisions with the present earth's surface and the erosional processes responsible for its form. Certainly the possibility of completely overturned beds should be kept in mind, especially in areas of complicated structure. In such cases, the down-dip view yields a picture of the strata in a completely upside-down orientation, but such a view may actually help the interpretation of such overturning if other obvious evidence is lacking.

3.2 EXPOSURES IN AREA OF TOPOGRAPHIC RELIEF

In areas of sloping terrain, additional factors are involved in determining the character of outcrop patterns, and these include slope angle and direction relative to the attitude of the strata, and on variations in slope angle and direction. In other words, in addition to thickness an dip, the map pattern also depends on the details of the topography. The relationships between dip and topography have been formalized into a series of statements, collectively called the Rule of V's, by which the direction of dip can be determined directly from their outcrop patterns. Wherever the trace of a plane crosses a valley, the resulting pattern is characteristic of the attitude of the plane. There are several distinct types of patterns (Fig. 3.4):

1. Horizontal planes: Topographic contour lines can be thought of as the surface traces of imaginary horizontal planes. The outcrop traces of real horizontal planes therefore exactly follow the topographic contours. Such patterns are completely controlled by the topography; the outcrop trace faithfully reflects the local contour lines in every detail. Therefore, the outcrop pattern V's upstream, just as the contour lines do (Fig. 3.4a).

2. Planes inclined upstream: As the attitude departs from the horizontal, with the dip direction in the upstream direction, the pattern made by the traces of the structural planes is progressively modified into a blunter V, still pointing upstream (Fig. 3.4b). With steepening dip, the outcrop pattern is an increasingly subdued reflection of topographic detail.

3. Vertical planes: In the special case of a 90° dip, the outcrop traces are straight and parallel to the strike direction, regardless of topo-

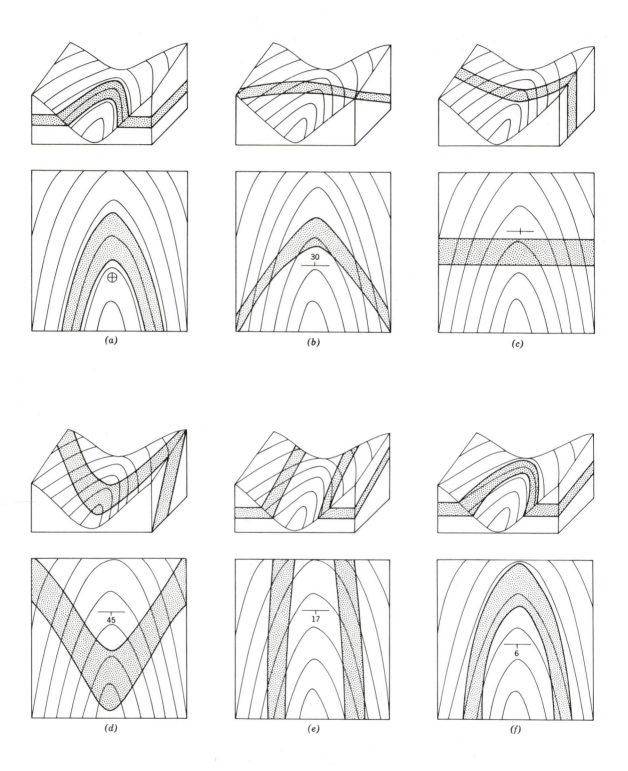

FIGURE 3.4 Outcrop patterns illustrating the rule of V's: (a) horizontal layer; (b) layer dipping upstream; (c) vertical layer; (d) layer dipping downstream; (e) layer and valley axis with equal inclinations; (f) layer dipping downstream at an angle less than valley gradient.

graphic detail. There is no V at all, and thus absolutely no control on the pattern by the topography (Fig. 3.4c).

4. Planes inclined downstream: There are three subcases, depending on the dip angle relative to the stream gradient.

a. With dip greater than valley gradient, the pattern V's downstream (Fig. 3.4d).

b. If the dip angle and valley gradient are exactly equal, the outcrop trace will not cross the valley axis, and there is no V (Fig. 3.4e). However, stream gradients generally headward, and a continuous planar structure must therefore cross somewhere upstream.

c. If the dip is less than the valley gradient, but still in a downstream direction, the pattern will V upstream (Fig. 3.4f).

As the valley gradient and dip determine whether the pattern will V up- or downstream, the boundary case occurs when the two are parallel.

These examples illustrate clearly the general principle that map patterns depend on the relation between slope and dip, and not directly on actual slope and dip angles alone.

Another and opposite set of rules could be formulated for traces which cross ridges. However, such ridges tend to have a greater variety of shapes, depending on the area's geomorphic history, and the patterns are correspondingly more varied. As stated, the strike direction was assumed to cross the valley axis at approximately a right angle; as a result, the V patterns were more or less symmetrical. With other strike directions, asymmetrical V's are produced, with the limiting case occuring when the valley and strike are parallel.

There is a simple, easily remembered single statement which summarizes these relations is: the V of the outcrop trace points in the direction in which the formation underlies the stream (Screven, 1963).

Better yet, however, is to visualize the geometrical relationship between the planar structures and to topography in three dimensions. In an area of topographic relief the outcrop pattern of uniformly dipping beds is irregular, yet if these same beds were viewed from an airplane in an oblique, down-dip direction, they would appear in edge view (the block diagram of Fig. 3.4b is very nearly in this orientation). The irregularities due to the topography are then eliminated and the traces of the inclined planes are straight, and true thickness appears directly. This same relation would, of course, hold true for a scaled topographic relief model with the outcrop pattern included. By learning to perceive the earth's surface, as depicted by the topographic contour lines on the map, as a relief model, the mind's eye can accomodate the influence of the topography of the outcrop pattern. This is not always easy to learn; practice is the key. Once the ability is attained, however, it is an enormously powerful aid in map interpretation, for even in areas of considerable and varied relief, and therefore of highly irregular map patterns, the structure can be viewed on the map in a down-dip direction with a great visual simplification.

3.3 DIP AND STRIKE FROM GEOLOGIC MAPS

So far, the examples have treated the attitude of inclined planes in semiquantitative terms only. However, the actual values of the dip and strike angles can be found from the outcrop trace of a structural plane. All that is required is the knowledge of the position in space of three points on the plane. In the simplest case, the strike is determined by joining two points on the plane of equal elevation. In Fig. 3.5, the trace of the lower bounding plane of the inclined stratum cuts the 1620 m contour at point A and B; the line joining these two points is therefore a line of strike.

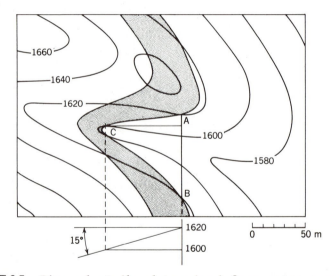

FIGURE 3.5 Dip and strike determined from outcrop pattern.

The dip angle is measured in the vertical plane normal to the strike. By drawing a line perpendicular to the line of strike AB to intersect the 1600 m contour at point C, a third point is established on the inclined plane. The dip angle can then be found from the horizontal or map distance h and the vertical distance v, which is equal to the difference in elevation of the line of strike and point C, using

(3.1) $\tan \delta = v/h$

Alternatively, a triangle may be constructed from this same data. First, a folding line is established perpendicular to the strike, and then, locating the relative elevations of both the line of strike and the third point using the map scale, the line of dip can be drawn, and the dip angle measured (Fig. 3.5). In either approach, chosing widely spaced points improves accuracy.

3.4 THREE-POINT PROBLEM

A more general form of this problem involves determining the dip and strike of a plane from three known points with different elevations.

PROBLEM

From the map location and elevation of three points located on a plane, determine its attitude.

APPROACH

A line joining two points of equal elevation establishes the strike direction. Chosing one of these to be the point of intermediate elevation, a second point with the same elevation is located along the line connecting the highest and lowest points. With the strike direction known, the inclination of the plane in the dip direction is then found by the methods of the previous section.

CONSTRUCTION (Fig. 3.6)

1. Join the three given points on the map to form a triangle, and label the highest point A, the intermediate point B, and the lowest point C.

2. With FL1 parallel to side AC, construct a vertical section showing points A and C in their correct vertical positions by plotting their elevations using the map scale. Draw in the inclined line AC in this section.

3. On this same section, draw a horizontal line whose elevation is that of the intermediate point B. Projecting the point of intersection of this horizontal line with the inclined line AC back to the map view locates the point B' on the side AC whose elevation is the same as point B. The line BB' is then a line of strike.

4. With FL2 perpendicular to this line of strike, construct a vertical section showing the correct elevations of points B and C. On this section the inclination of line BC is the dip angle.

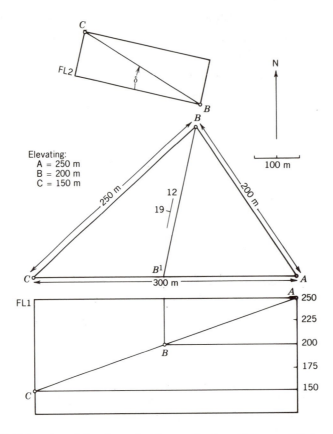

FIGURE 3.6 Graphical solution of the three-point problem.

Alternatively, the three-point problem may be viewed as an exercise in finding the true dip and strike from two apparent dips. First, apparent dips in each of two directions are found from the map distance h and elevation difference v using corresponding pairs of points, either by a graphical construction, or with

$$(3.2) \qquad \tan \alpha = v/h$$

The dip and strike can then be found with the construction of Figs. 1.9 or 1.10, or by using Eqs. 1.8 and 1.4.

3.5 PREDICTING OUTCROP PATTERNS

The reverse operation of locating the outcrop trace of a structural plane from its attitude at one known point is also possible. The earth's surface is represented on a map by topographic contours. Structural surfaces can be similarly represented by <u>structural</u> <u>contours</u>, that is, by lines of equal elevation drawn on the surface at a fixed vertical interval, each line representing a vertical distance from an established datum. If two different surfaces are both represented by contours with a common datum and interval, the line of their intersection is defined by the points of intersection of pairs of contours of equal elevation. For a structural plane it is then only necessary to construct these structure contours; this is particularly easy because they are a series of equally-spaced straight lines parallel to the strike direction.

The technique for accomplishing this can be illustrated with the aid of a block diagram (Fig. 3.7). Knowing the attitude of the structural plane at a single point A, a vertical section perpendicular to the strike is established, and topographic contours are added. Starting at the known point on this section, the trace of the dipping plane is then drawn. The intersection of this inclined line and the topographic contours fixes the locations of each structural contour (points 1, 2, 3, 4). Projecting these contours in the strike direction then locates points of intersection with the topographic contours on the earth's surface. The outcrop trace is completed by connecting these points. Note that not all of the structure contours are used; the contour associated with point 4 remains totally underground. In other cases, the contours may be completely in the air.

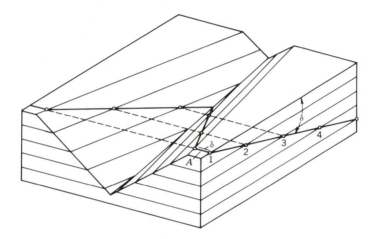

FIGURE 3.7   The basis for predicting the outcrop trace of a plane; the structure contours on the inclined plane are shown dashed.

PROBLEM

Given a topographic map and a single outcrop point Z on a structural plane whose dip is 20° due east, construct the outcrop trace of the plane across the map area.

CONSTRUCTION (Fig. 3.8)

1. With a FL perpendicular to the strike direction establish a vertical section. On this section:

   a. Draw a series of horizontal lines spaced at the interval of the topographic contours, and plotted using the map scale. Label each line with its elevation.

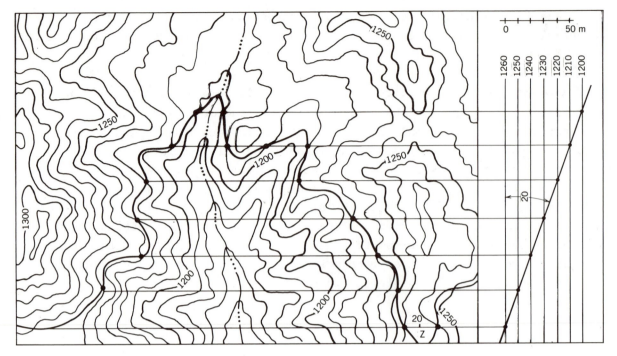

FIGURE 3.8 Outcrop trace of an inclined plane exposed at point Z.

   b. Using these elevation lines, locate the correct vertical and horizontal position of the outcrop point (in this example, the known point Z lies exactly on a topographic contour, but this will not usually be the case).

   c. Through this outcrop point draw a line in the dip direction inclined downward at the dip angle.

2. Each point where the dip line crosses an elevation line fixes the position of the corresponding structure contour on the structural plane. These contours are projected to the folding line and then back to the map view.

3. Each intersection of a structure contour with its matching topographic contour is an additional outcrop point, and these should be marked distinctly.

4. Complete the outcrop trace by joining successive outcrop points. This trace must cross at and only at these established points. If the contour spacing is wide, the outcrop trace can usually be sketched across the gap.

   If both the upper and lower bounding planes of a layer are to be shown it is a simple matter to plot the layer, using its scaled thickness, on the cross section, and then repeating the exercise for this second trace.

Instead of constructing the cross section, it may be easier and faster, especially if the dip angle is small, to determine the spacing s of the structure contours on the map using

$$s = i \cot \delta$$

where i is the contour interval.

In any case, the construction of an outcrop trace should be something more that an exercise in connecting points, as in a child's work book. Especially at breaks in topographic slopes, in valley bottoms, and on ridge crests, it may be necessary to interpolate both the structure contours and the topographic contours, at least mentally, in order to achieve the desired sensitivity to the effects of topography on the outcrop pattern.

EXERCISES

1. Determine the attitude of the mapped unit of Fig. X3.1. With this result view the map in a down–dip direction and, in combination with a visualization of the topography, try to see the unit as a layer in edge view. The topography of Fig. X3.2 is identical. With your visualization as a reference, now try to look down the dip of this layer, and estimate its attitude and thickness. Check your results.

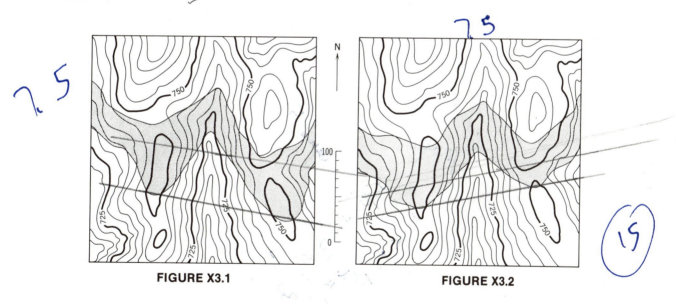

FIGURE X3.1   FIGURE X3.2

2. With the following information and the topographic map of Fig. X3.3 (see the end of the book for a full-scale version), construct a geologic map. The base of a 100 m thick sandstone unit of lower Triassic age is exposed at point A; its

attitude is N70°W, 25°S. Point B is on the east boundary of a 50 m thick, vertical diabase dike of Jurassic age; its trend is N20°E. At point C, the base of a horizontal Cretaceous sequence is exposed and at point D the base of a conformable sequence of Tertiary rocks is present.

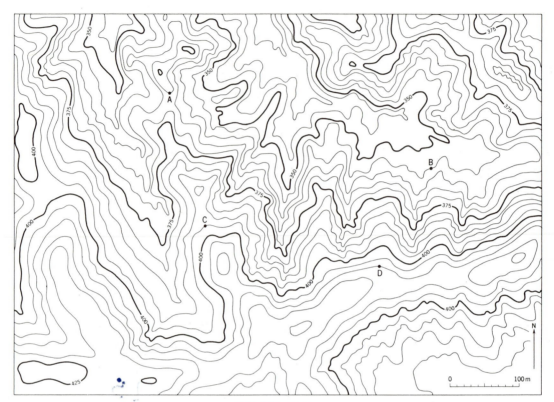

FIGURE X3.3

4
Lines and Intersecting Planes

4.1 DEFINITIONS

Line: a geometric element that is generated by a moving point; it has only extension along the path of the point. Lines may be rectilinear (straight) or curvilinear (curved). Only rectilinear elements are considered here.

Plunge: the vertical angle measured downward between a line and the horizontal (Fig. 4.1a). Plunge is a measure of inclination, and is analogous to the dip of a plane.

Pitch: the angle between a line and the horizontal measured in a specified plane (Fig. 4.1b). Rake is synonymous, but is not now widely used.

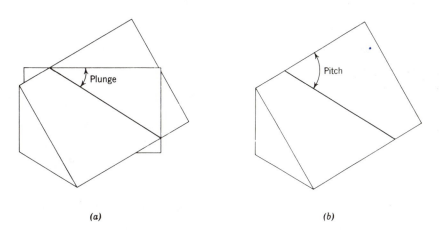

(a)   (b)

FIGURE 4.1   Inclination of a line: (a) plunge is measured in the vertical plane containing the line; (b) pitch is measured in the inclined plane containing the line.

4.2 LINEAR STRUCTURES

Structural lines may be of two types. They may exist in their own right, such as the long axes of mineral grains or streaks of mineral aggregates; elongate rock bodies and drill holes may also be considered essentially linear for some purposes.

More commonly, lines occur in conjunction with structural planes; examples include striations on slickensided fault surfaces, mineral lineation on foliation planes and lines formed by the intersection of planes.

The orientation of a line in space is specified by its trend and plunge. As with planes, there are a set of map symbols for lines, also with three parts:

1. a trend line,

2. an arrowhead in the direction of downward inclination,

3. a plunge angle written near the arrowhead.

The most common symbols are shown in Fig. 4.2. Special symbols may be invented when needed, and all symbols used must be explained in the map legend.

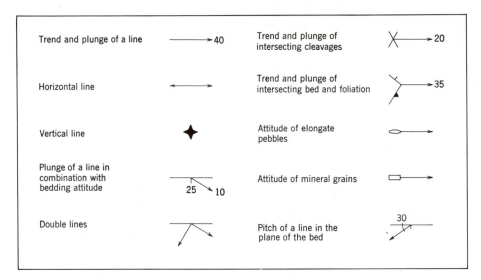

FIGURE 4.2 Map symbols for structural lines.

The attitude of a line may also be written out. The usual form gives the plunge angle followed by the trend as a bearing, as in 30°, S45°W, meaning that the line plunges 30° toward S45°W. Alternatively, this same attitude may be given with the azimuth of the trend, as in 30°/225°. Again, the difference depends of the type of compass dial used, and on personal preference. The azimuth form is particularly useful for computer processing of orientational data.

4.3 PLUNGE OF A LINE

Most of our concern in this chapter is with lines lying in structural planes. For such a line the plunge angle is, in effect, an apparent dip (see Fig. 4.3), and all the techniques of §1.5 are directly applicable to problems involving the relation of the plunge and trend of a line and the dip and strike of the plane.

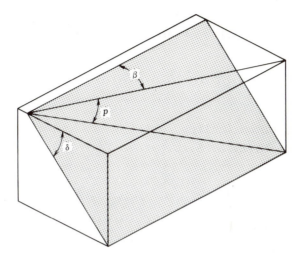

FIGURE 4.3 Block diagram showing the plunge of a line in an inclined plane.

For the same reason, the analytical solution for the plunge is obtained directly from Eq. 1.4, which we rewrite as,

$$(4.1) \qquad \tan p = \tan \delta \sin \beta$$

4.4 APPARENT PLUNGE

If a linear feature is to be depicted on a vertical section not containing the line, the angle of <u>apparent</u> plunge, rather than true plunge must be used. One need for such an angle arises if non-vertical drill holes and the rock units they penetrate are to be shown on a vertical section. The geometry of the situation is illustrated in Fig. 4.3. It should be noted that such projections may introduce errors; other things being equal, it is better to project over small distances and small angles. If the error become large, it may be necessary to realign the section.

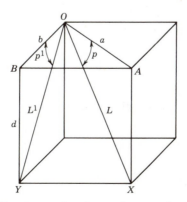

FIGURE 4.4   Block diagram showing the angle of apparent plunge.

Apparent plunge is analogous to apparent dip. However, in contrast, the angle of apparent plunge p' is greater than the angle of true plunge. When the section line is parallel to the trend of the plunging line, the true plunge is shown, and this then is the minimum value of the apparent plunge. If the section line is perpendicular to the trend of the line, then the apparent plunge angle is 90°, and this is its maximum value.

PROBLEM (Fig. 4.5a)

Show the projection of an inclined drill hole whose attitude is 30°, S45°E on a north-south vertical section.

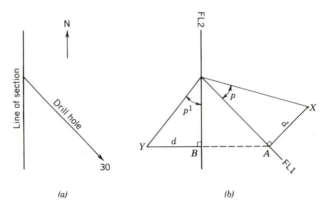

(a)   *(b)*

FIGURE 4.5   Construction of angle of apparent plunge: (a) sketch map of the problem; (b) orthographic construction.

CONSTRUCTION (Fig. 4.5b)

1. Draw the line of section and trend of the inclined drill hole in map view. The section may include the surface end of the hole O, as here; if not, then the point O must itself be projected orthogonally to the section line.

2. Using the trend line as FL1 draw a vertical section showing the angle of true plunge. Then either

    a. locate depth d at some convenient horizontal distance OA, or

    b. use the actual depth at the end of the hole to determine OA.

3. Project this surface point back to the section line to give point B.

4. With the line of the section as FL2, locate point Y at the same depth d below point B. The angle BOY is the angle of apparent plunge.

ANSWER

The apparent plunge of the inclined drill hole in the plane of the north-south section is p' = 39°.

The trigonometric expressions for solving this same problem can be derived from Fig. 4.4; three relationships, each involving plane triangles, are

$$a = d/\tan p, \quad b = d/\tan p', \quad \text{and} \quad \cos \beta = b/a$$

Substituting the expressions for a and b into the third equation then gives,

(4.2) $\tan p' = \tan p/\cos \beta$

When projecting to a vertical plane of section, any original length ℓ along the plunging line is shortened to ℓ'. Again from Fig. 4.4,

$$d = \ell \sin p \quad \text{and} \quad d = \ell'\sin p'$$

Equating these two and rearranging resulting expression then yields,

(4.3) $\ell' = \ell \sin p/\sin p'$

4.5 PITCH OF A LINE

The pitch angle r is measured in the plane containing the line. It may there-fore range in value from r = 0° when the line is horizontal, to r = 90° when the line is in the dip direction. In describing pitch it is necessary only to give the angle and the direction in which the acute angle faces; for example, 35°N means that the pitch angle is measured downward from the northern end of the strike line of the plane containing the line.

In the field, the pitch angle may be determined on an exposed plane by first marking a horizontal line and then with a protractor measuring the angle between this line and the linear structure. If, however, the plane is not well exposed, the pitch angle may have to be determined from indirect measurements.

PROBLEM

What is the pitch angle of a line with a structural bearing β = 35° on a plane with a dip of δ = 40°?

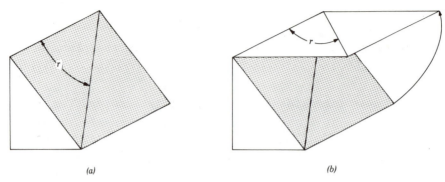

(a)   (b)

FIGURE 4.6 Elements of the pitch problem: (a) block diagram of line on inclined plane; (b) rotating the inclined plane to horizontal.

APPROACH (Fig. 4.6)

In the map view of an inclined plane containing a line we see directly the
angle β, which is the projection of the pitch angle   r.   In order to determine this angle, we must obtain a normal view of the inclined plane and
the line it contains. This is done by rotating the structural plane into
a horizontal position about an axis parallel to the strike direction.
This procedure can be usefully illustrated by unfolding a small paper
model.

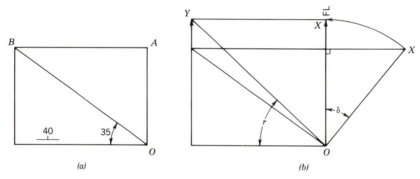

(a)   (b)

FIGURE 4.7 Finding pitch: (a) map view of problem; (b) orthographic construction involving rotating plane to horizontal.

CONSTRUCTION

1. Draw a map view showing a line of strike, the trend of the inclined
   line, and a line in the direction of the true dip (Fig. 4.7a).

2. With the dip line as FL1 draw a vertical section showing the angle of
   true dip (Fig. 4.7b). Then using a convenient length OA determine the
   depth d to point X directly beneath the surface point A.

3. In this section and using a compass, rotate the trace of the inclined plane to horizontal through the dip angle, thus locating point B such that OB = QX.

4. Construct two lines perpendicular to the dip line, the first through A to intersect the trend line at A'.

5. Through A' draw a line parallel to the dip line to intersect the second perpendicular at B'. The angle between the strike line and OB' is the pitch angle.

ANSWER

The pitch of the line in this plane is r = 42°.

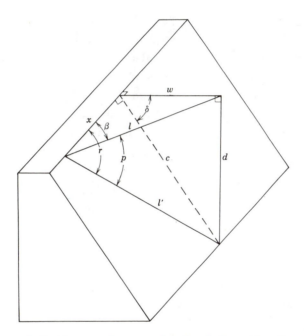

FIGURE 4.8 Diagram with the angle use in deriving the formulas for pitch.

There are two useful formulas which can be used to solve for the pitch angle. The first relates pitch to the structural bearing and the plunge angle. From Fig. 4.8,

$$x = \ell \cos \beta, \quad \ell' = \ell/\cos p, \quad \text{and} \quad \cos r = x/\ell'$$

Combining these gives,

(4.4) $$\cos r = \cos \beta \cos p$$

The second formula relates the pitch to the structural bearing and the dip of the plane. Again from Fig. 4.8,

$$w = c \cos \delta, \quad x = c/\tan r, \quad \text{and} \quad \tan \beta = w/x$$

Hence

(4.5) $\tan r = \tan \beta / \cos \delta$

4.6   INTERSECTING PLANES

Two non-parallel planes intersect in a line, and in many structural situations it is important to determine the attitude of this line.

Before attempting to solve this type of problem, it is important to visualize the geometry of such intersecting planes, and it helps to practice with the aid of flattened hands held parallel to the two planes. Folded paper models are also useful. There are three down-structure views which should be adopted: down-dip views of each of the planes, and especially a down-plunge view of the line of intersection. From these views and the previously established relationship between true and apparent dip, we may then clearly see that the line of intersection, being common to both planes, cannot have a plunge angle greater than the smaller dip angle. The trend of the line of intersection exactly bisects the angle between the two strike directions in the special case of two identical dip angles. Otherwise the trend of the line will be closer to the strike of the steeper plane.

PROBLEM

Find the plunge and trend of the line of intersection of the two planes whose attitudes are N21°W, 50°E and N48°E, 30°NW.

APPROACH

Down-structure views of the map (Fig. 4.9a) show that the trend of the line of intersection is northerly, and the plunge angle is less than 30°. The two lines of strike may be taken to represent the outcrop traces of the two planes on a flat surface, and the intersection of these two lines at point 0 is the outcrop point of the line. A second point on the line at depth will then fix its attitude.

CONSTRUCTION (Fig. 4.9b)

1. Plot strike lines for each of the two planes in map view. These two lines then represent structure contours on the two planes which intersect at point 0.

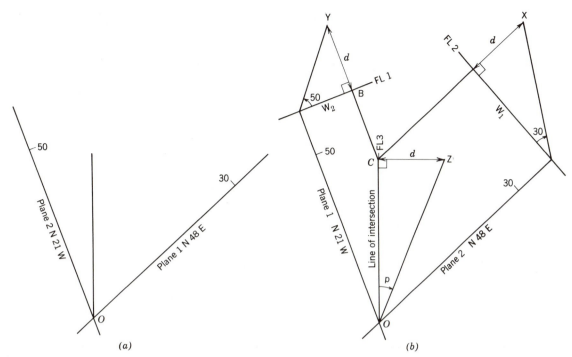

FIGURE 4.9 Line of intersection: (a) sketch of the problem; (b) construction using folding lines.

2. To obtain a second set of intersecting structure contours:

   a. With FL1 perpendicular to the strike of Plane 1, construct a vertical section showing the true dip. Then with a convenient length w_1 determine the depth d to point X below the surface point A.

   b. Similarly, with FL2 perpendicular to the second strike line, construct a vertical section showing the true dip of Plane 2, and using the same depth d to point Y, determine the length w_2 and locate the surface point B.

3. Through points A and B draw a second pair of line parallel to the strike directions of the respective planes intersecting at point C. These two lines are a second pair of structure contours on the planes at a depth d below the first pair, and the direction OC represents the trend of the line of intersection.

4. With this trend line as FL3, construct a vertical section using the same depth d at the surface point C. The vertical angle COZ is the plunge.

ANSWER

The line of intersection of these two planes trends due north, and plunges 23° in that direction.

This same type problem may also be solved using the semi-graphical cotangent method.

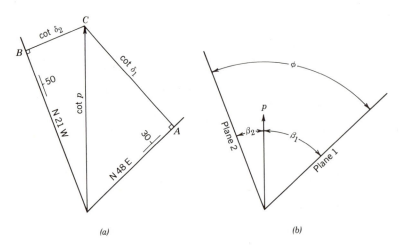

FIGURE 4.10 Line of intersection: (a) cotangent method; (b) labeled angles used in the derivation of the equation for the trend of a line of intersection.

CONSTRUCTION (Fig. 4.10a)

1. As before, plot the two strike lines radiating from a point O.

2. Draw a line perpendicular to each strike line equal in length to the cotangent of the corresponding dip angle.

3. This second pair of structure contours intersect at point C. The trend of the line of intersection is OC, and the length of the segment OC is equal to cot p.

In the derivation of the formula for solving this problem, we first seek the structural bearing of the line of intersection with respect to one of the two strike directions. Eq. 4.1 gives the plunge angle in terms of the dip of one plane and the corresponding structural bearing. Because the line is common to both planes, we then have, in the notation of Fig. 4.8b,

$$\sin \beta_1 \tan \delta_1 = \sin \beta_2 \tan \delta_2$$

or

(4.6)
$$\frac{\sin \beta_2}{\sin \beta_1} = \frac{\tan \delta_1}{\tan \delta_2}$$

Labeling the horizontal angle between the two strike directions of the intersecting planes $\phi = \beta_1 + \beta_2$, then,

$$\beta_2 = \phi - \beta_1$$

and we can eliminate β_2 from Eq. 4.6, giving

$$\frac{\sin(\phi - \beta_1)}{\sin \beta_1} = \frac{\tan \delta_1}{\tan \delta_2}$$

Substituting the identity for the sine of the difference of two angles gives

$$\frac{\sin \phi \cos \beta_1 - \cos \phi \sin \beta_1}{\sin \beta_1} = \frac{\tan \delta_1}{\tan \delta_2}$$

and this simplifies to

$$(4.7) \qquad \tan \beta_1 = \frac{\tan \delta_2 \sin \phi}{\tan \delta_2 \cos \phi + \tan \delta_1}$$

With the structural bearing relative to the strike of Plane 1, together with the dip angle, the angle of plunge is found by using Eq. 4.1.

4.7 ACCURACY OF TREND DETERMINATIONS

As with dip and strike, measurements of the angle of trend and plunge cannot be made without error, and as before, situations where small measurement errors may become magnified are of special concern.

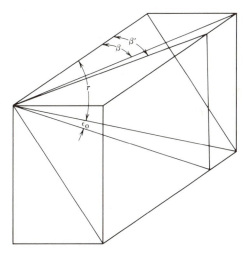

FIGURE 4.11 Geometry of errors in measuring the trend of a line.

In measuring the trend of a line on a structural plane it is common practice to align the compass in the direction of the horizontal projection of the line. If an error is made in this alignment as measured by the angle ε in the structural plane containing the line, a significant error of trend may result. To find this trend error we compare the true trend of a line as given by β with β' found from the pitch angle $(r+\varepsilon_o)$ on the same plane. The maximum trend error will then be

$$\varepsilon_t = \beta - \beta'$$

To find an expression for this error, we obtain from Eq. 4.5, and using Fig. 4.9,

$$\tan \beta = \tan r \cos \delta \quad \text{and} \quad \tan \beta' = \tan(r+\varepsilon)\cos \delta$$

Using these in the equation for the tangent of the difference of two angles,

$$\tan (\beta'-\beta) = \frac{\tan \beta' - \tan \beta}{1 + \tan \beta \tan \beta'}$$

we have (after Woodcock, 1976, p. 352),

(4.8)
$$\tan \varepsilon_t = \frac{[\tan(r+\varepsilon) - \tan \delta]\cos \delta}{1 + [\tan r \tan(r+\varepsilon)]\cos^2\delta}$$

A graph of this equation is given in Fig. 4.12; for a combination of a steep plane and a large pitch angle, a large trend error may result.

It is also of interest to note that the maximum trend error associated with the pitch angle $(r-\varepsilon_o)$ is less, and as a result, repeated measurements will not be symmetrically distributed about the true trend. The formula in this case is

(4.9)
$$\tan \varepsilon_t = \frac{[\tan \delta - \tan(r-\varepsilon)]\cos \delta}{1 + [\tan r \tan(r-\varepsilon)]\cos^2\delta}$$

To avoid such errors it is advisable to measure the pitch of the line directly and then determine its trend either graphically, or with the aid of Eq. 4.5.

A second type of error magnification occurs in determining the attitude of the line of intersection of two plane. Because both sets of dip and strike measurements are subject to their own errors, the derived attitude of the line of intersection will correspondingly be in error, and this error may be large if the angle between the two planes is small; a detailed treatment of this situation is given in Chapter 5.

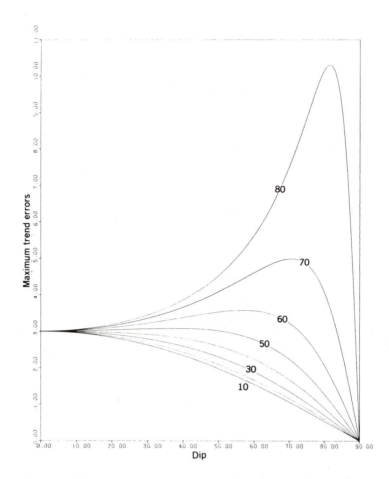

FIGURE 4.12 Maximum trend error as a function of pitch and dip.

Using the map of Fig. X4.1, structural plane A is a narrow shear zone whose attitude is N66°E, 50°S, and plane B is the top of a limestone bed whose attitude is N22°W, 40°W. Determine the orientation of the line of intersection of these two planes, the pitch of this line in plane B, the surface outcrop point of the line, and the depth at which the line would be found by drilling in the bed of Boulder Creek. (Note that A and B do not have the same elevation).

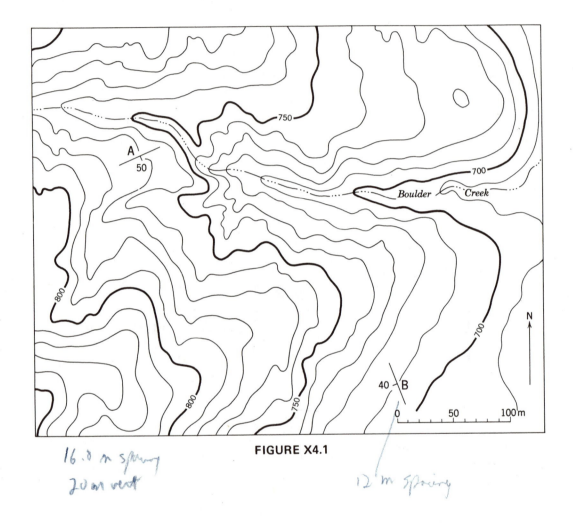

FIGURE X4.1

16.0 m spring
20 m vert

12 m spring

5

Graphic Solutions with the Stereonet

5.1 STEREOGRAPHIC PROJECTION

The solution of problems by the methods of descriptive geometry requires the construction of at least two orthographic views, and this is a time consuming process. One can always resort to a mathematical solution, but this requires that the appropriate equations are readily available; also, the lack of a visual check involves some risk. Fortunately, there is an alternative graphical approach by which the angular relationships of lines and plane can be found quickly and efficiently.

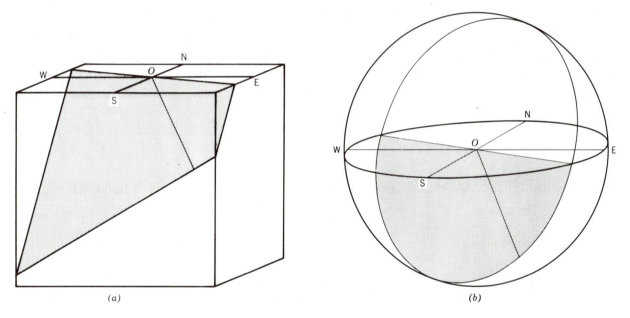

FIGURE 5.1 An inclined plane containing a line: (a) block diagram with point O on the outcrop trace; (b) a sphere with center at point O.

Imagine a sphere centered on the outcrop trace of an inclined structural plane containing a structural line (Fig. 5.1a). This plane will intersect the sphere as a great circle, and the line will be a point on this circle (Fig. 5.1b). The angular relationships in this simple situation are now represented in spherical projection. We now require some way of reducing this three-dimensional representation to the two dimensions of a plane. Our choice is dictated by the desire to preserve angular relationships, and this is best accomplished with the method of stereographic projection.

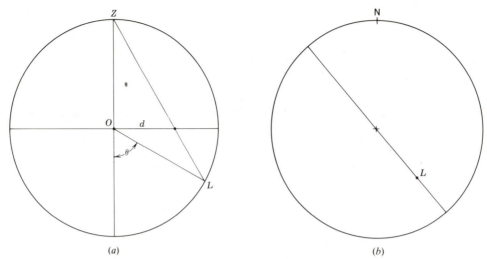

(a)   (b)

FIGURE 5.2   Stereographic projection of the inclined line of Fig. 5.1: (a)
the vertical diametral plane containing the structural line; the
point of intersection with the lower hemisphere is projected to
the horizontal equatorial plane using the projection point Z; (b)
the corresponding stereogram showing the same vertical plane and
the projection of the line.

Consider first a structural line. In the vertical plane containing the line,
the point of its intersection with the sphere is projected to the horizontal diametral projection plane using the projection point Z at the zenithal position on the
sphere (Fig. 5.2a). The distance d from this point to center O on the projection
plane is easily found to be

(5.1) $d = R \tan(\theta/2)$

where R is the radius of the sphere and θ is the complement of the plunge of the
line.

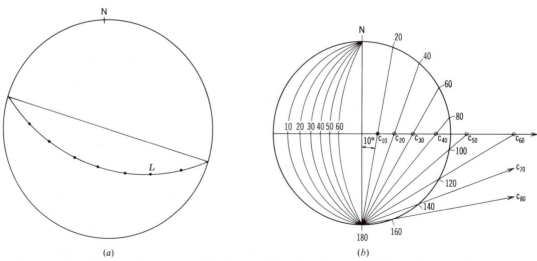

(a)   (b)

FIGURE 5.3   Stereographic projection of the inclined plane of Fig. 5.1: (a)
the locus of points representing a series of lines on the structural plane defines a circular arc; (b) the construction of great
circular arcs.

With this result, we may draw a direct view of the projection plane showing the line as a point. This representation plotted inside the horizontal great circle, called the primitive, together with the cardinal compass directions is a <u>stereogram</u> (Fig. 5.2b).

With this method of locating points we may easily represent a series of lines of apparent dip on a single structural plane. As will be seen, these fall on a circular arc (Fig. 5.3a). One of the important properties of the stereographic projection is that great circles on the sphere project as circular arcs on the stereogram. This permits the representation of any structural plane to be constructed easily. The distance of the center of such a circular arc from point O may be found graphically (Fig. 5.3b), or from the formula,

$$(5.2) \qquad d = R \tan \delta$$

where δ is the angle of dip of the plane.

Planes not passing through the center of the reference sphere cut the surface of the sphere as small circles. Such small circles may also be thought of as the intersection of the surface of the sphere and a cone with vertex at O. One easy way of generating such a cone is by plotting the locus of a line of fixed pitch angle on a series of dipping planes. These also fall on a circular arc (Fig. 5.4a). A second, and closely related property of the stereographic projection is that small circles also plot as circular arcs on the stereogram (that all circles project as circles is proved in §17.4). The distance to the centers of such small circles can be found graphically (Fig. 5.4b), or from

$$(5.3) \qquad d = R/\cos r$$

where r is the pitch angle.

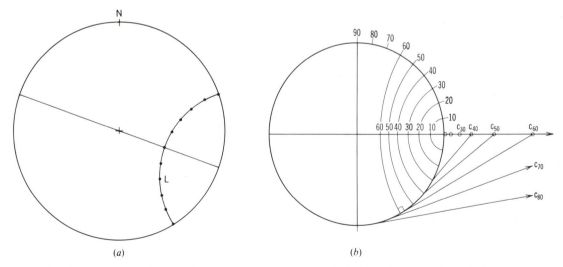

(a)   (b)

FIGURE 5.4 Stereographic projection of a cone: (a) the locus of points representing a line of constant pitch on a series of dipping planes; (b) the construction of a small circular arc.

So far only the lower hemisphere has been used, and there is a reason for this. Consider again the structural line of Fig. 5.1. Usually, there will be no directional properties associated with this line. That is, there is no difference between its upper or lower end. We are, therefore, free to represent it by either end, and we choose the end which intersects the lower hemisphere. This is done in order to maintain the correspondence between the representation of the line and the convention for giving its trend, which was previously established to be the direction associated with it <u>downward</u> inclination. For example, the structural line of Fig. 5.1 plunges toward the SE, and its representation on the stereogram of Fig. 5.2 is in the SE quadrant. For similar reasons, a plane is represented only by the projection of the arc of the great circle in the lower hemisphere in order to maintain correspondence between its representation and the convention for specifying the dip direction. For example, the structural plane of Fig. 5.1 dips toward the SW, and its great circle representation on the stereogram of Fig. 5.3 cuts the SW quadrant.

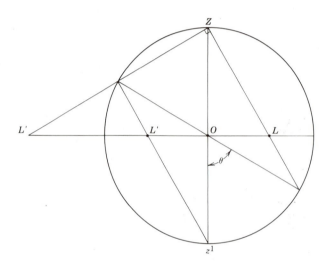

FIGURE 5.5 The construction of the opposite of a point in the lower hemi-
           sphere.

There are, however, some situations where it is necessary to distinguish the two ends of a line. With the same construction used to locate the point representing the intersection of a line with the lower hemisphere, we may also project points of intersection with the upper hemisphere. This is done joining Z and this point of intersection with a line to find the location of the point on the projection plane (Fig. 5.5). As can be seen, the projectors of the two ends of the line form a right angle at Z, so the <u>opposite</u> of any line can be easily and simply found. The distance d of this opposite from the center of the net can also be found from

(5.4) $d = R/\tan(\theta/2)$

This construction is not be used in this chapter, but it will be needed later.

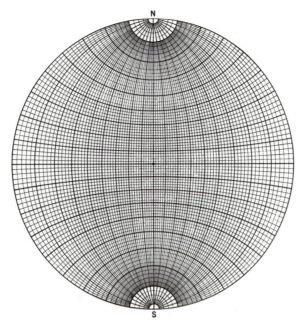

FIGURE 5.6 The Meridional Stereographic or Wulff Net.

By constructing a series of great and small circles on the lower hemisphere, we then have the Meridional Stereographic, or Wulff Net, or simply stereonet, in which the two families of curves are usually drawn every 2° (Fig. 5.6; see the full-scaled version end of the book). This net is a great aid in many constructions. Once the technique is learned, greatest benefit is gained if a net is always at hand. This is easily accomplished if the printed form is permanently mounted on a rigid backing and its surface protected with a sheet of clear plastic. In use, structural data are plotted and problems solved on an overlay sheet of tracing paper. This overlay is affixed to the net by a single pin placed exactly at its center in order to allow the sheet to revolve freely. A small piece of clear plastic tape on the back side of the tracing sheet to reinforce the pin hole will prevent tearing or enlarging of the point of rotation.

5.2 TECHNIQUES OF PLOTTING

When using the stereonet, it is important to visualize the net as if you were looking into a hemispherical-shaped bowl, and to imagine the circular arcs inscribed on its inner surface. Illustrations such as Fig. 5.1b and 5.2b may help in achieving this mental picture. The various structural elements to be plotted can then be visualized as passing through the center of the sphere and intersecting its lower surface. The importance of this of this visualization can not be emphasized enough. Not only does it make the plotting easier, but it also serves as an important check on the proper location of the plots, and on the general correctness of the various manipulations. For example, there are four different positions which satisfy the numerical components of dip and strike, but three of these are incorrect. The practice of visualization will quickly show which of these is the right one. The following examples should be worked through in every detail. Once the plots can be visualized in their three-dimensional setting, a variety of short cuts will suggest themselves by which the whole process can be made more efficient.

PROBLEM

Given a single structural plane with attitude N30°E, 40°E, plot its great circle representation in stereographic projection.

VISUALIZATION

With the net before you (oriented as in Fig. 5.6), hold the left hand, palm upward, over the center of the net with the fingers pointing toward N30°E, and the plane of the hand inclined 40° to the southeast. The plane of the hand in this position can be readily imagined to extend into the lower hemisphere and intersect its surface (Fig. 5.7a). Its trace cuts the southeast quadrant, and this is where the final plot must also be.

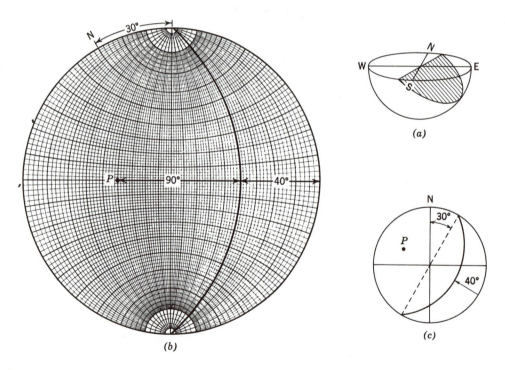

FIGURE 5.7 Stereographic plot of a plane and its pole: (a) perspective view of the inclined plane and its intersection with the lower hemisphere; (b) the position of the overlay and the net for the actual plot of the great circle and its pole, (c) the overlay as it appears after the plot.

PLOTTING A PLANE (Fig. 5.7b)

1. With an overlay in place, make a small tick mark over the north point of the net and label it N. This is the first step in all work on the stereonet, and it should soon become automatic.

2. To locate the line of strike, count off 30° clockwise from north, and make a second small mark on the primitive at this point.

3. No great circle on the net passes through this direction, so it is necessary to revolve the overlay until one does. Therefore, turn the overlay until the strike mark is exactly over the north point of the net, that is, revolve the overlay 30° anticlockwise.

4. To locate the particular great circle representing a plane dipping 40° eastward, count off from the primitive on the right side of the net inward along the east-west diameter of the net. Carefully trace in the arc of this great circle.

5. Revolve the overlay back to the original position and check the result by visualization (Fig. 5.7c). Note that is would have been easy to revolve the overlay in the opposite direction, or to plot from the left, or both, with erroneous results.

In common with most other projections, the dimensions of the object are reduced by one. The hemisphere is reduced to a plane, a plane to a curve and a line to a point. A further advantage of this particular projection is that a plane can be represented as a point, thus reducing the dimensions of the plot by one more. For every plane, there is a unique line normal to the plane, called its pole. To visualize, hold the left hand oriented as before, but with a pencil between the fingers so that it is perpendicular to the plane of the hand. The line of the pencil will pierce the lower hemisphere at a point in the northwest quadrant. This point is everywhere 90° from the plane. Therefore, from the great circle trace count off 90° from the right to left along the same east-west diameter used in Step 4 above, and mark P, the projection of the pole of the structural plane (Fig. 5.7b). Note that we can easily recover the great circle from its pole by reversing this procedure.

The line which is the pole of the plane is projected as a point; this point therefore represents the plane. Any line can be similarly represented by a direct plot, but when a pole is used to represent a plane it is a reciprocal plot.

PROBLEM

Represent a line with attitude 30°, S42°E on the stereonet as a point.

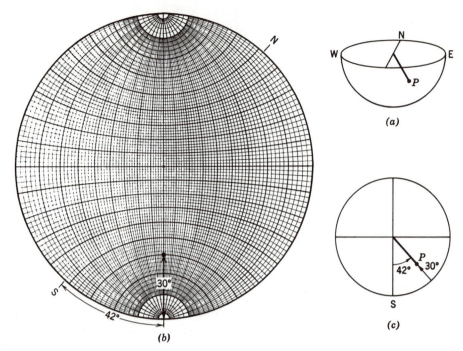

FIGURE 5.8 Stereographic plot of an inclined line: (a) perspective view of the line and its intersection with the lower hemisphere; (b) the position of the overlay and net for the actual plot; (c) the overlay as it appears after the plot.

VISUALIZATION

Hold a pencil in the given orientation over the center of the stereonet and visualize its intersection in the southeast quadrant of the lower hemisphere.

PLOTTING A LINE (Fig. 5.8)

1. With the overlay in place, and the south index marked S, locate a point on the primitive representing the trend of the line by counting 42° anticlockwise from S.

2. Revolve this trend mark to the south point of the net.

3. Count off 30° from the primitive inward along the north-south diameter, and mark the point P.

4. Restore the overlay to the starting position and recheck by visualization.

In this particular exercise, the graduations along the north-south diameter marked by the small circles were used for the first time. However, the trend mark made in Step 2 above could just as easily been move to the east point of the net, and the point plotted by counting off along the east-west diameter. In order to assure yourself that this is so, revolve your plotted point to the east-west diameter and check that the vertical angle measured here is also 30°. Thus we see that in some routines there is a choice of plotting positions. This may confuse some beginners, and it is advisable to stick closely with the listed steps until confidence develops. Once the process becomes familiar, however, it will be found that the use of these alternatives increases the ease and speed of plotting.

Just as structural lines and planes often occur together, so too can they be combined in a single, simple plotting routine.

PROBLEM

Given a plane (N10°W, 45°W) and a line in that plane (31°, N46°W) plot both on the stereonet.

VISUALIZATION

The flattened hand with a pencil held against it in the proper orientation helps one see the three dimensional aspects of the problem more clearly.

PLOTTING A PLANE AND A LINE (Fig. 5.9)

1. To plot the plane:

   a. Mark the direction of the strike on the primitive.

   b. Count off 45° from the left on the east-west diameter.

   c. Draw in the great circle (dashed arc on Fig. 5.9).

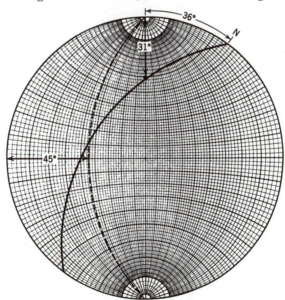

FIGURE 5.9   Stereographic plot of a plane containing a line; the overlay is in position for locating the line on the plane.

2. To plot the line:

    a. With the overlay oriented north, make a tick mark on the primitive indicating the trend of the line.

    b. Revolve this mark to the north point of the net, and count off 31° from this point inward along the north–south diameter and plot the point.

3. Just as the line lies on the plane, so too must the point representing the line lie on the great circle (solid arc of Fig. 5.9). If it does not, then an error has been made, either in the plot or in the original data.

5.3 ATTITUDE PROBLEMS

Problems dealing with the angular relationships of lines and planes solved by orthographic methods in Chapters 1 and 4 can be solved directly on the stereonet. We note here that analytical solutions of these problems can be easily obtained by spherical trigonometry, as outlined in Appendix B. Several additional applications of this method are given here.

PROBLEM

Given an inclined plane (N50°E, 50°SE) find the apparent dip in the N80°E direction.

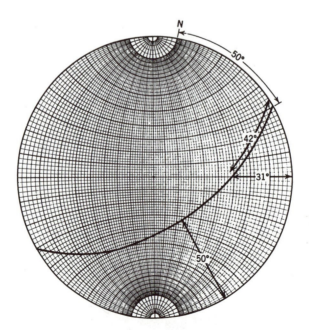

FIGURE 5.10 Apparent dip from true dip and strike; the overlay is in position for measuring the apparent dip angle.

CONSTRUCTION OF APPARENT DIP (Fig. 5.10)

1. To plot the plane:

    a. Revolve the north mark on the overlay 50° anticlockwise.

    b. From the east point of the net, count off 50° along the east–west
       diameter, and trace in the great circular arc.

2. Revolve the overlay back to the starting position, and make a tick mark
   at N80°E on the primitive.

3. Revolve this mark to the east point, and read off the angular position
   where the great circle crosses the east–west diameter.

ANSWER

The apparent dip in the N80°E direction is 31°.

PROBLEM

Given two apparent dips: (1) 28°, N56°W, and (2) 22°, N14°E, find the true
dip and strike.

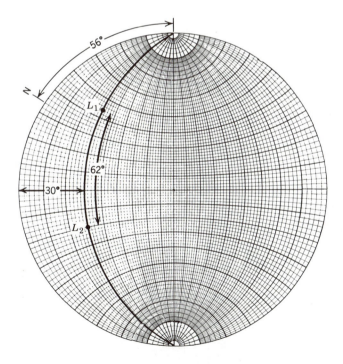

FIGURE 5.11 True dip and strike from two apparent dip angles; the overlay is
in position for locating the great circle common to the two apparent dips.

CONSTRUCTION OF TRUE DIP (Fig. 5.11)

1. Plot the two apparent dip lines:

    a. Line 1: revolve the north mark 45° clockwise and count off 30° from north along the north-south diameter. Mark this L_1.

    b. Line 2: revolve the north mark 14° anticlockwise and count off 22° from north. Mark this L_2.

2. Revolve the overlay until L_1 and L_2 lie on the same great circle. Trace in this arc. The true dip of the plane is then read directly using the EW diameter of the net. Note also that in this position the angle between the two lines in the plane can be read off. The strike of the plane is easily determined by restoring the overlay to the original position and then noting the angle between the point of intersection of the great circular arc at the primitive and north.

ANSWER

The true attitude is N56°E, 30°N. The angle between the two lines is 62°.

PROBLEM

Given two plane whose attitudes are: (1) N40°E, 60°SE, and (2) N60°W, 30°S, find the plunge and trend of the line of intersection.

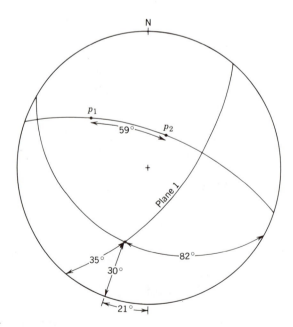

FIGURE 5.12 Two intersecting planes: (a) line of intersection and the angle between the two planes; (b) pitch of the line in each plane.

FINDING THE LINE OF INTERSECTION (Fig. 5.12a)

1. Plot the two great circles representing the planes:

    a. Plane 1: revolve the overlay 50° anticlockwise from north and count off 60° from the east point along the east-west diameter. Trace in the great circle.

    b. Plane 2: revolve the overlay 70° clockwise from north and count off 20° from the west point on the east-west diameter, and complete the great circle.

2. The point of intersection of the two great circular arcs represents the line of intersection of the two planes. Revolve this point to the north-south diameter, and make a trend mark on the primitive. To determine the plunge count off from the primitive along this diameter. For the trend, return to the original position, and count off the angle between the trend mark and the south point.

ANSWER

The attitude of the line of intersection is 30°/201°.

PROBLEM

Find the angle between these two intersecting planes.

DETERMINING THE DIHEDRAL ANGLE (Fig. 5.12b)

1. Plot the poles P_1 and P_2 of the two planes.

2. Trace in the arc of the great circular between these points, and count off the number of degrees between the two points along it.

ANSWER

The dihedral angle is 59°. By convention, the dihedral angle is acute, so if the measured angle is greater than 90°, its supplement is reported.

PROBLEM

Using the same data, determine the pitch of the line of intersection in each of the planes.

MEASURING PITCH (Fig. 5.12b)

1. Revolve the net so that the great circle representing one of the planes containing the line is in the plotting position.

2. Count off the angle between the primitive and the point along this great circle trace. Repeat for the other plane.

ANSWER

The line of intersection has a pitch of 35° SW in Plane 1, and a pitch of 82°E in Plane 2.

5.4 ROTATIONS

In a number of situations it is necessary to geometrically <u>rotate</u> structural elements in space. Every rigid body rotation can be defined by an angle and sense of rotation about a specified axis. The simplest rotation to perform on the stereo-net is when the rotational axis R is vertical. Fig. 5.13a illustrates a plane originally dipping 45° due north rotated 45° clockwise about such an axis to a new orientation (N45°E, 45°SE). Either the great circular arc or the pole of the plane may be rotated with equivalent results. As is evident from this figure, to find the new position one simply revolves the overlay sheet through the required angle but in reverse sense--a now familiar manipulation. Yet there is an important difference. Before, the process of turning the overly about the center of the net was one of

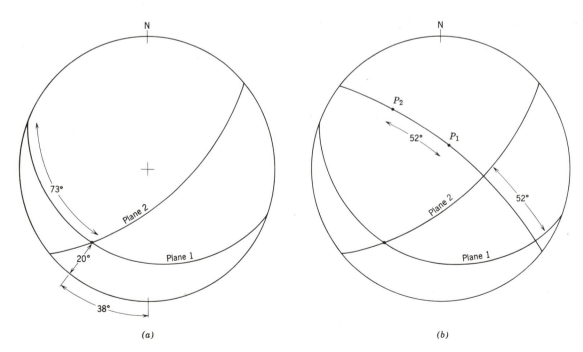

(a)   (b)

FIGURE 5.13 Simple rotations: (a) about a vertical axis; (b) about a horizontal axis.

convenience in plotting and measuring angles, but the overlay always carried with it the N mark, so that the original orientations were never really changed. The term revolve has been used specifically to describe this maneuver. In contrast, after rotation a plane or line has an entirely new orientation relative to a fixed coordinate direction.

Rotation about a horizontal axis can also be performed readily on the stereonet. First, the overlay is revolved so that the horizontal axis R coincides with the north-south diameter of the net. In this position, a rotation moves points along small circle paths. In Fig. 5.13b, a plane dipping 60° is rotated anticlockwise, as viewed from the south end of R. Although either poles or great circles may be rotated, working with poles is much easier.

It is sometimes necessary to rotate a structural element to horizontal and beyond. Fig. 5.14 illustrates how this is handled. A line (30°, N29°E) is rotated anticlockwise 100° about a horizontal axis. After just half the rotation, the point lies on the primitive--the line is horizontal. With an additional increment of rotation, the other end of the line moves into the lower hemisphere at a point diametrically opposite, and then proceeds along the same small circle.

Two methods for rotating about an inclined axis are available. The first depends on previous methods, and consists of first rotating R to a horizontal orientation, performing the require rotation, and then returning R, together with the structural elements, to the original orientation. The second method, more direct but not quite as easy to visualize, is the one we will illustrate here.

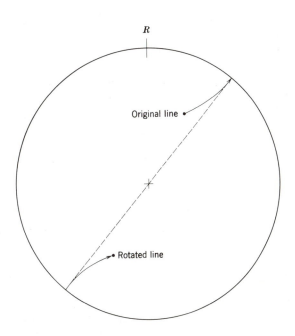

FIGURE 5.14 Rotation of a line to the primitive and beyond.

Rotate the plane (N83°E, 52°S) 80° clockwise, as viewed looking toward the northeast, about an axis plunging 30° toward N42°E.

APPROACH

As a pole rotates about the axis R with constant angle, it will describe a small circle on the surface of the sphere and on the stereonet. While it is useful to sketch in this circle as an aid to visualization (Fig. 5.15), it is not a necessary part of the construction. The basic technique consists of rotating the plane containing both the axis of rotation and the pole of the given plane about the rotational axis.

CONSTRUCTION (Fig. 5.15; after Turner and Weiss, 1963, p. 69)

1. Plot the rotational axis R and the pole P of the plane to be rotated.

2. Trace in the great circle representing the plane normal to R.

3. Draw in the great circle containing P and R; this intersects the great circle of Step 2 at L'. The angle between P and R can be easily read off, and it is 41°

4. As P rotates about R, so too will the line of intersection L rotate in the plane perpendicular to R. To find the final position of this line

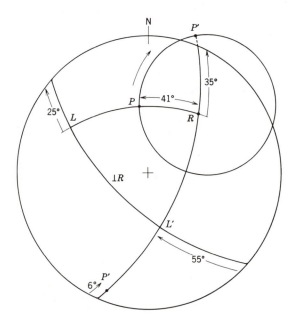

FIGURE 5.15 Rotation of a plane about an inclined axis.

L count off the required 80° from L' going clockwise. In this example, the line passes through the primitive so that the total rotation of 80° is measured in two segments (25°+55°).

5.5 ROTATIONAL PROBLEMS

It is often important to determine the orientation of a given feature as it existed before tilting. Simple examples include the restoration of primary sedimentary features such as current lineation or cross bedding, and the pretilt attitude of structures below an angular unconformity. The most common type of tilting movement occurs during folding, but may also be associated with faulting. For folds, it is a simple matter to unfold the structure and thus restore the beds to their original horizontal position. Provided there is no distortion within the plane of the bedding, the various features are thereby also returned to their original orientation.

To restore the beds of a horizontal fold, the bedding is rotated about an axis parallel to the strike of the beds through an angle equal to the dip.

PROBLEM

An inclined sandstone bed (N70°W, 70°S) contains cross beds (N40°W, 50°SW). Determine the original current direction.

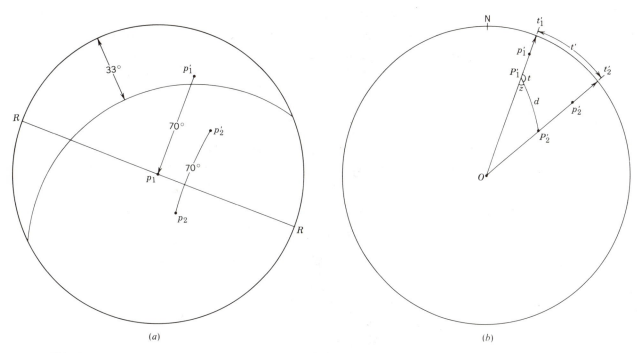

(a)   (b)

FIGURE 5.16 The two tilt problem: (a) stereographic solution; (b) analytical solution (after McIntyre, 1963).

CONSTRUCTION (Fig. 5.16a)

1. Plot the poles of both the tilted sandstone bedding P_1' and the cross beds P_2'.

2. The rotational axis R is parallel to the strike of the sandstone beds, that is, N70°W, which may be added to the stereogram as an aid to visualization. Rotate the overlay so that this line coincides with the north-south diameter of the net.

3. To restore the sandstone bed to horizontality, pole P_1' moves 70° inward along the east-west diameter to the center of the net. At the same time, the pole of the cross beds P_2' moves 70° in the same direction along its small circle to point P_2.

4. From this new pole position, the great circle representing the restored cross bedding can then be drawn in (dashed arc), and its attitude read.

ANSWER

The original attitude of the cross beds was N65°E, 33°N, and the associated current moved toward N25°W. Note that if the uncorrected orientation of the dip direction of the tilted cross beds is assumed to reflect the original current direction, an error of 115° is introduced.

In restoring current directions this procedure may have to be repeated many times, and an analytical solution would be helpful. The only requirement is to find two unknown parts of the spherical triangle whose corners are the points P_1', P_2' and 0 (Fig. 5.16b), and this can be done quite easily. Adapting Eq. B.8, and using the identity $\cos(90°-x) = \sin x$, we have,

$$(5.5) \qquad \cos d = \sin p_1' \sin p_2' + \cos p_1' \cos p_2' \cos t'$$

where p_1' and p_2' are the plunges of the two poles and $t' = t_2' - t_1'$ is the angle between the trends of these two poles. The original plunge of the pole of Plane 2 is then

$$(5.6) \qquad p_2 = 90° - d$$

where d is the angle between the two poles P_1' and P_2'. Using Eq. B.9,

$$(5.7) \qquad \cos z = \frac{\sin p_2' - \sin p_1' \cos d}{\cos p_1' \sin d}$$

where z is the supplement of the angle giving the trend of P_2 with respect to P_1.

Then

(5.8) $$t_2 = t'_1 + (180° - z)$$

where t_2 is original trend of the pole of Plane 2. Using the appropriate values in Eq. 5.5, $d = 32.5°$, and the original plunge $p_2 = 90° - 32.5 = 57.5°$. With this and the other given values in Eq. 5.7, $z = 45.4°$, and the trend of the restored pole of the cross beds is

$$t_2 = 20° + (180° - 45.4°) = 154.6°$$

If the plane returned to horizontality by this method was overturned, then the resulting orientation of the second plane will be incorrect. There are two ways of proceeding in this case.

METHOD I (Fig. 5.17a)

Using the same general set-up as before, rotate the pole P_1 in the opposite direction: first 20° to the primitive, and then using its opposite 90° back to the center of the net. The P_2 pole moves along its small circle 110°.

METHOD II (Fig. 5.17b)

Start with the results based on Fig. 5.17a; then continue the rotation with an additional 180° so that the opposite of P_1 is at the center of the net. The point P_2 now bears a simple relation to its position found by the

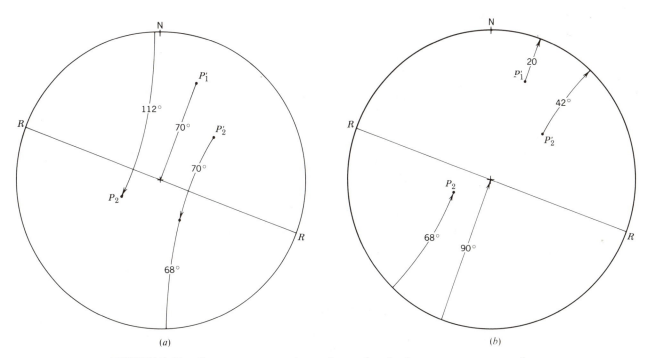

FIGURE 5.17 The construction when the beds are overturned.

earlier method. If the diameter normal to R is extended, then the correct position of P_2 makes the same angle with this diameter as before, but on the opposite side. This allows a simple correction to be applied: the plunge is the same, but the trend $t_2 = 157° + 2(53°) = 263°$.

ANSWER

The pole of the cross beds is now 57°/263°, which corresponds to a pretilt attitude of N17°W, 33°N.

In cases where the folds plunge, the tilted beds can be considered to have two rotational axes: one of them is the fold axis, and the other is a horizontal axis perpendicular to the fold axis. Reversing the angular rotations on both these two axes then unrolls the folded beds to their original orientation.

PROBLEM

On an anticline plunging 30° due north, beds on the east limb (N19°W, 60°E) contain sole markings which trend due east. Determine the pre-folding orientation of this sedimentary lineation.

CONSTRUCTION (Fig. 5.18; after Ramsay, 1961)

1. Plot the elements of the problem: Plane 1 = bedding plane, ℓ'' = lineation within the twice tilted bedding, and F = fold axis.

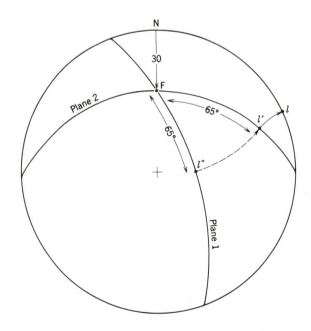

FIGURE 5.18   Unfolding of the beds of a plunging fold.

2. The rotation of Plane 1 and ℓ'' about the inclined axis F = R could be constructed as in Fig. 5.13b, but there is a simpler approach. If the plane of the bedding is unrolled about the plunging fold axis, the result will be plane dipping 30° due north. During this rotation, the angle between ℓ'' and F will remain constant. Thus after removing the first tilt, Plane 2 and its lineation ℓ' can be plotted directly.

3. During the second rotation the plane rotates about its line of strike, and the lineation moves along a small circle to its original position ℓ on the primitive.

ANSWER

The original trend of the sedimentary lineation was N65°E. Again, if this correction is ignored, a considerable error may result.

5.6 DIP AND STRIKE ERRORS

As we saw in §1.7 (especially Fig. 1.13), an error in placing the compass on a structural plane generally results in a strike error. Because of this, a series of repeated measurements would be expected to show a scatter which could be depicted by plotting the pole representing each measurement. There is another factor which also contributes to this scatter--local departures of the attitude of the plane from some mean value. Assuming that these measured poles are symmetrically distributed about a mean pole, and that the probability of finding a pole of given angular distance from this mean pole decreases with increased angular distance, a simple expression for the maximum strike error can be obtained.

By this assumption, the actual position of the pole must lie within a cone whose angular radius is the maximum observer error ε_0, and whose axis is the mean pole. Fig. 5.19a shows the vertical diametral plane of a sphere and the intersection with the lower hemisphere of the structural plane s with dip δ, together with its pole P and the error cone of semiapical angle ε_0. In Fig. 5.19b the intersection of this cone with the projection plane is shown as a small circle (the method of constructing such small circles is given in §17.4). If the maximum error in determining the orientation of P is ε_0, then the maximum error in the plunge of P and the dip of s is also ε_0. The maximum strike error in the determination of s is ε_s, which is also the error in the trend of P. This error is represented by the angle COF between the radius through the center of the small circle and the radius tangent to it in Fig. 5.19b. Then

$$\sin \varepsilon_s = CF/OC$$

Expressions for the distance OC and the radius CF are given by Eqs. 17.1; using these, together with $\varepsilon = \phi$ and $\cos p = \sin \delta$, we obtain

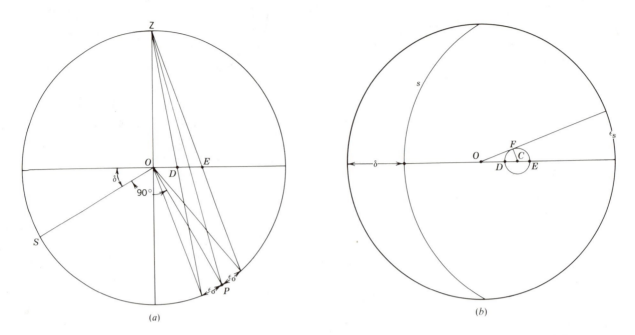

FIGURE 5.19 Cone of maximum error in determining the pole of a plane: (a) a vertical diametral section of the sphere containing the trace of the dipping plane s, its pole P, and the maximum error cone with semiapical angle ε_0; (b) the error cone projected as a small circle, with the maximum strike error ε_s being the angle between the two radii tangent to this circle (after Cruden and Charlesworth, 1976, p. 977).

$$(5.9) \qquad \sin \varepsilon_s = \frac{\sin \varepsilon_0}{\sin \delta}$$

A graph of this equation, first obtained by Pronin (1949; quoted in Vistelius, 1966, p. 51), is shown in Fig. 5.20. Cruden and Charlesworth (1976) tested this result against field measurements. Making the reasonable assumption that the maximum angle of error ε_0 is small, they found that Eq. 5.9 gave better results than the Müller hypothesis (see Eq. 1.9), and can be considered an adequate description of the data. They also noted that lithology, and presumably the character of the surfaces of the structural planes, effected the scatter of the dip measurements more than the strike measurements. They also found that for 10 measurements, at 95% confidence, the values of the total error in measuring a bedding pole ranged from ±3.4° to ±9.5°.

5.7 INTERSECTION ERRORS

Because errors are inevitably introduced when measuring the attitude of planes, it follows that when two measured planes intersect the attitude determined for the line of intersection will also be in error.

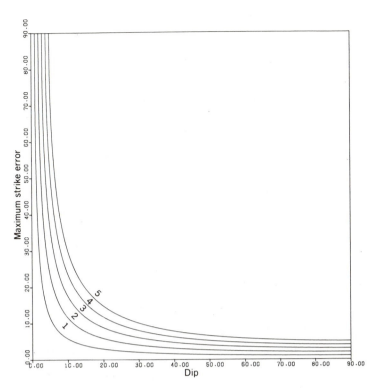

FIGURE 5.20 Maximum strike error as a function of dip for errors $\varepsilon_o = 1° - 5°$.

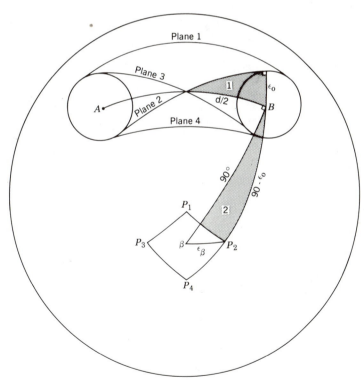

FIGURE 5.21 Error in the orientation of the line of intersection: (a) stereographic determination of the point of intersection β; (b) construction of the area of maximum error of the poles of the planes, and the area of maximum error of β (in part after Ramsay, 1967, p. 13).

The basic construction of an intersection is illustrated in Fig. 5.21a. However, the actual positions of the poles of the two planes lie within cones whose angular radius is the maximum observer error ε_o. As before, these cones project as small error circles. Correspondingly, the line of intersection actually lies within an area in the vicinity of the constructed intersection β, and we wish to find an expression for the maximum error ε_β associated with this location.

In Fig. 5.21b, the small error circles about the two measured poles are shown. For these, there will be four limiting cases represented by the four great circles tangent to each of the two circles and these are labeled Planes 1-4. The poles of each of these planes P_1-P_4 bound the area within which the β-axis actually lies, and we wish to determine its size. To do this, we construct two right spherical triangles. For Triangle 1, using the methods spherical trigonometry (see §B.2),

$$(5.10) \qquad \cos B_1 = \tan \varepsilon_o / \tan(d/2)$$

where B_1 is the angle of Triangle 1 associated with the pole at point B, and $d/2$ is half the dihedral angle between the two planes. Similarly, for Triangle 2,

$$(5.11) \qquad \cos \varepsilon_\beta = \cos \varepsilon_o \cos B_2$$

where B_2 is the angle at B associated with Triangle 2. Because the two angles at B form a right angle, $\sin B_1 = \cos B_2$, and Eq. 5.11 can be written as

$$(5.12) \qquad \sin B_1 = \cos \varepsilon_\beta / \cos \varepsilon_o$$

We now combime Eqs. 5.11 and 5.12 to eliminate B_1. First square both equations and add the resulting expressions. Then, using the identity $\sin^2 x + \cos^2 x = 1$, and rearranging slightly,

$$\frac{\tan^2\varepsilon_o}{\tan^2(d/2)} + \frac{\cos^2\varepsilon_\beta}{\cos^2\varepsilon_o} = 1$$

or, with some further rearranging,

$$\cos^2\varepsilon_\beta = \cos^2\varepsilon_o - \frac{(\cos \varepsilon_o \tan \varepsilon_o)^2}{\tan^2(d/2)}$$

With the identities $\sin x = \cos x \tan x$ and $\cos^2 x = 1 - \sin^2 x$, this becomes,

$$\sin^2\varepsilon_\beta = \sin^2\varepsilon_o \left(1 + \frac{1}{\tan^2(d/2)}\right)$$

Finally, with the identities $1/\tan^2x = \cos^2x$ and $1+\cot^2x = \cosec^2x = 1/\sin^2x$, and taking the square root, we have,

(5.13)
$$\sin \epsilon_\beta = \frac{\sin \epsilon_o}{\sin(d/2)}$$

A graph of this equation for values of the maximum error ϵ_o from 1°-5° is given in Fig. 5.22, and as can be seen, when the dihedral angle between the two planes is small, the maximum error in the location of β may be large. If the line of intersection between two gently dipping planes is desired, a special effort will be required to determine the attitudes of the planes as accurately as possible, or to measure the attitude of the line directly.

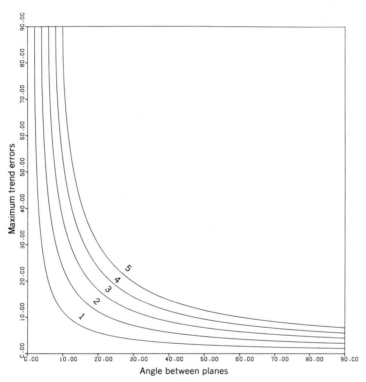

FIGURE 5.22 The maximum β error as a function of the dihedral angle for maximum errors ϵ_o = 1°-5° (see also Ramsay, 1967, p. 14).

EXERCISES

1. Construct a stereogram 15 cm in diameter showing the 45° small and great circles graphically or with the aid of Eqs. 5.2 and 5.3. Compare your result with the printed Wulff net.

2. Repeat the dip and strike problems of exercise 1 in Chapter 1 and the line of intersection problem of Chapter 4. Compare the methods of orthographic and stereographic projections for speed and accuracy.

3. A plane contains two lines: (1) 30°/320°, (2) 20°/020°. What is the attitude of the plane, and what is the angle between the two lines measured in the plane?

4. The beds below an angular unconformity have an attitude of N30°W, 40°W. The strata above the unconformity have an attitude of N20°E, 30°E. What was the attitude of the lower beds before the tilting of the younger bed occured?

5. An anticlinal fold axis plunges 24°/040°. On the east limb where the beds have an attitude of N5°W, 32°E, the crest line of current ripple marks pitches 70°N in the plane of the bedding. What was the pretilt orientation of these marks? Compare your result with the assumption that the tilted orientation of the lineation adequately represents the original direction.

6. An anticline plunges 50°/025°. The eastern limb is overturned, and at one point the attitude is N45°E, 50°W. At this same locality a sedimentary lineation plunges due west. What was the orientation of the lineation before folding.

7. Rotate the plane whose attitude is N10°E, 30°E, fifty degrees anticlockwise as viewed down the plunge of an axis whose attitude is 30°/340° in two ways: (1) as a series of steps involving rotation of the axis to the primative, rotating the line about the now horizontal axis, and then returning the axis to its original orientation; (2) as a single rotation about the inclined axis.

8. Two intersection planes have attitudes N5°E, 15°W and N15°E, 10°W. Determine the orientation of the line of intersection and the angle between the two planes. If the maximum operator error in measuring attitude is 2°, what is the maximum strike error for each of the planes, and what is the maximum β error for the line of intersection? How do these errors effect the calculated dihedral angle?

6
Faults

6.1 DEFINITIONS

Fault: a fracture surface along which appreciable displacement has taken place.

Fault zone: a tabular zone containing a number of parallel or anastomosing faults.

Shear zone: a zone across which two blocks have been displaced in faultlike manner, but without development of visible faults.

Slip: the relative movement on a fault, measured from one block to the other, as the displacement of formerly adjacent points (Fig. 6.1a). Slip is often represented by a vector--the relative slip vector--giving the direction and amount of the displacement of one block relative to the other.

Separation: the distance between the two traces of a displaced marker plane measured in a specified direction. Dip separation is measured parallel to the dip of the fault, and strike separation is measured parallel to the strike of the fault (Fig. 6.1b). Stratigraphic separation is measured perpendicular to the displaced planes, not in the plane of the fault.

Hanging wall block: the body of rock immediately above a nonvertical fault; the hanging wall itself is the bounding surface of this block.

Footwall block: the body of rock immediately below a nonvertical fault.

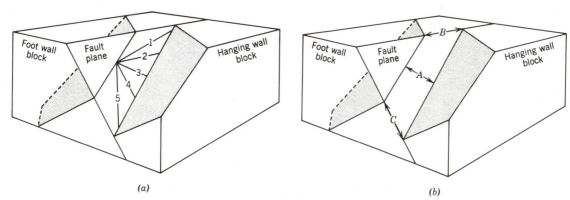

FIGURE 6.1 Fault displacing a marker plane (after Hill, 1959, 1963): (a) five possible slip vectors; (b) A = separation, B = strike separation; C = dip separation.

6.2 FAULT CLASSIFICATION

There are several geometrical approaches to the classification of faults. For example, there are descriptive schemes based on the relationship of the faults to other structures (longitudinal, transverse, or bedding-plane faults), and on patterns of groups of faults (radial, parallel, or en echelon faults).

The most important aspect of fault geometry is the relative displacement along it. Slip is the measure of this displacement, but unfortunately it is only occasionally ascertainable. Separation is more commonly measurable; in fact, observed separation is commonly the evidence for the existence of a fault. However, a clear distinction must be maintained between these two terms because an observed separation may result from many possible orientations of the slip vector (Fig. 6.1b). In order to emphasize this important distinction, two parallel classifications have evolved; one based on slip and the other on separation (Hill, 1959, 1963). However, the separation-based scheme is not really a classification of displacement at all. This will be clear when it is realized that for a given slip, the sense of separation depends entirely on the attitude of the disrupted marker plane (Fig. 6.2). Thus any classification based on separation is inherently misleading (see also Gill, 1971). If the separation is to be described, it should be spelled out in terms of the attitude of the disrupted plane and the direction in which the separation is measured. The sense of the separation can be described by the use of several commonly used terms (see Table 6.1).

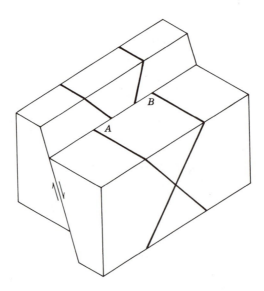

FIGURE 6.2   Fault with opposed senses of strike separation.

TABLE 6.1   Terms describing the sense of separation

Strike separation

    Left separation   Standing on the trace of a displaced marker on one
                          block and facing the fault, the trace of the marker
                          is found to the left on the opposite block (Fig. 6.1
                          shows left separation).

    Right separation   The trace of the marker is found to the right across
                          the fault.

Dip separation

    Normal separation   Viewing the fault in a vertical section, the trace
                          of the marker in the hanging wall is found below the
                          trace of the same marker in the footwall (Fig. 6.1
                          shows normal separation).

    Reverse separation   The trace of the marker in the hanging wall is found
                          above the trace in the footwall.

   Mackin's down-structure method of viewing a geologic map allows one to see directly the separation. In Fig. 6.3, three faults displace inclined strata.
Adopting a down-dip view of the beds reveals their true thickness, and also the
stratigraphic separation. Fault I, which is vertical, has a stratigraphic separation equal to the thickness of Bed 2. As an aid to this visualization, it is useful, especially for beginners, to hold the two flattened hands, fingers in the dip
direction of the beds, to represent the two sides of the fault. Moving the hands

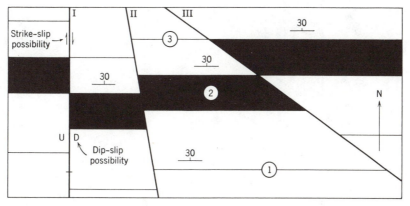

FIGURE 6.3   Diagramatic map illustrating the down-structure method of viewing
               map as applied to faults (after Mackin, 1950).

parallel to the plane of the fault then reproduces the observed separation for the various slip possibilities.

For example, it is easy to imagine a slip which would account directly for the stratigraphic separation--the left (west) block moved relatively upward and north- ward at an angle of 60°. It can then also be seen that many other slips are also possible. The west block could have moved relatively northward in pure strike slip, or it could have moved upward in pure dip slip, or any combinations of these two components.

For Faults II and III, the stratigraphic separation can also be seen in a down-dip view. In this view, Faults II and III may appear to be high- and low-angle reverse faults, but both are actually vertical.

A view down the dip of the displaced marker planes allows the correct sense of dip and strike separation to be determined directly in many cases. However, it fails in others, and this leads to a modification of Mackin's simple rule. Rather than looking down the dip of the faulted beds, one should look directly down the plunge of the line of intersection of the fault and the displaced plane (Threet, 1973b). In fact, because the strike of Fault I and the strike of the displaced layer are perpendicular, this view and the down-dip view of the faulted strata coin- cide.

For Fault II, this line plunges 28° and for Fault III, it plunges 18°. For each fault, a line of sight along these directions give views which are similar to those obtained in the down-dip direction.

It is when the fault plane is inclined that difficulties may be encountered. Fig. 6.4 shows two maps of inclined faults each displacing a single marker plane.

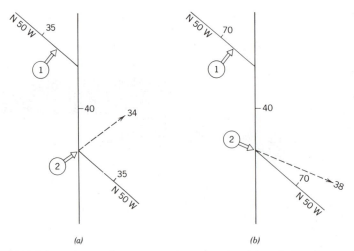

FIGURE 6.4 Visualization of separation on inclined faults: (a) view along
line of intersection of fault and displaced plane shows right and
normal separation; (b) view along line intersection shows right
and reverse separation.

In both cases a down-dip view of the marker planes leads to the conclusion that both faults exhibit right and normal separations. In order to obtain the modified down-structure view, it is necessary to determine the attitude of the line of intersection of each fault and its displaced plane; these are also shown in the figure. In the case of the moderately dipping marker plane (Fig. 6.4a), the down-plunge view is essentially the same as that obtained in the down-dip view, and confirms that the separations are right and normal. In the case of the steeply dipping marker plane (Fig. 6.4b), however, a down-plunge view gives quite a different picture. The strike separation is the same, but the dip separation is now seen to be reverse. Again, these relationships may be perceived more clearly if the displaced plane is modeled with the hands while viewing in the down-plunge direction.

Clearly it is <u>slip</u>, not separation, which most fundamentally describes the displacement on the fault, and a classification based on it is the only meaningful way of categorizing this information.

For completeness, it is also necessary to distinguish between <u>translational</u> faults, where the magnitude and orientation of the slip vector is constant and <u>rotational</u> faults, where the magnitude or orientation of the slip vector is variable (Fig. 6.5). Table 6.2 gives the classification of translational faults, and the associated terminology for describing the sense of slip.

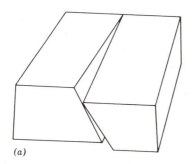
(a)

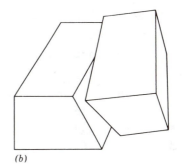
(b)

FIGURE 6.5 Rotation on planar faults (from Donath, 1962): (a) hinge fault; (b) pivotal fault.

TABLE 6.2 Classification of translational faults*

Dip slip	Normal-slip fault (hanging wall relatively downward). Reverse-slip fault (hanging wall upward).
Strike slip	Right-slip fault (opposite block to the right). Left-slip fault (opposite block to the left).
Oblique-slip	Dip- and strike-slip terms in combinations, for example, normal right-slip fault.

*Horizontal and vertical faults require special treatment.

During faulting, the displaced beds or the fault or both may rotate. In the case of a fixed, planar fault the relative rotational movement may be described by the terms given in Table 6.3. If the fault surface is curved, the beds in at least one of the blocks must rotate; this case, as well the case where the fault itself rotates are discussed in greater detail in later sections.

TABLE 6.3   A classification of rotation on planar faults

Clockwise rotation	Clockwise rotational fault (opposite block rotated relatively clockwise.
Anticlockwise rotation	Anticlockwise rotational fault (opposite block rotated anticlockwise).

The attitude of the fault plane is a second important factor in describing the relatively displacement of the two fault-bounded blocks. For example, the motion on a reverse-slip fault with a dip of 20° differs considerably from that on a reverse-slip fault with a dip of 70°. It is, therefore, necessary to introduce dip into the classification of fault displacements.

Rickard (1972) has suggested a useful way of combining the dip angle with the pitch of the net slip on a triangular diagram, which lead to a graphical classification scheme. Each possible dip-pitch pair is assigned a unique index; for example, for a fault dipping 60° on which the net slip pitches 80° the index symbol would be $D_{60}R_{80}$ (R, for rake, to avoid confusion with plunge). This is then represented by a point on the triangular grid (see point x, Fig. 6.6a). Four triangles are necessary to represent normal and reverse, and right and left slips (Fig. 6.6b). In this way all possible translational faults can be plotted. Further, the main catagories of faults can be added to the full diagram as an aid to classification. Rickard also suggest that the special cases of dip and strike slip be restricted to faults with pitch angles of 80°-90° and 0°-10°, respectively.

By including the dip of the fault plane in the classification, several additional catagories are needed. A reverse fault which dips less than 45° is called a thrust. The term overthrust is commonly used for a low angle thrust (dip = 0°-10°). The prefix over is used to emphasize the essentially horizontal character of the relative displacement of the upper part of the structure, and not to imply a direction of absolute movement.

There is no widely accepted term for a normal fault with dip less than 45°. Some normal faults have a pronounced curvature with the result that the angle of dip at depth may be less than 45°; these are called listric (or shovel-shaped) normal faults (see Fig. 6.23).

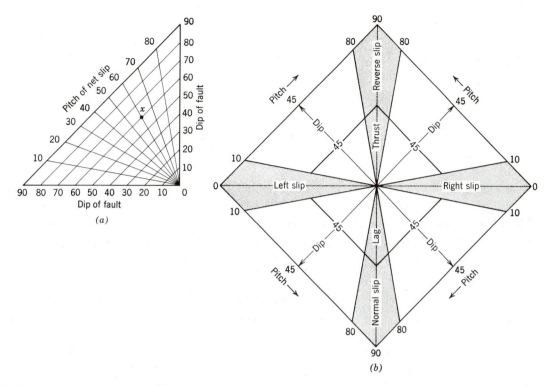

FIGURE 6.6   Classification of translational fault: (a) dip–pitch grid; (b) fault categories; the special cases of dip and strike slip faults are shown shaded, while oblique-slip categories are blank (after Rickard, 1972).

Frequently, it is not possible to determine the slip direction with enough precision to make Rickard's scheme practical.   Further, oblique-slip faults are somewhat neglected.   Threet (1973a) has suggested a simple alternative scheme for naming faults which remedies this flaw.   Fig. 6.7 is a view of the fault plane on the footwall side.   The slip vector which defines the relative displacement of the hanging wall block will fall into one of the eight sectors, and this determines the slip-based name.

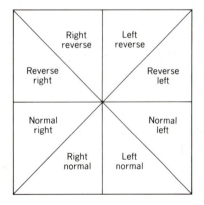

FIGURE 6.7   A simplified classification of faults (after Threet, 1973a).

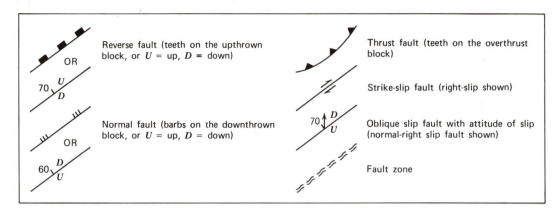

FIGURE 6.8   Suggested map symbols for faults (in part after Hill, 1963). Note that a heavy line is used for the trace of the fault; alternatively, a wiggly line is used for faults.

To depict faults and their slip directions on geologic maps special symbols are used. Fig. 6.8 shows a number of such map symbols for the combinations of slip and dip which cover most types of faults.

6.3   TRANSLATIONAL FAULTS

It is sometimes possible to find along faults polished rock surfaces, called slickensides, on which there are scratches or grooves. These linear features, which Fleuty (1975) suggests be called slickenlines, give the direction of slip, although if there has been several differently oriented episodes of movement, only the last will be recorded. There are also sometimes minute step-like features on these polished planes, called slickensteps by Fleuty, which are approximately perpendicular to the slickenlines. The direction in which these steps face has been used as an indication of the sense of slip, but counter examples are known.

These terms should not be applied to the fiberous coatings found on some movement planes. These are formed by deposition, commonly of calcite or quartz, from aqueous solutions during sliding, and can also be used to identify the latest slip direction, and in some cases also the sense of slip.

The determination of both the orientation and amount of the slip requires the recognition of two originally adjacent points on the fault surface. Strictly, geologic examples of such points are nonexistent, and, therefore, other features from which points can be derived must be found. In practice, lines may be recognized in several situations. These lines intersect or pierce the fault plane to give the required points; these are called piercing points (Fig. 6.9). Such lines may be represented by physical structures:

1. Intersecting planes (e.g., intersection of dike and marker bed).
2. Trace of one plane on another (e.g., beds truncated against an unconformity).

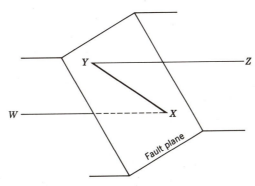

FIGURE 6.9 Piercing points: X = piercing point of line WX, and Y = piercing point of line YZ; the segment XY represents the slip (after Crowell, 1959).

3. Linear geologic bodies (shoe-string sands, linear ore veins, etc.).
4. Stratigraphic lines (pinch-out lines, ancient shore lines, etc.).

Or the required lines may be constructed from field data:

1. Isopachous or isochore lines.
2. Lithofacies lines.
3. Hinge lines of folds.

In order to introduce the methods used in the geometric analysis of fault displacement, we start with a situation where the direction and amount of the slip on a fault is already known.

PROBLEM

A normal fault displaces a marker bed which is exposed only in the hanging wall side (Fig. 6.10a). If the net slip is 100 m, find the location of the bed on the footwall.

APPROACH

As always, it is advisable to thoroughly visualize the elements of the problem. First, turn the map so that a down-dip view is obtained (View 1 of Fig. 6.10b). At the same time hold the hands together to represent the plane of the dipping bed. Now move the left hand upward in the direction of the known dip slip. It will then be clear that the displaced part of the bed is to be found to the east of its position on the hanging wall block, that it, it must show right separation. The more accurate view down the plunging line of the fault-bed intersection (View 2 of Fig. 6.10b) does not change this conclusion. To determine the location of the displaced bed, a direct view of the fault plane is required, and usually a view of the hanging wall is chosen.

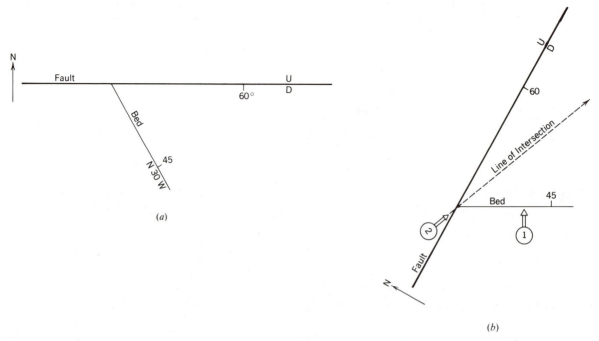

FIGURE 6.10 Locating a displaced plane across a fault: (a) map with the elements of the problem; (b) 1 = down-dip view, and 2 = down-plunge view of the displaced plane.

CONSTRUCTION

1. Plot both the fault and displaced bed on the stereonet as great circles, and determine the pitch of the trace of the bed in the plane of the fault (= 38° in Fig. 6.11a).

2. With this result, plot the line of intersection directly on the plane of the fault using a direct view. The slip is scaled off and the the separation measured.

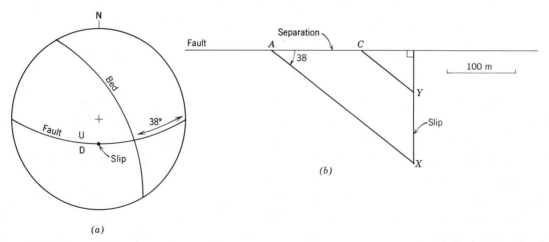

FIGURE 6.11 Solution of the problem of fault with known slip: (a) stereographic construction of the line of intersection; (b) direct view of the fault to find of the strike separation.

ANSWER

The measured separation is 128 m. With this, the location of the bed on the footwall side can be found (see Fig. 6.11b).

This problem is not very realistic because the direction and amount of slip was given. More often it is not known, and must be found from an analysis of displaced planes, or other related features.

PROBLEM

Given a fault dipping 55° due south, and displacing a bed and a vein as shown in Fig. 6.12a, determine the amount and orientation of the slip, and classify the fault accordingly.

APPROACH

Visualize! First, estimate the orientation of the lines of intersection of the fault and the displaced planes. Then turn the map so that a down-plunge view of this line is obtained. With the aid of the hands, note that both normal slip and left slip satisfy the stratigraphic separation of the vein. Repeat for the marker bed. The opposed senses of strike separation rules out the possibility of a sigificant strike-slip

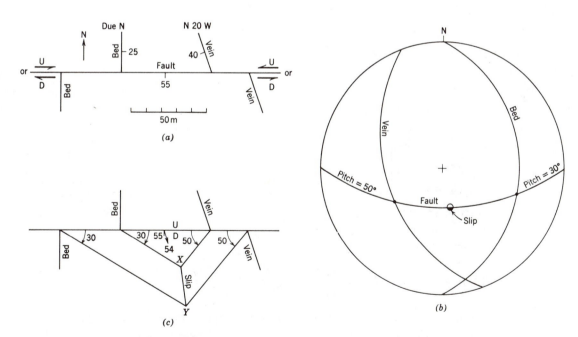

FIGURE 6.12 Slip from two displaced planes: (a) map of the fault and the planes; (b) sterogram with the pitch of a bed and a vein on the fault plane; the slip direction is shown by a small split circle with the dark half indicating the down side; (c) direct view of the fault giving the slip.

displacement, and a strong component of dip slip is indicated. Again, the determination of the slip requires a direct view of the fault plane.

CONSTRUCTION

1. A stereogram of the various planar elements yields the pitch of the bed (30°E) and of the vein (50°W) on the fault (Fig. 6.12b).

2. Using these pitch angles, the traces of the displaced planes on the fault are then drawn, giving the piercing point from the footwall side X, and from the hanging wall side Y (Fig. 6.12c).

3. The line XY is the slip. The pitch of the slip line is measured from the trace of the fault.

ANSWER

The net slip is 25 m, and the pitch of the slip line on the fault is 82°E. Plotting the slip line on the stereogram allows its plunge and trend to be read; it is 54°, S13°E. With this, the complete attitude symbol can be added to the map. It is a normal-slip fault, with a small left-slip component.

6.4 FAULT TERMINATIONS

The slip which accompanies the first increment of movement on a new fault, or the renewed slip on an established fault has only finite extent. The region of slip can be pictured as an area inside a closed curve drawn on the fault surface which may or may not intersect the earth's surface (Fig. 6.13a).

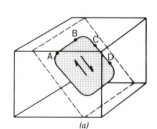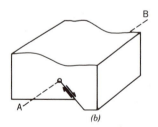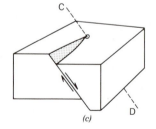

FIGURE 6.13 Slip area (from Hobbs and others, 1976, p. 310).

There is a geometric problem in accomodating the change from slip inside to no-slip outside this boundary. For small slips over large areas the accomodation may be by suitably distributed small ductile strain (Figs. 6.13b,c). Another way of accomodating misfits is by developing branch faults near the terminations (Fig. 6.14a). For large displacements, subsidiary structures developed by folding or secondary or faulting or both become necessary. Figs. 6.14b and 6.14c illustrate the termination of strike-slip faults by two types of transverse structures (note, however, that in such cases it maynot be clear which structure developed to accomodate the movement on the other).

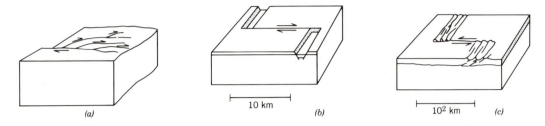

FIGURE 6.14 Fault terminations (from Hobbs and others, 1976, p. 311).

The mode of termination gives important information on the nature of the mechanism of faulting, including the slip rate (Williams and Chapman, 1983) and should be closely observed if at all possible. Unfortunately, the ends of faults are not often seen in the field. Sometimes the trace of the fault extends well beyond the particular area being mapped. In other cases, the exposures, lithologic contrasts or geomorphic expressions are insufficient to permit the faults to be traced onward. This presents the field geologist with a dilemma (Gage, 1979). On one hand, it is not desirable to show a fault where it can not be found. On the other hand, it is equally undesirable to suggest that the fault termination was observed at the place where the map trace ends. One way of resolving the matter is with the use of question marks in the general locality where the fault "disappears", together with an appropriate comment in the legend.

6.5 ROTATIONAL FAULTS

For the case of a fixed planar fault, the rotational component can be determined if two differently oriented planes have been displaced.

PROBLEM

Determine the angle and sense of rotation for the planar fault shown in the map of Fig. 6.15a.

METHOD

The traces of the displaced plane from the footwall and hanging wall sides can be plotted from pitch angles determined on the stereonet, or more simply, the difference in pitch angles may be read directly.

ANSWER

It is a clockwise rotational fault, and the rotational angle is 30°.

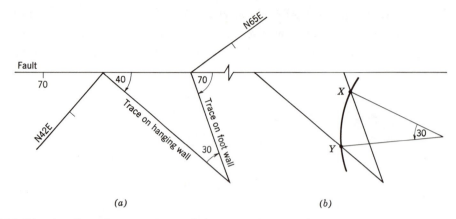

FIGURE 6.15 Angle of rotation: (a) map view with superimposed construction of the traces of a single displaced plane; (b) center of rotation constructed from the displaced points X and Y.

If two piercing points can also be located (points X and Y of Fig. 6.15b), a center of rotation can also be found. It lies at the apex of the isosceles triangle whose base in the straight line XY, and opposite angle is the rotational angle. If rotation and translation are combined, this construction will always result in the location of a center which would account for the motion by rotation alone. Thus if a rotational component is present, and only two piercing point are known, the slip is indeterminate. Despite this limitation, the angle of rotation and two originally adjacent points will still allow the position and location of all other structures to be predicted from one block to the other. If sets of point can be found at two different locations, and the angle of rotation is constant, the rotational centers will coincide in the case of rotation only, but will differ if a component of translation is involved. It should then be possible to separate the two rotational and translational components.

6.6 OVERTHRUSTS

Low angle thrust have a number of special and important features which deserve additional treatment.

The terms hanging and foot wall apply to such thrusts, but two additional terms are in common usage. Autochthon denotes the rock unit not separated or displaced from the site of its origin, and still rooted to its basement. The allochthon is the rock unit transported from its original site. The mechanism of transport may be by thrusting or overfolding or both. Use of allochthon most often implies considerable displacement with the result that the rocks which compose it differ greatly in lithologic character or structure from the underlying autochthon.

Because of the low angle, the outcrop pattern of the thrust plane is strongly influenced by erosion. One result is that an erosional outlier, called a <u>klippe</u>, may be produced. Similarly, erosion may also expose the autochthon in a <u>window</u> or <u>fenster</u>. These two features are illustrated in Fig. 6.16.

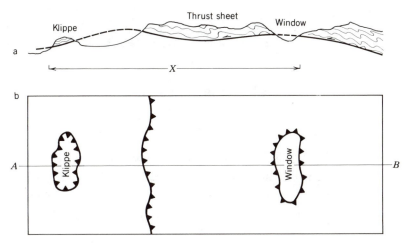

FIGURE 6.16 A klippe and a window (in part after Ramsay, 1969, p. 63): (b) cross section showing two features; (b) map showing the two erosional features depicted by a closed curve.

Overthrusts are found in a wide variety of settings and involve a variety of rocks. One particularly important habitat of overthrusts is in the relatively undeformed sedimentary sequence of the external parts of mountain belts. Weak units within such sequences, especially shales or evaporites, are often utilized as movement planes or zones for some distance by the thrust. The rocks above this detachment horizon may deform by folding or faulting, or both.

Commonly, the thrust remains in the weak material for some distance, and then cuts abruptly upward at a steeper angle, either to the surface or to a higher weak zone. The result of this behavior is that the thrust plane has a series of steps, the parts of which are called <u>flats</u> and <u>ramps</u> (Fig. 6.17).

There may also be a number of major or minor thrusts within the allochthon. Such additional faults commonly display either one of two distinct styles. One of these involves the presence of minor thrusts which root in the main basal thrust and curve upward to the surface; these are called <u>listric</u> thrusts. If a number of such

FIGURE 6.17 A schematic section of a thrust sheet showing a ramp separating two flats.

thrusts are present, the effect is to break the allochthon into a series slabs, and this result is described as imbricate or schuppen structure (Fig. 6.18).

FIGURE 6.18 Imbricate structure.

A body of rock within the allochthon bounded on all sides by minor thrusts is termed a horse. A duplex is an imbricate family of horses--"a herd of horses" (Boyer and Elliott, 1982, p. 1202). The faults bounding a duplex are called roof and floor thrusts (see Fig. 6.21). The lowermost thrust is commonly referred to as the sole thrust.

The determination of the direction and amount of slip for an overthrust sheet is difficult. In principle, the methods which are used for other types of faults are also applicable to overthrusts. For example, the slip could be obtained from two originally adjacent points. Unfortunately, such a situation where this method could be used has apparently not yet been found.

As with other types of faults, if the slip direction can be determined, then the problem of estimating the slip becomes tractable. A number of independent ways have been used in order to gain a handle on this slip direction.

One guide is that thrust sheets move up the dip direction in their central portions. This direction can be established by an application of the bow and arrow rule (Elliott, 1976, p. 298). A straight line is drawn connecting the two ends of the outcrop trace of a single thrust. The perpendicular bisector of this line then gives an estimate of the slip direction, and the length of the line is an estimate of the displacement (Fig. 6.19).

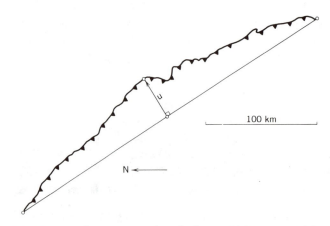

FIGURE 6.19 Bow and arrow rule (after Elliott, 1976, p. 298).

Additional techniques for estimating the slip direction use the orientation of folds and slaty cleavage in the thrust sheet (Pfiffner, 1980), and the orientation of the foliation and lineation in mylonites (Butler, 1982). The basis for these methods are treated in later chapters.

With the slip direction, the displacement may then be estimated in several ways. Perhaps the simplest approach is to draw a section parallel to this direction along a section across the exposed thrust plane, intersecting if possible a klippe and a window. The maximum width of the exposure of the thrust is then gives a measure of the mimimum displacement (see distance x in Fig. 6.16).

The fenster-to-klippe method of putting a limit on thrust displacement breaks down entirely if rock beneath the thrust are older than those above it, or if stratigraphic inversion occurred prior to thrusting by recumbent folding, or if a line drawn from fenster to klippe is not at least approximately parallel to the direction of displacement.

If a single displaced plane can be identified in both the foot wall and hanging wall, the amount of slip can be determined using the previous methods.

PROBLEM (Fig. 6.20a)

An overthrust sheet was emplaced by slip due west. A displaced marker plane is found in two places: at points A on the hanging wall block, and B on the foot wall block. Find the displacement.

CONSTRUCTION (Fig. 6.20b)

1. In a plan view of the thrust plane (not the plane of the map), determine the orientation of the trace of the displaced plane on the thrust plane from both the hanging wall and foot wall sides.

2. From point B on the foot wall side draw a line parallel to this trace to intersect the slip line at point C.

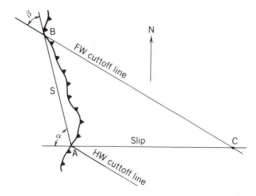

FIGURE 6.20 Net slip on a thrust from separation (after Elliott and Johnson, 1980, p. 77).

3. The length AC represents the magnitude of the slip.

The magnitude of the slip can also be calculated from these same measured quantities. From the map, the line AB is the oblique separation. This line makes an angle α with the slip direction, and an angle β with the trace of the displaced plane. In the oblique triangle ABC the angles at A and B are then known, as is the included side s which is the measured separation. From this triangle

$$C = 180° - (A+B) \quad \text{and} \quad b = c \sin B/\sin A$$

and also

$$A = (\alpha-\beta) \quad \text{and} \quad B = \beta$$

Then

$$(6.1) \qquad \text{slip} = \frac{s \sin \beta}{\sin(\alpha-\beta)} \qquad (\alpha > \beta)$$

There is a particularly powerful approach to estimating the minimum displacement on a basal thrust which requires an accurate cross section drawn parallel to the slip direction (Fig. 6.21). The method of constructing such <u>balanced</u> sections is described in detail in §12.3, and assumes that areas and lengths are conserved in the plane of the section. That is, the area A before and the area A' after thrusting are equal. If t is the stratigraphic thickness, then the original length ℓ of a trace of a disturbed bed can be found from,

$$(6.2) \qquad \ell t = A = A' \quad \text{or} \quad \ell = A't$$

Measuring the final length ℓ', we then calculate the shortening within the duplex,

$$(6.3) \qquad \Delta \ell = \ell' - \ell$$

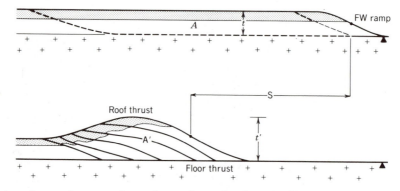

FIGURE 6.21 A schematic section through a duplex before and after thrusting; the duplex developed by progressive collapse of the footwall ramp under an advancing major thrust (after Elliott and Johnson, 1980, p. 72).

and this is also an estimate of the minimum displacement on the floor thrust. Williams (1984) has extended this analysis to include wedge-shaped thrust masses.

6.7 FOLDS AND FAULTS

Faults and folds are often found together. We are particularly interested in genetic relationships between faulting and folding. There are two important cases.

First, the folds may result directly from the fault movement. During slippage of one block pass the other, frictional drag on the fault plane may produce certain effects in the blocks themselves. One of these is the dragging of preexisting layers into folds. Such folds are termed <u>fault</u> <u>drag</u> <u>folds</u>. Where present, they represent displacement along the fault zone in addition to the slip. The total displacement, measured outside the zone of disturbance associated with the fault is termed <u>shift</u>.

The presence of such drag folds gives slightly more information concerning the fault movement than in cases where such effects are absent; the sense of fault separation can be determined from observations made on only one side of the fault. However, caution is necessary. The presence of fault drag folds tends to encourage an impulse to read into the pattern something more that the limited measurement of separation, that is, to use the curvature of the folds as evidence of the direction of slip. For example, the map pattern of Fig. 6.22a has been mistakenly used as an indication of strike-slip displacement. While this is certainly one possibility, there is no more evidence of such slip that the map pattern of Fig. 6.22b, where drag folds are absent.

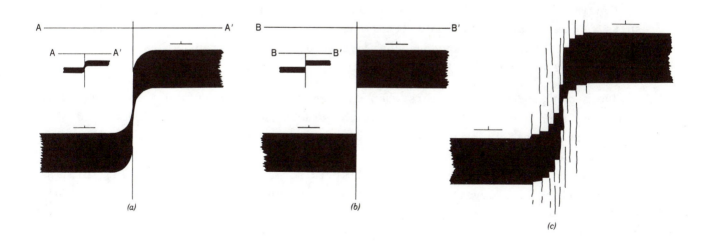

(a)   *(b)*

(c)

FIGURE 6.22 Examples of fault separation: (a) with drag; (b) without such drag; (c) distributive faulting.

Two considerations will make this limitation clear. First, a vertical section drawn perpendicular to the fault trace will always show these same folds (see section AA' of Fig. 6.22a). In fact, every general section that intersects the fault will reveal curvature. Clearly, the folds in each of these diversely oriented sections can not be taken as evidence of slip parallel to the plane of that section. Thus only separation can be determined--strike separation in map view, and dip separation in a vertical section (see section BB' in Fig. 6.22b).

The process of fault-drag folding is closely related to distributive faulting (Fig. 6.22c). It will then be evident that the orientation of fault drag folds is determined by the attitude of the fault and of the displaced layers; the folds are parallel to the intersection of the two intersecting planes. A down-plunge view of the folded beds should amply confirm this conclusion (see also Fig. 6.23).

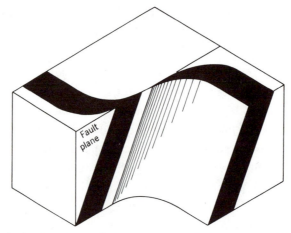

FIGURE 6.23 Axis of fault drag fold is parallel to the line of intersection of the fault and the plane of the displaced layer.

Considerable caution is needed in interpreting such folds because similar final results can be obtained from an initial shear zone which developed into a fault at a late stage. Further, simultaneous development of folds and faults has been demonstrated experimentally by Dubey (1980).

The second association occurs when the fault surface is curved. As a result of movement on such faults, at least one of the blocks must deform, and <u>fault-bend folds</u> result (Suppe, 1983). An example of such a fold is shown in Fig. 6.17. An interesting special case of such folding involves the bending of the upper part of the hanging wall block by movement on a listric normal fault; the resulting fold is called a rollover or "reverse drag" (though of course they have nothing to do at all with drag). A model of this case in shown in Fig. 6.28.

6.8 EXTENSIONAL FAULTS

A normal fault represents a local horizontal extension. The amount of this extension can be found from the horizontal component of the slip.

$$(6.4) \qquad s_h = s \cos p$$

where s is the magnitude of the slip and p is its angle of plunge. Where the oblique component is small, the dip angle may be used with little error.

Especially where more than one fault is involved, it is usually convenient to measure the extension as a fractional increase in length. Then, using Eq. 6.3, we have

$$(6.5) \qquad e = \Delta \ell / \ell = (\ell' - \ell)/\ell = \ell'/\ell - 1$$

If a tabular body of rock is extended in this way, rotation of the faults or beds will occur. Two separate cases arise. In the first, planar faults, together with the displaced beds, rotate. In the second, the faults are curved, and the beds rotate.

The geometry of the first case is shown in Fig. 6.24. Following Thompson (1960) we may obtain an expression for the combined effects of slip and rotation. In Fig. 6.24c, ω is the angle of rotation of the originally horizontal bedding, δ is

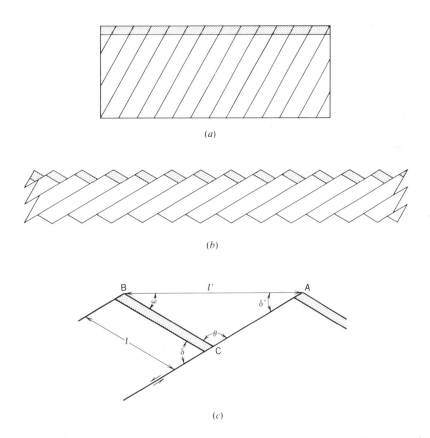

(a)

(b)

(c)

FIGURE 6.24   Geometry of extension by normal faults with rotation: (a) faults at the instant of failure; (b) after slip and rotation; (c) geometric elements in the derivation of Eq. 6.8 (after Wernicke and Burchfiel, 1982, p. 106).

the dip of the fault before and δ' is the dip after rotation. Then from the law of sines,

(6.6) $\qquad \ell'/\sin \theta = \ell/\sin \delta' \qquad$ or $\qquad \ell'/\ell = \sin \theta/\sin \delta'$

Also from this figure, $\theta = 180° - (\omega + \delta')$, hence $\sin \theta = \sin(\omega + \delta')$. Then Eq. 6.6 becomes

(6.7) $\qquad \ell'/\ell = \sin(\omega + \delta')/\sin \delta'$

and we may write Eq. 6.5 as

(6.8) $\qquad e = \dfrac{\sin(\omega + \delta')}{\sin \delta'} - 1$

Figure 6.25 is a graph of this equation.

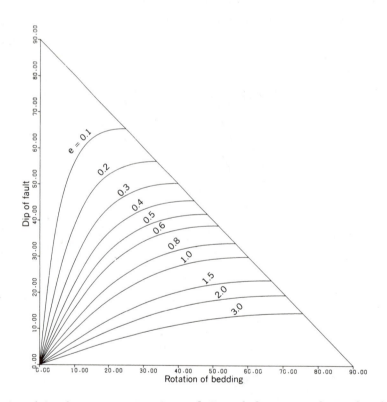

FIGURE 6.25 Graphical representation of Eq. 6.8; note that the limiting case occurs when $(\omega + \delta') = 90°$

In the second case, we may model a listric normal fault by assuming that the trace of both the fault and folded bedding in cross section are circular arcs, as shown in Fig. 6.26. In this illustration ℓ is represented by the arc length BQ, and

$$\ell = \omega r_2 \qquad (\omega \text{ in radians})$$

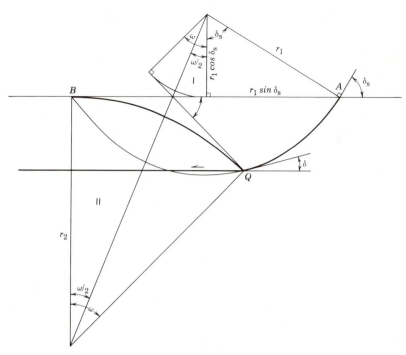

FIGURE 6.26 Model of a listric normal fault (after Wernicke and Burchfiel, 1982, p. 107).

The extended equivalent ℓ' is represented by the straight line segment AB, and

$$\ell' = 2r_1 \sin \delta_s$$

where δ_s is the dip at the surface point A. Then

(6.9) $$e = \frac{2r_1 \sin \delta_s}{\omega r_2} - 1 \qquad (\omega \text{ in radians})$$

We can also express r_2 in terms of r_1, and use the result to eliminate r_2. Because AB = 2(BX+XM), and utilizing relationships involving Triangles I and II,

$$2r_1 \sin \delta_s = 2[r_1 \cos \delta_s \tan(\omega/2) + r_2 \tan(\omega/2)]$$

Solving this for r_2 then gives,

$$r_2 = r_1 \left(\frac{\sin \delta - \cos \delta \tan(\omega/2)}{\tan(\omega/2)} \right)$$

Using this in Eq. 6.10, and noting that

$$\delta_s = \omega + \delta$$

where δ is the dip of the fault at point Q, we finally obtain,

$$(6.10) \qquad e = \frac{2}{\omega[\cot(\omega/2) - \cot(\omega+\delta)]} - 1 \qquad (\omega \text{ in radians})$$

Fig. 6.27 graphically illustrates the difference between the extension associated with rotated planar faults (as represented by Eq. 6.9) and listric normal faults (as represented by Eq. 6.10) in otherwise similar situations.

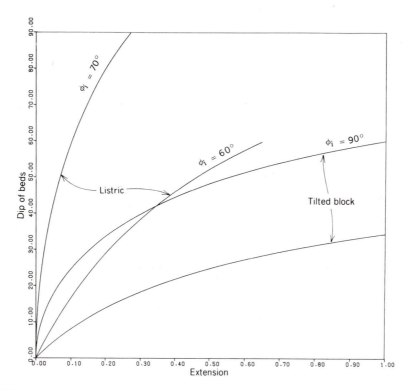

FIGURE 6.27   A comparison of the extension associated with listric and planar normal faults (after Wernicke and Burchfiel, 1982, p. 108).

To emphasize the importance of these differences in diagnosing fault geometry, Fig. 6.28 shows a hypothetical geologic map and accompanying cross section. A low-angle normal fault dips 5° to the east and offsets a section of Tertiary volcanics and underlying Precambrian basement rocks. Point Y represents one piercing point; we need a second. Because the exposed fault is planar and the bedding in the two blocks has the same attitude, it is attractive to project the fault as a straight line to find the point X; XY is then an estimate of the slip. However, if the exposed fault is actually the flat portion of a listric normal fault which originally steepened above the section, the estimated extension would be considerably less.

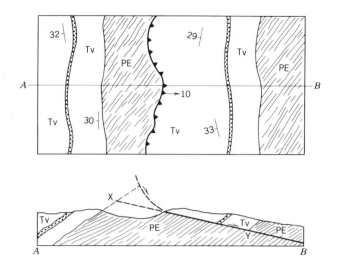

FIGURE 6.28 A geologic map and accompanying cross section illustrating the difference between treating the now eroded portion of a low-angle normal fault as a plane or curved surface (after Wernicke and Burchfiel, 1982, p. 107).

EXERCISES

1. The plane of a normal-slip fault strikes due north and dips 60° west. The fault displaces a structural plane whose attitude is N90°W, 30°N and which shows 100 m of left separation. What is the slip?

2. A fault whose attitude is N90°W, 60°N cuts two structural planes: Plane 1 has an attitude of N45°W, 30°NE; Plane 2 has an attitude of N50°E, 45°NW. The amounts and senses of separation are shown in Fig. X6.1. What is the orientation and magnitude of the relative slip vector, and what is the name of the fault?

3. A fault whose attitlude is N30°E, 60°W displaces two planes as shown in Fig. X6.2. What is the angle and sense of rotation? Locate the center of rotation which will account for the observed displacements.

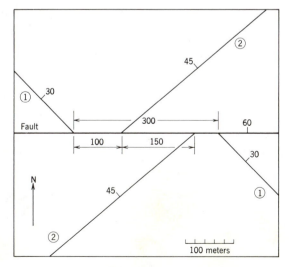

FIGURE X6.1

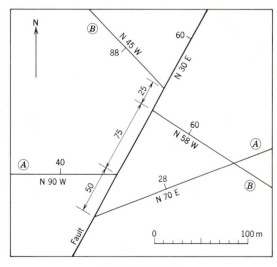

FIGURE X6.2

7
Stress

7.1 INTRODUCTION

By Newton's Second Law _force_ is the product of mass and acceleration. Having both magnitude and direction, force is a vector quantity. If a mass of one kilogram is given a linear acceleration of one metre per second per second, the force is one newton (= 1 N). Closely related is the moment of a force, or _torque_, which is also a vector quantity. If a force of one newton acts perpendicular to a moment arm one metre long, the torque is one metre-newton (= 1 m·N). Being vectors, forces and torques can be treated by the rules of vector algebra.

If two or more forces are balanced in such a way that the mass is at rest, it is said to be in a state of static _equilibrium_. For this condition to exist, the total force must vanish, or in other words, the vector sum of all the forces must be zero. The total torque must also vanish, that is, the vector sum of all the moments of the forces relative to some arbitrary point must be zero. These conditions are expressed by the two vector equations:

$$\mathbf{F}_1 + \mathbf{F}_2 + \mathbf{F}_3 + \ldots = 0$$

$$\mathbf{M}_1 + \mathbf{M}_2 + \mathbf{M}_3 + \ldots = 0$$

By resolving each force and each moment into its components we have the necessary and sufficient conditions for equilibrium in the form of six scalar equations:

(7.1) $\Sigma F_x = 0$ $\Sigma F_y = 0$ $\Sigma F_z = 0$

(7.2) $\Sigma M_x = 0$ $\Sigma M_y = 0$ $\Sigma M_z = 0$

The three relationships of Eqs. 7.1 express the fact that the components of the forces in each of the three coordinate direction are balanced. Similarly, Eqs. 7.2 express the fact that the moments of these forces about axes parallel to the three coordinate axes are also balanced.

7.2 SURFACE TRACTIONS

Consider a body subjected to several external or body forces (Fig. 7.1). Across any plane, real or imagined, within the body there will exist a field of forces equivalent to the loads exerted by the material on one side of the plane on the other, or in the case of a physical boundary, to the applied loads. In general, these forces will not be uniform in either direction or magnitude from point to point on such a plane. However, if the area in the neighborhood of a point 0 is contracted, the ratio of force to area tends to a finite limit. This limit is a vector called the <u>traction</u> **T** at that point. Its magnitude is

(7.3)
$$T = \lim_{\Delta A \to 0} \frac{\Delta F}{\Delta A}$$

There will also be an equal and opposite traction acting on the other side of the element of area.

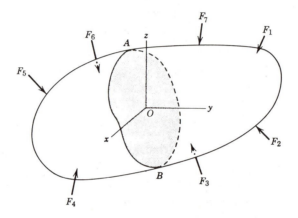

FIGURE 7.1 A body in a state of static equilibrium

The traction vector as defined above will have different directions and magnitudes for different positions of the element of area within the body. For any given ΔA the traction vector will, in general, be inclined to the plane of the element of the area (Fig. 7.2a), in which case it is convenient to resolve it into two components (Fig. 7.2b): a <u>normal</u> component, with the symbol σ (sigma), acting perpendicular to the plane, and a <u>shearing</u> component, with the symbol τ (tau), acting tangentially to the plane. These two components are also sometimes called, loosely, the normal stress and shear stress acting on the plane.

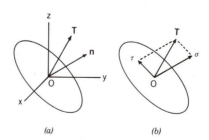

FIGURE 7.2 The normal and tangential components of the traction acting on an element of area; the orientation of the element of area is represented by the unit vector **n** normal to its plane.

A traction has the dimension of force per unit area; a force of one newton acting on one square metre is one pascal (= 1 N/m^2 = 1 Pa). One pascal is very small, so it is convenient in geology to use kilopascals (1 kPa = 10^3 Pa) or megapascals (1 MPa = 10^6 Pa); a megapascal is also equivalent to 1 N/mm^2, which increases the usefulness of this particular multiple. For larger magnitudes, the gigapascal is appropriate (1 GPa = 10^9 Pa = 1000 MPa). In the past, much of the geologic literature has expressed tractions in bars and kilobars; the factors for converting to pascals are easily remembered:

$$1 \text{ bar} = 10^5 \text{ Pa} = 0.1 \text{ MPa}$$
$$1 \text{ kilobar} = 100 \text{ MPa}$$

The traction, and therefore its normal and shearing components, will also have different orientations and magnitudes for different orientations of the elemental area ΔA. The totality of the tractions for all planes of all orientations constitutes the stress in the material at that point. It might seem that a great deal of information would be required in order to specify this state, but, in fact, if the tractions on any three non-parallel planes are known, the traction acting on any other plane can then be found.

7.3 STRESS COMPONENTS

For these three non-parallel planes it is most convenient to adopt the faces of an elemental cube whose edges are parallel to the three axes of a cartesian coordinate system (Fig. 7.3). On each of face of this cube the traction is resolved into three components, one in each of the coordinate directions. In all there are nine

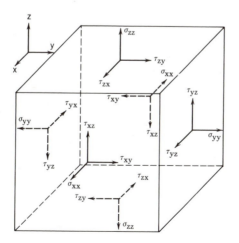

FIGURE 7.3 The full complement of stress components in three dimensions acting on the volume element ΔxΔyΔz.

components, called the cartesian <u>stress</u> <u>components</u>, which may be written out in the form of an ordered array or matrix.

$$\begin{bmatrix} \sigma_{xx} & \tau_{xy} & \tau_{xz} \\ \tau_{yx} & \sigma_{yy} & \tau_{yz} \\ \tau_{zx} & \tau_{zy} & \sigma_{zz} \end{bmatrix}$$

In this notation, the first subscript identifies the plane on which the component acts by noting the coordinate direction of the normal to the cube face, and the second subscript gives the coordinate direction in which the component itself acts. It can then be seen that each row contains the components which act on a face, and each column contains the components which act in one of the coordinate directions. There are several reasons for writing the components in this way. First, it represents an easy way of keeping track of them, and second, it emphasizes the important fact that stress is a single entity, albeit with nine components.

For this elemental cube to be in a state of static equilibrium certain restrictions must be placed on some of the components. In order to apply the condition of equilibrium, as expressed by Eqs. 7.1 and 7.2, stress components must be first converted to forces, and this is accomplished by multiplying each by the corresponding area on which it acts.

Only two pairs of tractions contribute to the moment about any particular coordinate direction. For example, the moment about a line through the middle of the

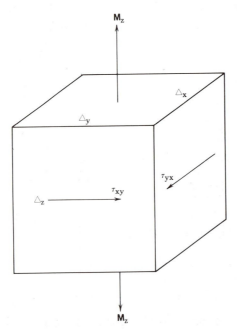

FIGURE 7.4 Torques due to the shearing components τ_{xy} and τ_{yx}.

cube parallel to the z-axis is due the tractions τ_{xy} and τ_{yx}. The tangential force on the front face of the cube in Fig. 7.4 is $\tau_{xy}\Delta y \Delta z$, and a tangential force with the same magnitude acts in the opposite direction on the back side of the cube. The magnitude of the torque about the center of the block due to these two forces is then

$$M_z = 2\tau_{xy} \Delta y \Delta z (\Delta x/2)$$

By the right-hand rule, the moment vector associated with this pair of forces points in the +z-direction.

In a similar way, the magnitude of the torque associated with the pair of tractions τ_{yx} is

$$M_z = 2\tau_{yx} \Delta x \Delta z (\Delta y/2)$$

and the associated moment vector points in the −z−direction. For the cube to be in a state of static equilibrium, the magnitudes of these two oppositely directed vectors must be equal; that is,

$$2\tau_{xy} \Delta y \Delta z (\Delta x/2) = 2\tau_{yx} \Delta x \Delta z (\Delta y/2)$$

and this reduces to

$$(7.4) \qquad \tau_{xy} = \tau_{yx}$$

Similar relationships hold between the two other pairs of components with different subscripts. Hence, only six of the nine stress components are independent.

7.4 STRESS IN TWO DIMENSIONS

If the stress components in one of the coordinate directions vanish, the body is in a state of <u>plane stress</u>. Strictly, this condition holds for thin plates only, but it is a convenient way to introduce the geometry of stress at a point, and the formulation has an additional meaning which makes it useful. Following conventional practice, we will assume that the components in the z-direction are all zero. This reduces the total number of components from nine to four, of which only three are independent.

$$\begin{bmatrix} \sigma_{xx} & \tau_{xy} \\ \tau_{yx} & \sigma_{yy} \end{bmatrix}$$

If a rectangular element with sides Δx and Δy and unit thickness is in a state of plane stress, we wish to determine the normal and shearing components of the trac-

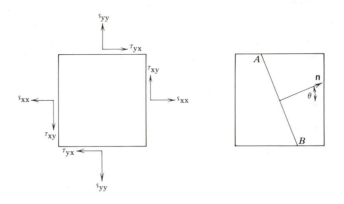

FIGURE 7.5 Stress in two dimensions: (a) stress components; (b) components of the traction acting on the inclined plane AB defined by the angle θ the normal to the plane makes with the x-axis.

tion acting on an inclined plane parallel to the z-axis and whose <u>normal</u> makes an angle θ with the x-axis (Fig. 7.5). Our approach will be to analyse several simple situations, and then combine these separate results.

UNIAXIAL STRESS

If the rectangular element is subjected to a uniform normal stress acting in the x-direction, what are the components of the traction which act on the inclined plane AB shown in Fig. 7.6a? To find these, we first imagine a triangular <u>free-body</u> cut from the element (Fig. 7.6b). To restore this body to equilibrium, we must re-place the action of the originally adjacent material with equivalent forces. If a is the area of the inclined plane, the area of side OA is equal to (a cos θ). The magnitude of the force acting on side OA is then

$$(7.5) \qquad F_x = \sigma_{xx} (a \cos \theta)$$

and an equal and opposite force must also act on the inclined side AB. We are par-ticularly interested in the components of this force which are normal and tangential

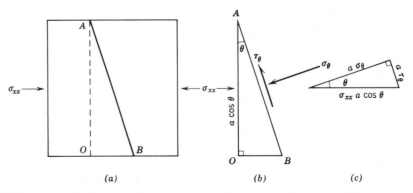

FIGURE 7.6 Uniaxial normal stress acting parallel to the x-axis.

to the inclined plane AB. From the vector triangle involving the uniaxial force and these component (Fig. 7.6c), we then find that

$$F_N = F_x \cos \theta$$

$$F_T = F_x \sin \theta$$

By substituting the expression for F_x from Eq. 7.5, and dividing by the area, we obtain expressions for the magnitudes of the two components of the traction acting on this inclined plane in terms of the applied tractions.

(7.6a) $$\sigma = \sigma_{xx} \cos^2\theta$$

(7.6b) $$\tau = \sigma_{xx} \sin \theta \cos \theta$$

BIAXIAL STRESS

To find the tractions acting on the inclined plane when normal components act in both coordinate directions, we first consider the effects of the uniaxial stress in the y-direction alone (Fig. 7.7a). Proceeding just as before, the force on the side AB of the free body (Fig. 7.7b) is

$$F_y = \sigma_{yy} (a \sin \theta)$$

Then from the triangle of forces (Fig. 7.7c), the normal and shearing tractions are

(7.7a) $$\sigma = \sigma_{yy} \sin^2\theta$$

(7.7b) $$\tau = \sigma_{yy} \sin \theta \cos \theta$$

By the <u>principle of superposition</u>, the total traction acting on the inclined plane

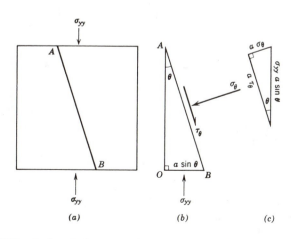

(a)   *(b)*   *(c)*

FIGURE 7.7 Uniaxial normal stress acting parallel to the y-axis.

is the sum of the tractions due to the two separate unixial cases. Adding the corresponding pairs of expressions for σ and τ from Eqs. 7.6 and 7.7, assuming

$$\sigma_{xx} \geqslant \sigma_{yy}$$

and observing that the contributions of the applied stress components make to the shearing tractions are opposite, we obtain

(7.8a) $\sigma = \sigma_{xx} \cos^2\theta + \sigma_{yy} \sin^2\theta$

(7.8b) $\tau = (\sigma_{xx} - \sigma_{yy})\sin\theta \cos\theta$

PURE SHEAR STRESS

Finally, we seek the effects of applied shear tractions alone (Fig. 7.8a). Again, the forces on the free body corresponding to these tractions are determined first (Fig. 7.8b). Considering each pair of tractions separately, two triangles of forces are constructed (Fig. 7.8c). From triangle of forces 1, the contributions from the shearing traction are

$$\sigma = \tau_{xy} \sin\theta \cos\theta$$

$$\tau = \tau_{xy} \cos^2\theta$$

and from triangle of forces 2, the contributions are

$$\sigma = \tau_{yx} \sin\theta \cos\theta$$

$$\tau = \tau_{yx} \sin^2\theta$$

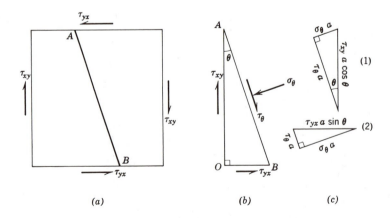

FIGURE 7.8   The state of pure shear stress.

Adding the pairs of equations for σ and τ, paying attention to the different senses of shear, and using the symmetrical relationship of Eq. 7.4, we then have

(7.9a)
$$\sigma = 2\tau_{xy} \sin \theta \cos \theta$$

(7.9b)
$$\tau = \tau_{xy} (\cos^2\theta - \sin^2\theta)$$

GENERAL TWO-DIMENSIONAL STRESS

The contribution of the full complement of applied components to the normal and shearing tractions acting on the inclined plane can now be found by adding the expression found in the biaxial case of Eqs. 7.8 and pure shear case of Eqs. 7.9. The results are

(7.10a)
$$\sigma = \sigma_{xx} \cos^2\theta + \sigma_{yy} \sin^2\theta + 2\tau_{xy} \sin \theta \cos \theta$$

(7.10b)
$$\tau = (\sigma_{xx} - \sigma_{yy})\sin \theta \cos \theta + \tau_{xy} (\sin^2\theta - \cos^2\theta)$$

We are particularly interested in the extreme values of σ, and these can be found from the condition $d\sigma/d\theta = 0$. Differentiating Eq. 7.10a and setting the result equal to zero gives,

(7.11)
$$2[(\sigma_{xx} - \sigma_{yy})\sin \theta \cos \theta + \tau_{xy} (\cos^2\theta - \sin^2\theta)] = 0$$

Comparing this results with Eq. 7.10b, we see that when σ has extreme values, $\tau = 0$. This shows that when the plane element is oriented such that the normal traction σ has its maximum or minimum value, the shearing traction is zero. We can find the orientation of the planes on which these extreme normal tractions act by substituting the identities,

$$\sin \theta \cos \theta = \tan \theta/(1+\tan^2\theta)$$

$$\cos^2\theta - \sin^2\theta = (1-\tan^2\theta)/(1+\tan^2\theta)$$

$$\tan 2\theta = 2 \tan \theta/(1-\tan^2\theta)$$

into Eq. 7.11. After some manipulation we find that,

(7.12)
$$\tan 2\theta = \frac{2\tau_{xy}}{\sigma_{xx} - \sigma_{yy}}$$

Two angles satisfy this expression, 2θ and $2\theta+180°$, or θ and $\theta+90°$. Hence the maximum and mimimum values of σ occur on two mutually perpendicular planes. These extreme normal tractions are called the <u>principal stresses</u>, and they are labeled so that

$$\sigma_1 \geqslant \sigma_3$$

We denote these principal stresses as σ_1 and σ_3 for two reasons. We wish to emphasize that in nature we are dealing with a three-dimensional setting, and because we are especially interested in the greatest and least principal stresses, as compared with the intermediate principal stress σ_2.

We now choose our coordinate axes to coincide with these two principal direction. We may then make the following replacements in Eqs. 7.10,

$$\sigma_{xx} = \sigma_1, \quad \sigma_{yy} = \sigma_3 \quad \text{and} \quad \tau_{xy} = 0$$

The results are

(7.13a) $\sigma = \sigma_1 \cos^2\theta + \sigma_3 \sin^2\theta$

(7.13b) $\tau = (\sigma_1 - \sigma_3)\sin\theta\cos\theta$

A graph of these two equations is shown in Fig. 7.9.

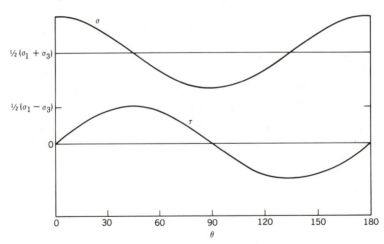

FIGURE 7.9 Graph of σ and τ as a function of the angle θ given by Eqs. 7.13 (after Jaeger, 1969, p. 8).

7.5 MOHR CIRCLE FOR STRESS

Eqs. 7.13 can be put into a much more useful form by substituting the double angle identities:

(7.14a) $\sin^2\theta = \dfrac{1}{2}(1 - \cos 2\theta)$

$$(7.14b) \qquad \cos^2\theta = \frac{1}{2}(1+\cos 2\theta)$$

$$(7.14c) \qquad \sin \theta \cos \theta = \frac{1}{2} \sin 2\theta$$

We then obtain the important results,

$$(7.15a) \qquad \sigma = \frac{\sigma_1 + \sigma_3}{2} + \frac{\sigma_1 - \sigma_3}{2} \cos 2\theta$$

$$(7.15b) \qquad \tau = \frac{\sigma_1 - \sigma_3}{2} \sin 2\theta$$

These have the form

$$x = c + r \cos \alpha \qquad \text{and} \qquad y = r \sin \alpha$$

which are the parametric equations of a circle with radius r, which is centered on the x-axis at a distance c from the origin. This property of Eqs. 7.15 allows a wide variety of problems involving σ and τ to be solved graphically with a <u>Mohr</u> circle for stress (Fig. 7.10). The main feature of this construction is a circle centered on the horizontal σ-axis at a distance from the origin and a radius equal to

$$c = (\sigma_1 + \sigma_3)/2 \qquad \text{and} \qquad r = (\sigma_1 - \sigma_3)/2$$

This circle is the locus of all permissible pairs of values of σ and τ for a given state of stress at a point.

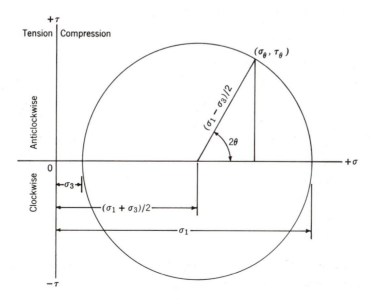

FIGURE 7.10 Mohr circle for stress in two dimensions.

The Mohr circle diagram is an excellent way of remembering some important relationships, and it is useful to commit its features to memory.

Mohr circle for stress can be used to solve a number of different types of problems. In order to construct a circle, we need a sign convention for the normal and shearing components of the traction. Following usual geological practice, compression is reckoned positive and tension is negative. If a pair of shearing tractions act in an anticlockwise sense they are taken positive, and if clockwise negative.

The simplest problem to solve graphically is the determination of the normal and shearing tractions acting on any specified plane given the principal stresses.

PROBLEM (Fig. 7.11a)

If σ_1 = +200 MPa and σ_3 = 50 MPa, what are the normal and shearing tractions on a plane whose normal makes an angle of 25° measured anticlockwise from the σ_1-direction.

CONSTRUCTION (Fig. 7.11b)

1. Draw a pair of orthogonal axes and locate point C on the horizontal σ-axis at $(\sigma_1 + \sigma_3)/2$ = +175 units from the origin using a convenient scale.

2. With point C as center, and with a radius of $(\sigma_1 - \sigma_3)/2$ = 75 units, complete the circle.

3. Draw a radius of this circle making an angle of 2θ = 50° measured in the same sense as θ, that is, anticlockwise from σ_1, thus locating point P on the circumference of the circle.

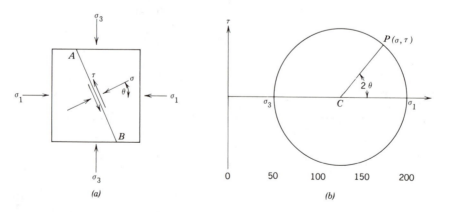

FIGURE 7.11   Mohr circle solution for the traction components acting on an inclined plane given the principal stresses.

4. The coordinates of P are the required values of the normal and shearing tractions acting on the specified plane.

ANSWER

Measuring the two coordinate distances gives σ = 162 MPa and τ = 65 MPa. Within the limits of the graphical method, these are the same results found by using Eqs. 7.15. The shearing traction is positive, and therefore acts in an anticlockwise sense.

Although the sign convention for the shearing traction give unambiguous results, the sense of shear on the plane can also be obtained directly by inspection, and this serves as a useful check. Imagine that the plane as a frictionless cut. The force equivalent to the maximum principal stress will then cause the two pieces of the block to slip in the correct sense.

We may also determine the magnitude of the principal stresses and their orientations given the four cartesian stress components.

PROBLEM

From the following stress components, determine the values of σ_1 and σ_3, and their orientations (see Fig. 7.12a):

$$\begin{bmatrix} \sigma_{xx} & \tau_{xy} \\ \\ \tau_{yx} & \sigma_{yy} \end{bmatrix} = \begin{bmatrix} 500 & 100 \\ \\ 100 & 150 \end{bmatrix} \text{ MPa}$$

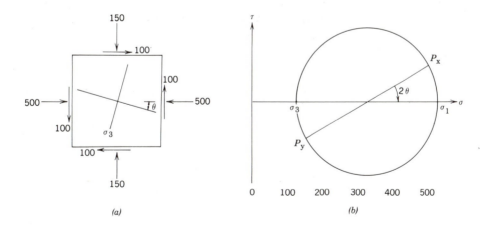

(a)   (b)

FIGURE 7.12   Determination of the principal stresses and their orientation from a Mohr circle construction.

CONSTRUCTION (Fig. 7.12b)

1. On a pair of orthogonal axes, plot the two points

$$P_x = (\sigma_{xx}, \tau_{xy}) = (+500, +100)$$
$$P_y = (\sigma_{yy}, \tau_{yx}) = (+150, -100)$$

2. A line joining these two points is the diameter of the circle. The intersection of this line with the σ-axis locates its center C and the circle can then be completed.

3. The two intercepts give the values of principal stresses. The line P_xC represent the x-coordinate direction; the orientation of maximum principal stress is found by measuring the angle $2\theta = 30°$ clockwise from P_x to σ_1.

ANSWER

The principal stresses are $\sigma_1 = 527$ MPa and $\sigma_3 = 123$ MPa, and the σ_1-direction makes an angle of $\theta = 15°$ measured clockwise from the x-axis.

It should be carefully noted that in these constructions, the angle 2θ is measured in the same sense on the $\sigma\tau$ plane as θ on the xy plane.

The orientation of the principal stresses in this type of problem can also be estimated by inspection. This is done by considering separately the set of normal tractions and the set of shearing tractions. Without the shearing tractions, σ_{xx} is σ_1 (Fig. 7.13a). With only the shearing tractions, two principal stresses make angles of 45° with the coordinate axes (Fig. 7.13b). These two separate principal stresses can not be combined directly because they act on different planes, but the orientation of the maximum principal stress for the total stress will lie between them.

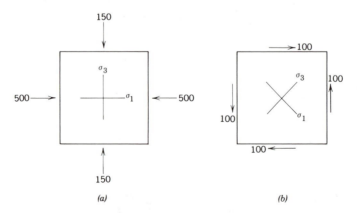

FIGURE 7.13 Estimation of the principal directions by inspection.

Finally, a closely related problem involves expressing the four cartesian stress components relative to a set of new, rotated set of coordinate axes.

PROBLEM

Express the stress components for a set of x'y'-axes which make an angle measured anticlockwise from the xy-axes of 50° (Fig. 7.12a)

CONSTRUCTION (Fig. 7.14b)

1. Draw the Mohr circle for this particular state of stress (as in Fig. 7.12b).

2. Locate a second diameter $P_{x'}P_{y'}$ by measuring an angle of $2(50°)=100°$ anticlockwise from the first diameter P_xP_y.

3. The coordinates of the points $P_{x'}$ and $P_{y'}$ are the stress components relative to the new set of axes.

ANSWER

These new stress components are:

$$\begin{bmatrix} \sigma_{xx} & \tau_{xy} \\ \tau_{yx} & \sigma_{yy} \end{bmatrix} = \begin{bmatrix} 195 & 154 \\ 154 & 455 \end{bmatrix} \text{MPa}$$

Note carefully that, while the components have changed, the state of stress they describe has not. In particular, the principal stresses have the same magnitudes and orientations.

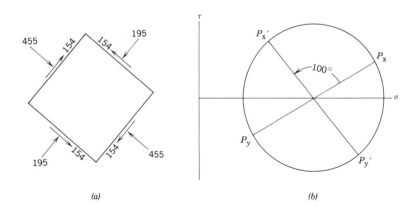

(a)   (b)

FIGURE 7.14   Stress components in a rotated coordinate system.

7.6 EFFECTS OF PORE FLUID PRESSURE

If a porous material contains a fluid under hydrostatic pressure p, the state of stress is altered. As shown in Fig 7.15a, this pressure in the pore fluid tends to counteract the effects of the principal stresses due to the applied load, which we label S_1 and S_3. Hence

$$\sigma_1 = S_1 - p \qquad \text{and} \qquad \sigma_3 = S_3 - p$$

Substituting these two expressions into Eqs. 7.15, and rearranging, yields,

(7.16a)
$$\sigma = \frac{S_1 + S_3}{2} + \frac{S_1 - S_3}{2} \cos 2\theta - p$$

(7.16b)
$$\tau = \frac{S_1 - S_3}{2} \sin 2\theta$$

In other words, the normal traction on any plane is reduced by the pressure p, while the shearing traction remains unchanged. This can be expressed as

(7.17)
$$\sigma = S - p$$

where S is the normal component of the traction acting due to the applied load. The term _effective stress_ is used to indicate that the pore fluid pressure has been taken into account in this way. This partitioning of stress and pressure in porous solids is generally called Terzaghi's relationship (see Gretener, 1981, for a good treatment of the subject).

In terms of a Mohr circle, the effect of a pore fluid pressure shifts the circle to the left by an amount p without changing its size (Fig. 7.15b).

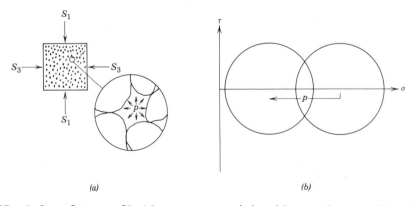

(a)

(b)

FIGURE 7.15   Role of pore fluid pressure: (a) effect of pore fluid pressure in a porous sandstone; (b) Mohr circle taking the pressure into account.

7.7 TRACTIONS VS FORCES

The description of stress given in the chapter was essentially established over 150 years ago by the French mathematician Cauchy and his predecessors. Historically, however, it has not always been applied correctly to geologic problems. The first clear statement of the principles in the geologic literature was given by Anderson (1942; see 1951), and reemphasized by Hubbert (1951), and by Hafner (1951). Unfortunately, errors persist and they most frequently root in the manipulation of stress components as force vectors.

An example from a recent text will illustrate how this vector approach goes wrong. Fig. 7.16a shows a low-angle thrust fault and a stress field in which the maximum principal stress is horizontal and the minimum principal stress is vertical (the details of this association will be given in the next chapter). Each of these principal stresses is treated as a force vector; in particular, the vector components of each principal stress in the plane of the fault is found. If $\sigma_1 = 40$ and $\sigma_3 = 28$, in arbitrary stress units, the magnitude of the two opposed "tangential force vectors" are 39.1 and 5.8. The vector sum of these is then equal to 33.3 and this, the author supposes, is the shearing force which drives the fault.

Fig. 7.16b shows a Mohr circle solution for the same problem. For the given principal stresses, the shearing traction on the plane of the fault is found to be $\tau = 2.4$ units. Not only is there a great discrepancy between these two results, but because the true maximum shearing traction is only 6.0 units, the values obtained by the vector treatment are not possible for any orientation of the fault plane. It should also be noted that this vector treatment would give a net tangential force on the plane of the fault even if the two principal stresses were equal, when in actuality, a shearing traction could not exist at all on any plane. An additional error is present in Fig. 7.16a; the supposed shearing components are either pointing in the wrong direction or are plotted on the wrong side of the trace of the fault.

Stress components are <u>not</u> vectors; they are a part of a more complicated physical entity which describes the relationship between the vectors **T** and **n**. To treat them simply as force vectors is a serious error, and conclusions based on such a treatment are invalid.

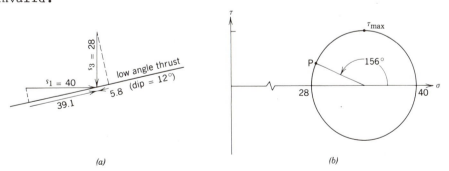

(a)   (b)

FIGURE 7.16 Stresses in the vicinity of a low angle thrust fault: (a) tractions as force vectors (after Badgley, 1965, p. 219); (b) Mohr circle solution.

EXERCISES

1. Reproduce the Mohr circle diagram of Fig. 7.10 from memory, and with it derive the two Mohr circle equations of Eqs. 7.15.

2. Four states of stress are illustrated below; each component is given in MPa. For each state determine the orientation and magnitude of each principal stress.

3. For the state (c), determine the components of stress for a coordinate system which is rotated from the vertical and horizontal 30° clockwise.

4. For the state (d), determine the orientation and magnitude of each principal stress if a pore fluid pressure of 7 MPa is present in the material.

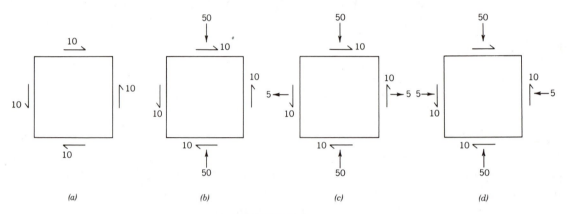

FIGURE X7.1

8
Faulting and its Causes

8.1 INTRODUCTION

An important goal of structural geology is to determine the nature of the stress field together with the mechanical properties of the rock material at the time of the formation of geological structures. In some situations, a part of this information may be obtained by combining data obtained from experiments with a detailed field study of the geometrical features of faults.

8.2 STATE OF STRESS AT DEPTH

In the earth's crust, the vertical component of stress is generally a function of depth z, and its magnitude is given by

$$(8.1) \qquad \sigma_{zz} = \rho g z$$

where ρ is the mean bulk density, and g is the acceleration due to gravity. For example, assuming a mean density of 2.4×10^3 kg/m^3, what is the vertical component of stress at a depth of 1 km?

$$\sigma_{zz} = (2400 \text{ kg/m}^3)(9.8 \text{ m/s}^2)(1000 \text{ m})$$

$$= 23.5 \times 10^6 \text{ Pa} = 23.5 \text{ MPa}$$

If this is rounded upward to 24 MPa an error of only 2% is introduced (this is equivalent to the commonly made assumption that g = 10 m/s^2). This approximation is more accurate than any in situ determination of the density of crustal rocks, hence quite acceptable. With this change, it will be seen that there is a close numerical relationship between σ_{zz} and ρ, which can be expressed in an easily remembered rule:

The overburden pressure is $\rho/100$ MPa per kilometre, ρ in kg/m^3.

In the absence of tectonic forces, past and present, what can be said about the horizontal components of stress at depth? Measurements of the state of stress made underground or in deep drill holes show that the horizontal stress components vary from place to place, and the reasons for this variation are not always clear.

One possibility is that the horizontal components are equal to the overburden pressure, that is, state of stress is "hydrostatic", and this is a simple and convenient starting point. This is expressed by

$$\sigma_{xx} = \sigma_{yy} = \sigma_{zz}$$

In rocks, this state is commonly referred to as being <u>lithostatic</u>, and is sometimes called Heim's rule. With the addition of tectonic forces, the state of stress at the point in question then becomes a more general, and more interesting type.

8.3 EXPERIMENTAL FRACTURES

Geometrically, there is no limit to the size of the circle which might be drawn or conceived. Physically, however, there is a definite limit, for if the differential stress becomes too large, the material will fail. To learn the details of rock behavior under various stress conditions it is necessary to resort to experiment. The usual practice is to apply a load to the ends of a carefully prepared cylinder of rock which is immersed in a pressurized fluid. In this configuration, the greatest principal stress is parallel to the axis of the cylinder, and the other two principal stresses are perpendicular to this axis.

Briefly, the results of such tests show that the conditions required to prouce failure are related to the stress level as measured by $(\sigma_1 - \sigma_3)$, but also that the higher the pressure in the confining fluid the higher the required differential stress to produce the same results. Further, they show that whether the failure occurs by fracture or flow is also stress dependent.

Two types of fractures may occur: extension fractures perpendicular to the least principal stress σ_3 (Fig. 8.1a), or shear fractures which are related to,

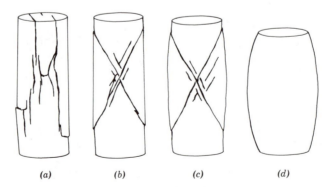

(a)   (b)   (c)   (d)

FIGURE 8.1 Types of failure in experimentally deformed limestone at varying confining pressures (see Verhoogen, and others, 1970, p. 458): (a) extension fractures (0.1 MPa); (b) brittle shear fractures (3.5 MPa); (c) semibrittle shear fractures (30 MPa); (d) ductile failure (100 MPa).

but not identical with the planes of maximum shearing stress. If fracture occurs before negligible permanent strain, the material is said to be <u>brittle</u> (Fig. 8.1b); if a small permanent strain (≤ 5%) preceeds fracturing, it is <u>semibrittle</u> (Fig. 8.1c). At higher confining pressures, the deformational mode is entirely <u>ductile</u> (Fig. 8.1d).

8.4 AMONTON'S LAW

Specifically, what conditions must exist before a brittle rock will fail in shear? Consider first the case where a fracture plane already exists in the rock mass. Renewed sliding will take place when the shear traction on the plane is sufficient to overcome the friction and the effect of the normal traction across the plane. This situation is analogous to the familiar problem of a block sliding on an inclined plane (Fig. 8.2a). The magnitude of the force acting on such a block of mass m is $F = mg$. This force can be resolved into two components, one tangential to the inclined surface, and the other normal to it (Fig. 8.2b). The following relationships then hold:

$$\cos \alpha = F_N/F \quad \text{or} \quad F_N = mg \cos \alpha$$

$$\sin \alpha = F_T/F \quad \text{or} \quad F_T = mg \sin \alpha$$

These components of force can then be converted by tractions by dividing each by the area of contact a,

$$\sigma = (mg/a)\cos \alpha \quad \text{and} \quad \tau = (mg/a)\sin \alpha$$

As the angle of inclination increases, a point is reached where the block will slide. This limiting angle is called the <u>angle of friction</u> ϕ. It is also convenient to define a <u>coefficient</u> of friction μ as the ratio of the shearing traction to normal traction at the point of sliding.

$$\mu = \tau/\sigma = \tan \phi$$

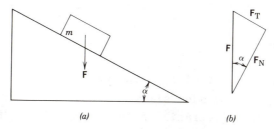

FIGURE 8.2 Geometry of a block sliding on an inclined plane: (a) force acting on the block; (b) components of force normal and tangential to the plane.

This expression is commonly written in an alternative form, and is known as Amonton's law.

(8.2) $\tau = \mu\sigma$

When the conditions expressed by this equation are met, the block will slide.

Now consider a body of rock at some moderate depth containing a plane whose physical property is completely determined by a coefficient of friction, which depends on the nature of the sliding surfaces. The angle of friction varies from about 20° to 40°, but 30° is a common value which we will use. As the differential stress increases from its assumed initial lithostatic state as shown by the the change from a point to a Mohr circle, a point will be reached when the magnitude of the tractions on the plane expressed by Eq. 8.2 will be satisfied, and renewed sliding will then occur. Thus the first possibility of renewed faulting occurs when the Mohr circle is tangent to the line whose slopes is tan ϕ (Fig. 8.3). However, it will occur then only on a plane whose normal measured from the σ_1-direction is given by

(8.3) $\pm\theta = 45° + \phi/2$

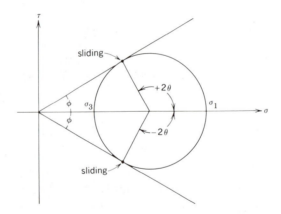

FIGURE 8.3   Mohr circle for renewed sliding on a plane of whose frictional properties given by $\phi = 30°$, and in the optimum orientation.

Or, alternatively, on the plane making an angle α with the σ_1-direction, where

(8.4) $\pm\alpha = 45° - \phi/2$

If the plane has some other orientation, a further increase in the differential stress is required in order to produce sliding (Fig. 8.4). In this figure we have adopted a common practice of illustrating only a half of the Mohr circle; because of the symmetry about the horizontal axis the other half can be easily completed if necessary.

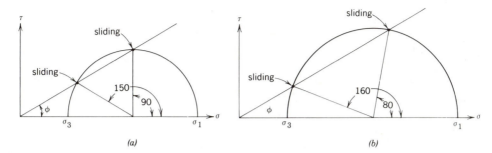

FIGURE 8.4 Conditions for renewed sliding when the plane is not in the optimum orientation: (a) $\theta = 45°$ or $75°$; (b) $\theta = 40°$ or $80°$.

8.5 THE COULOMB CRITERION

But what of a rock mass without such preexisting planes? About two hundred years ago, Coulomb (see Handin, 1969) suggested that two factors resist the shearing traction which tends to cause shear failure:

1. the normal traction across the potential shear plane,

2. the cohesive shear strength of the material.

This is now known as the Coulomb criterion of shear failure, and is usually written as

(8.5) $$\tau = \sigma_0 + \sigma \tan \phi_i \qquad \text{or} \qquad \tau = \sigma_0 + \mu_i \sigma$$

where σ_0 is the cohesive shear strength of intact rock when the normal traction across the potential plane is zero, the angle ϕ_i is by analogy with sliding friction the angle of _internal_ friction, and μ_i is the coefficient of internal friction. The cohesive strength is of the order of 10–20 MPa for most sedimentary rocks and 50 MPa for crystalline rocks. The value of the angle of internal friction is usually not the same as the angle of sliding friction, but the average is still close to $30°$.

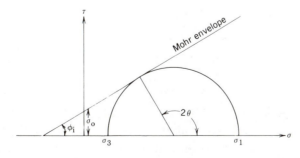

FIGURE 8.5 Coulomb criterion of shear failure.

The Coulomb criterion can also be depicted graphically on a Mohr circle diagram (Fig. 8.5). The only difference between this and the previous diagram illustrating Amonton's law is is that the line representing the Coulomb criterion, usually called the Mohr envelope, does not pass through the origin. Compressive tests made at low to moderate confining pressures confirm that this linear envelope describes the conditions of shear failure to a good approximation.

This criterion has two important limitations. First, the predicted tensile strength, as represented by the point of intersection of the envelope and the σ-axis is much too great; commonly it is only about half the cohesion. The envelope can be empirically modified as shown by the curve in Fig. 8.6 to take this into account (Price, 1966, p. 27). Second, at high confining pressures failure occurs by ductile flow. This flow is pressure insensitive, and the yield condition can be represented by a horizontal line which intersects the Coulomb envelope (Fig. 8.6; Goetze and Evans, 1979, p. 471).

The analogy between the Coulomb criterion and the conditions for sliding is useful both as a memory aid, and as a simple physical explanation. It is not entirely clear, however, what internal friction is. In loose, dry sand it is of the nature of external or sliding friction, but for cohesive material it can be neither be measured nor is there a simple physical explanation. Here, we will regard it as no more than the experimentally determined slope of the Mohr envelope. Despite these uncertainties and difficulties, it is still a very useful criterion.

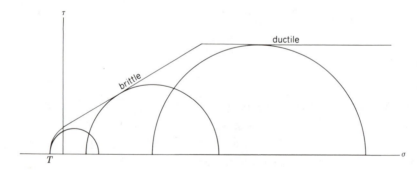

FIGURE 8.6 The Mohr envelope representing the Coulomb criterion of shear failure, modified to take failure in tension and by ductile flow into account; the circles illustrate the state of stress at tensile, shear and ductile failure.

8.6 CLASSIFICATION OF FAULTS

Clearly, no significant shearing traction acts along the air-earth interface, and therefore one of the principal stresses must be perpendicular to the earth's surface. In many areas this means that one of the principal stresses is approximately vertical. In combination with the geometrical relationship between fracture planes and principal stress directions, this leads to a dynamic classification of faults.

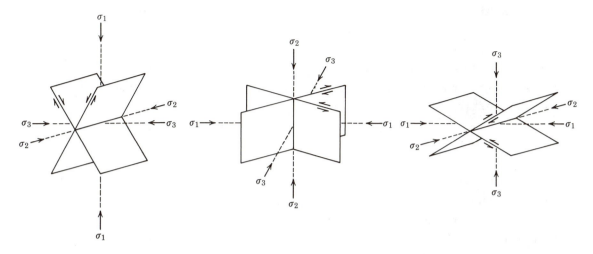

FIGURE 8.7 Dynamic classification of faults: (a) normal faults; (b) wrench faults; (c) thrust faults (after Anderson, 1951, p. 15).

1. If σ_1 is vertical, the dip of the fault plane is $\delta = 45° + \phi/2$, or about 60°. These are <u>normal</u> faults (Fig. 8.7a).

2. If σ_2 is vertical, the fault planes are also vertical, and the slip direction is horizontal. These are <u>wrench</u> faults (Fig. 8.7b).

3. If σ_3 is vertical, the dip of the fault plane is $\delta = 45°-\phi/2$, or about 30°. These are <u>thrust</u> faults (Fig. 8.7c).

8.7 FAULTS AND STRESSES

It is now possible to solve certain problems dealing with the geometrical relationship between shear fractures and principal stress directions. The pertinent features are:

1. The intersection of a pair of conjugate shear fractures defines the intermediate principal stress direction.

2. The acute angle between conjugate pairs is 2α, and this angle is bisected by greatest principal stress.

3. The slip direction is defined by the intersection of the fault plane and the plane containing the greatest and least principal stresses.

4. The sense of slip is such that the wedge of material in the acute angle moves inward in the direction of the greatest principal stress.

5. The angle of internal friction is related to the angle α by Eq. 8.3, or $\phi_i = 90°-2\alpha$

This relationship of conjugate pairs of shear planes is sometimes referred to as Hartmann's rule (Bucher, 1920, p. 709).

With these rules we can recover some important aspects of the orientation of the stress field, and of an important physical property of the rock material at the time of faulting.

PROBLEM

Given two faults with attitudes N24°W, 50°W, and N48°W, 76°NE, and assuming they are conjugate, determine the orientation of the principal stresses, the direction and sense of slip and the angle of internal friction.

METHOD (Fig. 8.8)

1. Plot the planes of the two faults as great circles on the stereonet. The point of intersection defines the direction of the intermediate principal stress.

2. Draw in the great circle for which this direction is the pole. This is the plane containing the greatest and least principal stresses, and its intersection with the two faults fixes the slip directions.

3. Bisect the acute segment of this great circle between the faults to locate the orientation of the greatest principal stress. The least principal stress direction is 90° along this same great circle.

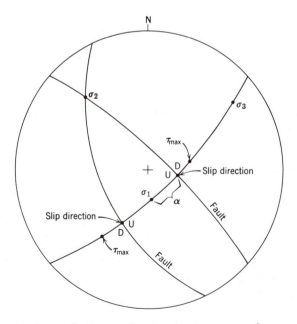

FIGURE 8.8 Orientation of the principal stresses from conjugate faults.

4. The angle between the greatest principal stress and one of the slip directions is the angle •, and the angle of internal friction can be determined.

ANSWER

The orientations of the principal stress directions are:

σ_1: 68°/006° σ_2: 21°/317° σ_3: 13°/054°

The angle of internal friction is 32°. Since the greatest principal stress direction is close to vertical, the displacement on the two faults must be dominantly dip slip. This sense can also be obtained by visualizing each fault with the flattened hand representing the plane of the fault and the index finger of the other hand representing the direction of the greatest principal stress.

If less information is available, then, of course, fewer firm conclusions concerning the stress directions can be obtained. However, an approximate answer may still be possible.

PROBLEM

Given a dominantly left-slip fault with an attitude of N30°E, 70°W, and a slip direction which pitches 15°N in the plane of the fault, estimate the orientations of the principal stress directions.

APPROACH

The direction of the intermediate principal stress and the plane of the other two principal stresses can be obtained by previous methods. An estimate of the value of angle of internal friction then fixes the direction of the greatest principal stress. Without additional information, use of the average figure of $\phi_i = 30°$ is probably the best that can be done.

METHOD (Fig. 8.9)

1. Plot the great circle representing the fault, and locate the slip direction as a point on this trace. The intermediate principal stress direction is found by counting off 90° from the slip line. Since this is the pole of the plane of the greatest and least principal stress directions, it is easy to add this great circle to the diagram.

2. With an estimated value of $\phi_i = 30°$, $\alpha = 30°$ (note that this is a special case; by Eq. 8.4, these two angles will generally differ).

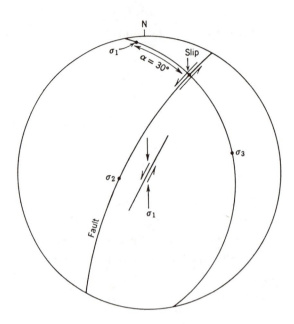

FIGURE 8.9 Orientation of the principal stresses estimated from a single fault with known slip direction.

Locate the direction of the greatest principal stress 30° from the slip direction along the great circle containing the greatest and least principal stresses. It is critical that σ_1 be plotted on the correct side of the slip line, and this depends on the sense of slip. In order to locate σ_1 it is necessary to visualize the relationship between the orientation of this greatest principal stress and the slip direction, and again, the use of a hand and a finger will help. It may also help to rotate the fault plane into some other orientation. Here, for example, it is easy to imagine the fault to be vertical, with the slip direction also horizontal. In this orientation, the direction of the greatest principal stress must be north-south (see inset Fig. 8.9), and it is then clear that the point representing it must be plotted on the north side of the fault trace.

ANSWER

The approximate orientation of the principal stresses are:

σ_1: 2°/357° σ_2: 65°/263° σ_3: 25°/088°

Other estimates of the angle of shear will, of course, give other orientations of the greatest and least principal stresses, though the probability of great error is small. For reasons which are discussed in later sections, caution is needed in this case. A reasonable level of confidence may be obtained if two conditions are met: the fault is a primary shear fracture, and not a reactivated plane of weakness, and the amount of slip is small.

Fractures, often filled with quartz or calcite, are associated with some faults. These are extension fractures (compare Fig. 8.2a), and their presence gives additional information about the state of stress and the displacement associated with the fault.

PROBLEM

A fault has an attitude of N10°E, 25°W. Subhorizontal quartz veins (N20°W, 10°E) are present adjacent to the fault plane (Fig. 8.10a). Determine the orientation of the principal stress directions, the angle of internal friction, and the direction and sense of slip.

METHOD (Fig. 8.10b)

1. Plot the plane of the vein-filled fracture as a great circle. The pole of this plane is the least principal stress direction.

2. Plot the fault plane as a great circle. The plane of Step 1 intersects the fault plane to locate the intermediate principal stress.

3. Using the intermediate principal stress direction as the pole, the plane of the greatest and least principal stresses can be drawn. This

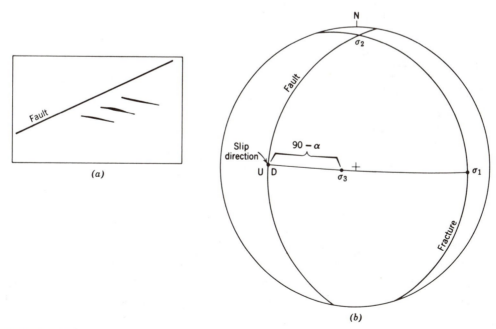

FIGURE 8.10 Orientation of the principal stresses from a fault and associated extension fracture.

great circle intersects the fault plane at the slip direction and the plane of the veins at σ_1.

4. The least principal stress direction is 90° away from the greatest principal stress direction. The angle between σ_1 and the slip line is α.

ANSWER

The orientation of the principal stresses are:

σ_1: 10°, 093° σ_2: 4°/002° σ_3: 80°/251°

The angle of internal friction is 20°; the sense of slip is reverse and the fault is a thrust.

8.8 MAGNITUDES OF THE STRESS COMPONENTS

It is also possible to get some feeling for the magnitudes of the principal stresses responsible for renewed movement on each of the three type of faults (the following presention is a graphical version of an analysis by Sibson, 1974).

First, we need to determine the relationship between the principal stresses for renewed sliding. From Fig. 8.3,

$$\sin \phi = \frac{(\sigma_1 - \sigma_3)/2}{(\sigma_1 + \sigma_3)/2}$$

and this can be rearrange to

(8.6) $$\frac{\sigma_1}{\sigma_3} = \frac{1 + \sin \phi}{1 - \sin \phi}$$

For the case of $\phi = 30°$, we then have the minimum ratio for renewed sliding,

(8.7) $$R_{min} = \sigma_1/\sigma_3 = 3.0$$

For a normal fault, σ_1 is vertical. If the mean bulk density of crustal rocks is such that the vertical component of stress at a depth of 1 km is 24 MPa, then with Eq. 8.7 the magnitude of σ_3 is 8 MPa. This situation is illustrated in Fig. 8.11, where it can be seen that, assuming conditions were originally lithostatic and that the fault is in the optimum orientation, only a modest reduction of the horizontal stress is required for renewed slip. In a similar way, we can

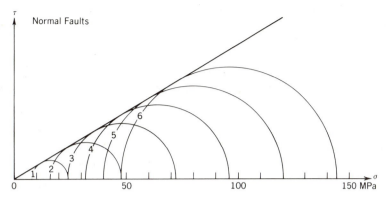

FIGURE 8.11 Estimation of the magnitude of the two principal stresses required for renewed slip on a normal fault at 1-6 kilometres.

determine the conditions for slip at depths of 2, 3, 4, 5 and 6 km. Note that the differential stress required for renewed slip increases with increasing depth.

For thrust faults, σ_3 is vertical, and at a depth of 1 km its magnitude is the same 24 MPa. From Eq. 8.7, the magnitude of σ_1 is then 72 MPa, and we see that a considerable increase in the horizontal stress component is required for renewed movement to occur. Similarly, we can find the stress conditions at greater depths (Fig. 8.12).

For a wrench fault σ_2 is vertical, and there are a whole range of possibilities. All that is required is that

$$\sigma_3 \leqslant \sigma_{zz} \leqslant \sigma_1$$

Hence the conditions for renewed slip on a wrench fault are intermediate between the conditions for normal and thrust faults.

An independent source of information concerning the state of stress associated with episodes of slip on faults comes from a study of earthquakes. It seems

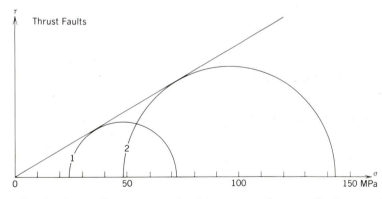

FIGURE 8.12 Conditions for renewed slip on a thrust fault at 1 and 2 kilometres.

probable that the maximum differential stress within the crust rarely exceed about 100 MPa, and it may commomly be a good deal less (Raleigh and Evernden (1981). This immediately presents a problem. With 100 MPa as an upper limit, our analysis shows that normal faults can not slip below about 6 km, thrust faults below about 2 km, and wrench faults between these two depths, and yet it is known that earthquakes are generated on all three types of faults at considerably greater depths. Admittedly, our analysis contains a number of approximations. In particular, the frictional sliding rock surfaces is a good deal more complicated than a block sliding on an inclined plane. Yet it is unlikely that a more detailed approach could change our results by more than a relatively small amount, and therefore can not account for the large discrepancy. Something must be missing from our consideration, and it is this--the pressure in the fluid contained in porous crustal rocks.

Using Eq. 7.17 we can rewrite Amonton's law in terms of the effective normal traction as,

$$(8.8) \qquad \tau = (S - p)\mu$$

The ratio of the pore fluid pressure to the overburden pressure at a given depth is give the symbol λ, where,

$$(8.9) \qquad \lambda = p/\sigma_{zz}$$

This pore fluid factor expresses the fraction of the vertical load which is born by the fluid. If $\lambda = 0$ the entire load is supported by the rock, and if $\lambda = 1$ it is entirely supported by the fluid. If the water in the pore spaces is connected to the atmosphere and if the groundwater table is essentially at the earth's surface, then the pore pressure at any depth can be calculate from

$$(8.10) \qquad p = \rho_w gz$$

where ρ_w is the density of water. When the pore fluid pressure has this value it is referred to as being _normal_. At a depth of 1 km, and using the same rule of thumb for the overburden pressure, p = 10 MPa. Hence in the situation under consideration, the pore fluid factor has a value of

$$\lambda = 10 \text{ MPa}/24 \text{ MPa} = 0.42$$

Because the deep water is usually salty and the density of crustal rocks are usually a bit less, a more representative value is $\lambda = 0.45$. This relationship between p and σ_{zz} can also be illustrated graphically (Fig. 8.13a).

However, abnormal values are regularly observed in deep oil wells, where they may be as high as $\lambda = 0.9$ (Fig. 8.13b). Further, there is compelling geologic

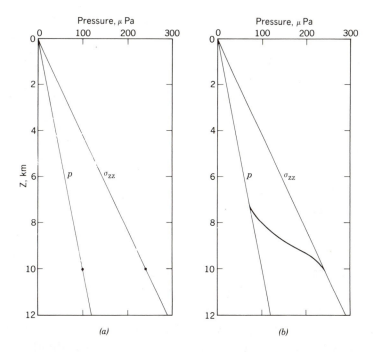

FIGURE 8.13 Fluid and overburden pressure as a function of depth in a sedimentary sequence: (a) normal conditions; (b) abnormal conditions.

evidence that at least briefly they may reach values of 1.0 or greater. The main reason for such high pore fluid pressures is that the permeability of clay shale is so low that the water can not escape as it is compacted during burial, and the pore fluid pressure continues to rise.

It is in this way that a differential stress of realistic magnitude can produce slip on a fault in the deeper parts of the crust. For example, consider the case of renewed thrusting at a depth greater than 2 km. From Eq. 8.7, the condition for slip is then,

$$(8.11) \qquad \frac{\sigma_1 - p}{\sigma_3 - p} = 3 \quad \text{or} \quad p = \frac{3\sigma_3 - \sigma_1}{2}$$

If the depth is 10 km, then $\sigma_3 = 240$ MPa. For a differential stress of 100 MPa, $\sigma_1 = 340$ MPa. With these values the pore pressure must be p = 190 MPa, and the corresponding value of the pore fluid factor is $\lambda = 0.79$ (see Fig. 8.14). In effect, the fault "thinks" that it is at a much shallower depth.

Although additional parameters are involved, a similar problem is encountered in the primary shear fracture of rocks. This too indicates that the Coulomb criterion should be modified to take pore pressures into account.

$$(8.12) \qquad \tau = \sigma_0 + (S - p)\mu_i$$

and this is commonly referred to as the Coulomb-Terzaghi criterion.

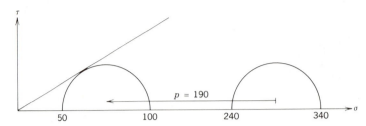

FIGURE 8.14 Conditions for slip on a thrust fault at a depth of 10 km, assuming the differential stress is 100 MPa.

Finally, at high confining pressures ductile flow may dominate over fracture (see Fig. 8.1d), in which case mylonites may form along the zone of shear. The condition favorable for such flow would be in the deeper parts of the crust and a high differential stress (and presumably relatively dry conditions which would permit this build up). This in turn suggest that mylonites should be found most commonly along thrust fault and rarely along normal faults, and this is just the case.

8.9 FAULTS IN ANISOTROPIC ROCKS

If a rock mass is anisotropic, the geometrical relationship between the principal stress directions and shear fractures is more complicated. Two type of structures may influence fracturing in this way: pervasive planar fabrics, such as cleavage, and planes of discontinuity with low or zero cohesion, such as joints.

Donath (1961, 1964) experimentally investigated the role of cleavage in shear fracturing. By loading to failure a series of cylinders of slate cut at different angles, a varied relationship between the attitude of the plane of fracture and the cleavage was found (Fig. 8.15). If an isotropic rock had been used in these experiments, a conjugate pairs of fractures inclined to the greatest principal stress direction at approximately 30° would have been expected. In the slate, such fractures were obtained in only two situations: in one experiment at higher confining pressure with the cleavage parallel to the long axis of the cylinder, and to σ_1, and in all experiments with the cleavage perpendicular the cylinder axis and to σ_1. In these two orientations, the slate was effectively isotropic. In all others, only one fracture developed, and its attitude was controlled, directly or indirectly, by the cleavage. This is most clearly demonstrated by the essentially 1:1 relationship between cleavage and fracture attitude on the left side of the graph. Although for larger angles of cleavage inclination the fractures are no longer parallel the cleavage, fracture angles greater than 45° also indicates a continuing by the planar weakness.

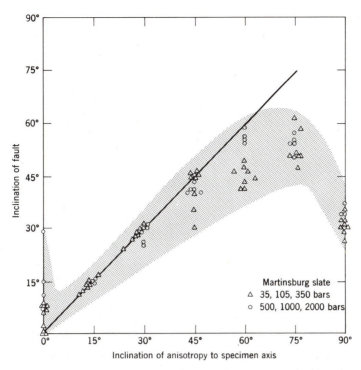

FIGURE 8.15 Relationship between shear fracture orientation and slaty cleavage (after Donath, 1963).

From the point of view of interpreting field examples, however, the uncertainties introduced by the effect of anisotropism makes the determination of the principal stress directions from a single fracture difficult or impossible. This can be illustrated more clearly by considering the second type of structure--a single plane of weakness. In three dimensions, the direction of the resolved shearing stress on a plane inclined to the principal stress direction depends on the magnitude of the principal stresses, and is not, in general, simply related to the principal directions. It follows, then, that if this shear stress exceeds the resistance on the plane, the direction of slip will also not be related directly to the principal stress directions. Therefore, the orientation of the stress field responsible for renewed movement on such a single plane of weakness is also unsolvable from geometric considerations alone.

There is one situation where stress orientation can still be estimated from a knowledge of renewed movement on preexisting planes. If a rock mass is cut by abundant, variously oriented fractures, renewed slip will occur on those planes with approximately the orienton of the conjugate pair that would have formed in intact rock. Given such a collection of fractures, the analytical procedure consists simply of plotting the poles of the preexisting planes which show evidence of movement. Barring a preferred orientation of these planes, the poles will form two clusters which are symmetrical to the principal stress directions (Fig. 8.16). Compton (1966, p. 1370) has applied a similar technique with interesting results.

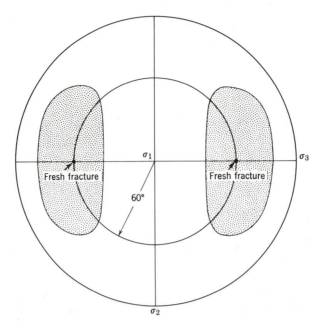

FIGURE 8.16 Assuming an average angle of friction, planes of weakness with poles within the shaded zones will show renewed movement (after Jaeger, 1969, p. 161).

8.10 LIMITATIONS

The Coulomb criterion describes conditions at the instant of failure. As a result of fracture and displacement, the magnitude and orientation of the principal stresses will no longer be the same as before failure. As the state of stress builds up for an additional increment of slip, it may or may not be the same as the original state, though it is probably unlikely that any drastic changes will occur between small increments of slip.

For large displacements, not only may the state of stress change, but also the geometry of the fault may be altered. Thrusts, in particular, are subject to such changes. The 30° dip of a primary thrust cannot be maintained for large displacements for an overhang at the surface would result. A consequence of continued slip would be a flattening of the thrust plane. The state of stress during the continued motion of such an overthrust sheet would be different, and possibly quite different from that responsible for the primary thrust.

For large-scaled faults another difficulty arises. For example, regionally extensive wrench faults may well extend to depths beyond the zone of even semi-brittle fracture. The slip then may be related to deep crustal or subcrustal flow, and for such flow the relatively simple relationships on which the fracture analysis is based does not hold. Deep flow would, of course, set up stresses in the overlying brittle rocks, but these would not necessarily be uniform along the entire length of the fault. Such nonuniformity, as interpreted from local fracture and fold analysis, has been demonstrated along parts of the San Andreas fault of California (Dickinson, 1966). The picture which emerges is one of a series of irregular blocks or "tectonic rafts" along and adjacent to the fault. The stress states within adjacent rafts are, at least partly, local. This nonuniformity, together with the long history of movement, makes it difficult to interpret such faults in terms of a regionally developed stress field. Probably all such major faults have a more complex origin and history.

Finally, as a result of the movement on a pair of conjugate faults the body of rock lengthens in the σ_1-direction, shortens in the σ_3-direction, and essentially remains of constant length in the σ_2-direction. If, however, the strain field is three-dimensional, then three or more intersecting faults may develop and the procedure used for two cannot be applied (Reches, 1978; Aydin and Reches, 1982).

EXERCISES

1. Two faults have attitudes of N10°W, 60°E and N30°E, 70°W. What were the orientations of the principal stresses at the time of faulting, what is the angle of internal friction, and what are these senses of slip on the faults.

2. Vertical extension fractures as associated with a fault whose attitude is N63°E, 70°S. Find as many of mechanical features as possible.

3. A fault with slickensides (N32°E, 60°SW) and striations (54°, N23°W) shows normal separation. Find the orientations of the principal stresses.

4. Of a large number of fractures, the following orientations showed evidence of renewed slip. Estimate the orientations of the principal stresses.
    a. N30°E, 50E   e. N84°E, 16°S   i. N50°E, 49°SE
    b. N5°W, 15°W   f. N33°E, 63°SE   j. N34°E, 28°NW
    c. N30°W, 30°W   g. N4°E, 43°°E   k. N12°E, 30°SW
    d. N54°E, 66°SE   h. N80°E, 20°N   l. N14°E, 56°E

5. A rock has the following characteristics: σ_0 = 38 MPa, ϕ_i = 33°. If remains a constant 100 MPa, what will σ_3 be at failure?

6. Under the same conditions, a plane of zero cohesion with ϕ = 30° is inclined 45° to σ_1. Which will occur first, fracture or slip?

7. At a depth of 3.5 km what would the magnitude of the horizontal component of stress be for renewed slip on a thrust fault in the ideal orientation. If the differential stress is 50 MPa, what pore fluid pressure is required.

9
Concepts of Deformation

9.1 INTRODUCTION

As a result of forces acting within the earth at various times and places, bodies of rocks are displaced from the positions of their origin. These displacements can involved any one, or more, of four distinct pattern:

1. translation: change in position,
2. rotation: change in orientation,
3 distortion: change in shape,
4. dilatation: change in volume.

Examples in which one type dominates are the emplacement of an essentially intact thrust sheet, the tilting of a fault block, the flattening of a pebble, and the volume change accompanying a reduction in porosity. Generally, however, all four types of displacements are involved in producing the geologic structures we see in the field. As a result of these displacements, a body of material with one configuration is transformed into another configuration, and the body is said to be deformed.

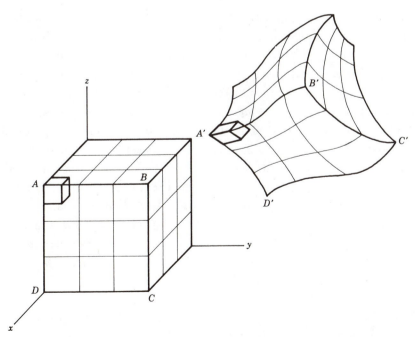

FIGURE 9.1 Translation, rotation, distortion and dilatation of a reference cube.

A complete geometric description of the deformed state for even a simple structure may be quite involved. Imagine a reference body embedded in a rock mass (Fig. 9.1). After deformation, this body will have a different location and orientation, and particularly a different form. In general, original lines will have become curvilinear, original planes curviplanar, and original parallel lines and planes will no longer be parallel. Such a deformation is termed inhomogeneous. The simplest way of proceeding with the description of such a structure is to consider a small part of the inhomogeneously deformed mass (for example, the small cube at corner A, Fig. 9.1). Compared to the larger body, the deformed equivalent of this much smaller portion has a high degree of regularity; original lines are still linear, original planes still planar, and originally parallel lines and planes are still parallel. This state is homogeneous.

Accordingly, one method of attack involves seeking such small parts which are effectively homogeneous, and then building up an overall pattern, either by comparing a series of homogeneously deformed parts, or by extrapolation based on continuity of structures which develop as a consequence of the deformation. In order to do this, however, it is necessary to develop some understanding of the geometry of homogeneously deformed bodies. Although it is ultimately necessary to treat deformation in a three-dimensional setting, many situations can be approached from a consideration of just two, and this also serves as a useful way to introduce the subject. A particularly fruitful way of exploring the geometry of deformation in two dimensions is with card-deck models.

9.2 CARD-DECK MODELS

The technique of modeling deformation with a card deck is quite simple. Various forms are drawn or printed on the edge of a reasonably thick deck of cards, and the deck is then sheared. This shearing can be accomplished in one of two different ways. The easiest way is to flex the deck with one end held firmly and then release the deck with the other end held firmly. This procedure may be repeated a number of time until the deck has the desired shape. Alternatively, a special box may be constructed and the card deck deformed with a variety of shaped endpieces or formers (Ramsay and Huber, 1983, p. 2). Strictly, deformation involves displacements which vary continuously, whereas the deformation of the card deck is simulated by a series of small slips, with each card remaining intact (Fig. 9.2) However, the thinner the cards, the closer continuity is approached. When forms of about one centimetre or larger are printed on a deck of thin cards, the distortions are for most practical purposes continuous. Used IBM computer cards are ideally suited to this purpose; they are uniform in size, thin but sufficiently stiff, and they are readily available from most computing facilities.

Even if the deck is deformed by hand it is still convenient to have an accurately square, three-sided tray to hold the cards in the sheared position for quantita-

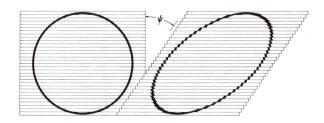

FIGURE 9.2 A card deck illustrating the formation of the strain ellipse by homogeneous simple shear; ψ is the angle of shear.

tive experiments (see Ragan, 1969, p. 135). The result can then be analyzed directly on the cards, or the patterns can be reproduced easily with a flat-window copying machine.

9.3 HOMOGENEOUS DEFORMATION BY SIMPLE SHEAR

The type of deformation simulated with the card deck is called simple shear. Rocks containing a pronounced planar structure, such as bedding or cleavage, may deform in a manner similar to that of a card deck, that is, by distributive slip on planes of weakness. It should be emphasized, however, that material may also deform by simple shear without utilizing such planes, either because they are inappropriately oriented, or absent altogether.

Simple shear is a very special type of deformation, yet despite this special character, the card-deck models portray a number of properties of more general types of deformation, including a component of rotation.

In the models, the condition of homogeneity is met by insuring that the edge of the sheared deck is straight; this is easy to do with the flexing mechanism. The simple character of simple shear lies in the fact that a single parameter describes a homogeneous deformation. This parameter is the angle of shear ψ (psi), defined as the change of an original right angle. This angle is easily measured by starting with a square-ended deck (see Fig. 9.2).

The most basic experiment is to homogeneously deform a deck on which a circle has been drawn. An ellipse results. If the original circle had a unit radius (Fig. 9.3a), the result is the strain ellipse (Fig. 9.3b), although as we will see, all the same information can be obtained from a circle of any known radius. The term strain refers to the changes in length of lines and the changes of angles between lines, as expressed by the strain ellipse; it describes the distortional and dilatational parts of the deformation.

The emphasis in the model is on the plane edge of the card deck; the deformation is, however, taking place throughout the deck, that is, in three dimensions.

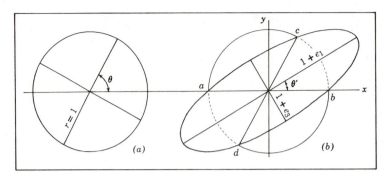

FIGURE 9.3 Geometric properties of a unit circle transformed into an ellipse by simple shear: (a) reference circle--the angle θ defines the orientation of the principal axes in the undeformed state; (b) strain ellipse--the angle θ' fixes the orientation of the principal axes in the deformed state.

The original circle can, therefore, be thought of as the diametral plane of a sphere embedded in the deck. As a result of the deformation, this sphere becomes an ellipsoid, and the ellipse seen on the edge of the deck is a principal plane of this ellipsoid.

After producing a pronounced and reasonably large ellipse the following observations can be made:

On the deformed deck, locate and draw in the two mutually perpendicular axes of the ellipse. The long axis can be located more accurately and should be drawn in first; the short axis can then be constructed as its perpendicular bisector. These two axes were diameters of the original circle, and therefore represent the directions of greatest lengthening and greatest shortening. All other lines have changed length by intermediate amounts. A measure of these changes is the <u>extension</u> e, defined as the change in length per unit length,

(9.1) $\qquad e = (\ell'-\ell)/\ell = \ell'/\ell - 1$

where ℓ is the initial length and ℓ' is final length. Note that an increase in length is reckoned positive, and a decrease is negative. The maximum and minimum, or <u>principal extensions</u> are denoted by subscripts so that

$$e_1 \geqslant e_3$$

The intermediate principal extension e_2 is perpendicular to the plane of the card deck, and is equal to zero in simple shear.

The lengths of the semi-axes of the strain ellipse are then S_1 and S_3. These are the <u>principal stretches</u>:

(9.2) $\qquad S_1 = 1+e_1 \qquad$ and $\qquad S_3 = 1+e_3$

The orientation of these two directions, called the principal axes in the deformed state, is established by the angle θ' which S_1 makes with the shear direction. The stretch associated with any other radius of the strain ellipse is, by Eq. 9.1,

(9.3) $S = 1+e = \ell'/\ell$

By superimposing a concentric circle on the final ellipse identical in size to the starting circle, two special lines are identified which have undergone no net change in length (lines ab and cd of Fig. 9.3b). These are the lines of no finite longitudinal strain. In simple shear one of these always remains fixed in the direction of shear; the orientation of the other is a function of the angle of shear.

Returning the card deck to its starting position, the lines which marked the axes of the strain ellipse are also perpendicular in the initial circle; these are the principal axes in the undeformed state (Fig. 9.3a). The orientation of these two axes is established by the angle θ which the line destined to become S_1 makes with the shear direction. In a general deformation, as here, this pair of axes has a different orientation before and after the deformation. This change in orientation is measured by the angle of rotation ω (omega), where

(9.4) $\omega = \theta' - \theta$

$\theta' \equiv \theta_d$
rufan   *Davis*

It is especially important to note that we are able to determine this angle of rotation only because we can measure θ and θ' from the shear direction, which we assumed remained fixed throughout the experiment. There could well be an additional component of rotation, simulated in the model by bodily rotating the deformed deck. We will see later that in certain types of folding, sedimentary bedding undergoes layer parallel shear while at the same time rotating bodily. We will return briefly to the problem of determining the rotational component of the deformation in the next chapter.

In the restored position, the circle used to define the lines of no finite longitudinal strain becomes an ellipse identical to the strain ellipse, but inclined in the opposite direction. This is the reciprocal strain ellipse, and its axes also coincide with the principal directions in the initial circle.

9.4 ANALYSIS OF SIMPLE SHEAR

The geometrical properties of a homogeneous deformation by simple shear illustrated with the card-deck experiment can calculated using a method first set out by Thomson and Tait (1867, 1962; see also Treagus, 1981). Before proceeding, we need an additional parameter--the shear strain γ (gamma) defined as

(9.5) $\gamma = \tan \psi$

Referring to Fig. 9.4a, for any given angle of shear there will always be a line before deformation OA which after deformation has the same length OB. This is a line of no finite longitudial strain. The orientations of the lines OA and OB are easily found; the angle δ which each makes with the normal ON to the shear plane is related to γ by

$$(9.6) \qquad \tan \delta = \gamma/2$$

The angle α' which the line of no finite longitudinal strain makes with the shear direction is then

$$(9.7) \qquad \alpha' = 90° - \arctan(\gamma/2)$$

Because the two lines of no finite longitudinal strain are symmetrical to the ellipse axes, the orientation of the principal axes on the deformed deck can be found by bisecting the two angles at B. This can be easily accomplished with an elementary construction: draw a semi-circle centered at O with radius OA = OB (Fig. 9.4b); the chord BC then bisects the acute angle at B and fixes the orientation of the long axis of the strain ellipse. Similarly, the chord BD bisects the obtuse angle at B and gives the orientation of this same line in the undeformed state.

As a result of the deformation line AC becomes line BC, which represents the long axis of the strain ellipse. The angle between these two lines is, therefore, the angle of rotation. Because the inscribed angle ACB and the central angle AOB intercept the same arc, $\omega = 2\delta$ and therefore by Eq. 9.5,

$$(9.8) \qquad \tan\left(\frac{\omega}{2}\right) = \gamma/2$$

An expression for the orientation of the principal directions can also be obtained. From Fig. 9.4b, $\theta' = \alpha'/2$. Then with Eq. 9.7,

$$\theta' = [90° - \arctan(\gamma/2)]/2 \quad \text{or} \quad \tan(90° - 2\theta') = \gamma/2$$

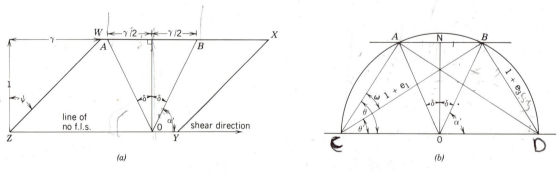

(a)   *(b)*

FIGURE 9.4 Simple shear and the principal axes: (a) location of the lines of no finite longitudinal strain; (b) construction of the principal axes from these lines (after Thomson and Tait, 1962, p. 124).

Introducing the trigonometric identities $\tan(90°-x) = \cot x = 1/\tan x$, we have

(9.9) $\qquad\qquad \tan 2\theta' = 2/\gamma$

Two angles, θ' and $\theta'+90°$, satisfy this equation, and give the orientations of the two principal axes in the deformed state. In a similar way the orientations of the principal axes in the undeformed state can be found from

(9.10) $\qquad\qquad \tan 2\theta = -2/\gamma$

The first solution of Eq. 9.9 gives θ', and the first solution of Eq. 9.10 gives $\theta+90°$.

The magnitudes of the principal stretches can also be found from angular measures. In the S_1-direction $\ell = AC$ and $\ell' = BC$. Then from Eq. 9.2,

$$S_1 = BC/AC$$

From Fig. 9.4b, and noting that $AC = BD$,

$\tan(\theta - \theta') = \tan \psi/2$

$$\tan \theta' = AC/BC = 1/S_1$$

Hence

(9.11) $\qquad\qquad S_1 = 1/\tan \theta'$

The area strain Δ (delta) is defined as the change in area per unit area, or

(9.12) $\qquad\qquad \Delta = (A'-A)/A = A'/A - 1$

where A is the initial area and A' is final area; by this definition an increase in area is positive, and a decrease is negative. In terms of a circle transformed into an ellipse, this may be written as

$$\Delta = [\pi S_1 S_3 - \pi r^2]/\pi r^2$$

For our reference circle $r = 1$, and this expression then reduces to

(9.13) $\qquad\qquad \Delta = S_1 S_3 - 1$

Simple shear does not change area; the parallelogram WXYZ of Fig. 9.4a has the same area as the original rectangle from which it was derived. Because no changes occur perpendicular to the plane of this parallelogram, there is also no change in volume in simple shear. Deformations which conserve volume are termed isochoric. Simple shear is one example of an isochoric deformation.

For the condition of no area change, the area strain $\Delta = 0$; using Eq. 9.13, we then have a necessary relationship between the principal stretches,

(9.14) $S_3 = 1/S_1$

Then with this and Eq. 9.11,

(9.15) $S_3 = \tan \theta'$

Knowing the orientation of the simple shear ellipse, the shear strain can be determined. From Eq. 9.9 we obtain directly

(9.16) $\gamma = 2/\tan 2\theta'$

The value of γ can also be obtained from the principal stretches. Using the identity for the tangent of a double angle in Eq. 9.16 we have

$$\frac{2}{\tan 2\theta'} = \frac{1-\tan^2\theta'}{\tan \theta'} = \frac{1}{\tan \theta'} - \tan \theta'$$

Using Eqs. 9.11, 9.15 and 9.16, we find that

(9.17) $\gamma = S_1 - S_3$

Because it immediately establishes the shape of the strain ellipse, a very useful measure is the strain ratio,

(9.18) $R_s = S_1/S_3$

With Eq. 9.14,

$$R_s = S_1^2 \quad \text{or} \quad S_1 = \sqrt{R_s}$$

and

$$R_s = 1/S_3^2 \quad \text{or} \quad S_3 = 1/\sqrt{R_s}$$

With these we may rewrite Eq. 9.16 as

(9.19) $\gamma = \sqrt{R_s} - 1/\sqrt{R_s}$

9.5 SUPERIMPOSED DEFORMATIONS

The card deck is also useful for illustrating what happens when two homogeneous deformations are superimposed. A simple experiment consists of drawing a circle and transforming it into an ellipse, which then represents the result of the first deformation D_1. A second circle is then added to the deck, and it in turn is also transformed into an ellipse which then represents the second deformation D_2. The first circle, now twice deformed, represents the total deformation D_T.

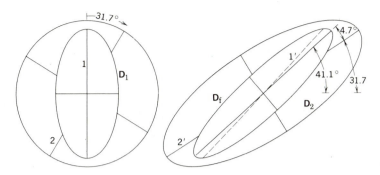

FIGURE 9.5 Card-deck experiment illustrating the superposition of two homogeneous deformations; the numbers 1 and 1', 2 and 2' identify material lines before and after deformation.

In describing the ways in which such ellipses are superimposed on ellipses, several ratios are defined in the same way as the strain ratio, that is, the ratio of the long to short axes of an ellipse, as measures of the distortions which take place.

Actually, we are not restricted to superimposed simple shear deformations, for any ellipse drawn on the deck can represent the first deformation. For example, in Fig. 9.5, D_1 is represented by an ellipse with an axial ratio $R_1 = 2.0$ and oriented with its long axis perpendicular to the shear direction. It and a concentric circle

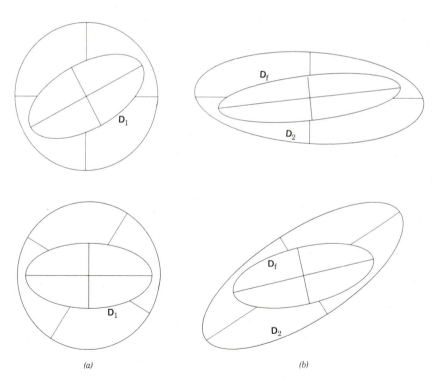

FIGURE 9.6 Apparent rotation: (a) the D_2 ellipse of Fig. 9.5 is now irrotational; (b) counter rotation of an ellipse with long axis originally parallel to the shear direction.

are then sheared through an angle of $\psi = 45°$. In the result the D_1, the D_2 and the D_T ellipses are all differently oriented, and the axial ratios R_1, R_2 and R_T are not simply related.

An additional feature of this type of superimposed deformations is brought out more clearly if the rotational component of D_2 is removed; this is easily accomplished by redrawing the two ellipses so that the D_2 axes before and after the second deformation have the same orientation (Fig. 9.6a). It can then be seen that the D_1 ellipse has undergone a change in orientation even though D_2 is now irrotational. This is an apparent rotation because the axes of the D_1 and D_T ellipses are marked by different material lines. This effect can be seen even more dramatically by drawing the D_1 ellipse with its long axis parallel to the shear direction (Fig. 9.6b). After deformation, the D_T ellipse appears to have rotated in a sense counter to that of the simple shear rotation.

These examples illustrate the fully general, or noncoaxial superposition of two strain ellipses. The special cases of coaxial superposition can also be illustrated. In Fig. 9.7, the D_1 ellipse is drawn with its axes coinciding with the corresponding D_2 axes in the undeformed state. In the result, the D_2 and D_T ellipses are also coaxial, and the ratios of the three ellipses are related by

(9.20) $$R_T = R_1 R_2$$

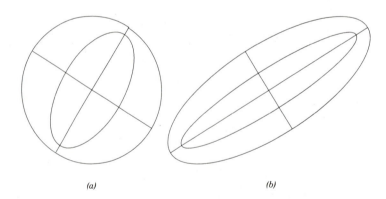

(a)　　　　　　　(b)

FIGURE 9.7   Coaxial superposition of strain ellipses where the two corresponding pairs of axes are parallel: (a) before deformation; (b) after deformation.

In a similar coaxial case, the D_1 ellipse is drawn with its axes coinciding with the opposite D_2 axes (Fig. 9.8). The axial ratios are then related by

(9.21) $$R_T = R_1/R_2 \qquad or \qquad R_T = R_2/R_1$$

depending on whether R_1 is greater or less than R_2.

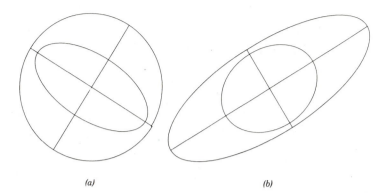

(a)   (b)

FIGURE 9.8 Coaxial superposition of strain ellipse where the two corresponding pairs of axes are perpendicular: (a) before; (b) after deformation.

In general, the axial ratios of the final ellipse representing the results of
the noncoaxial superposition of two homogeneous deformations lies between the ratios
of these two coaxial cases, that is,

$$(9.22) \qquad (R_2/R_1) \leqslant R_T \leqslant (R_1 R_2)$$

9.6 DEFORMED PEBBLE PROBLEM

The way pebbles or other elliptical grains deform is geometrically similar to
the results of superimposed deformations; the role of the strain ellipse after the
first deformation is replaced by the elliptical shape of the pebble. Given a collection of homogeneously deformed pebbles, subject to the conditions that:

1. the initial elliptical shapes were all the same, and
2. the pebbles were initially without preferred orientation,

we may easily determine the orientation of the principal axes in the deformed state,
the strain ratio and the initial shape ratio.

The graphical technique utilized here is aimed at identifying the two coaxially
deformed pebbles (as in Figs. 9.7 and 9.8). The axial ratios of these two extreme
shapes are related to the strain ratio R_s and the initial shape ratio R_i by

$$(9.23) \qquad R_{max} = R_s R_i \qquad \text{and} \qquad R_{min} = R_s/R_i$$

With the maximum and minimum ratios known, these two equations may then be solved
for the two unknown ratios, giving

$$(9.24) \qquad R_s^2 = R_{max} R_{min} \qquad \text{and} \qquad R_i^2 = R_{max}/R_{min}$$

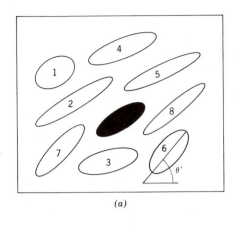
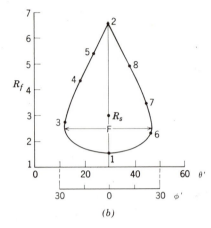

(a)   (b)

FIGURE 9.9   Homogeneous deformation of variously oriented ellipses: (a) the angle θ' fixes the orientation of the long axes of each elliptical form; the black ellipse is the strain ellipse; (b) graph of the final ratio R_f against θ' for each ellipse; R_s plots near the center of the tear-drop curve (after Ramsay, 1967, p. 211).

PROCEDURE

1. From a suite of deformed pebbles (Fig. 9.9a):

   a. Measure the angle θ' between the long axis of each pebble and an arbitrary reference direction.

   b. Measure the two axial length of each pebble and calculate the final shape ratio R_f = long/short axes.

2. Plot each pair of values (θ', R_f) as a point on a graph (Fig. 9.9b).

3. The resulting points lie on a closed curve, symmetric about a fixed value of θ', which defines the orientation of the S_1-direction of the strain ellipse. Having identified this direction, it is then convenient to adopt it as the reference direction and to relate the orientation of the deformed pebbles to it by the angle ϕ'.

4. The two points on the line $\phi' = 0$ represent the extreme shapes R_{max} and R_{min}. With these, the values of R_s and R_i are found with Eqs. 9.20.

It should be emphasized that without additional information there is no way of knowing the pre-deformational orientation of the reference direction of Step 1a, and therefore no way of determining the rotational component of the deformation.

As can be seen clearly from the R_f/ϕ' graph of Fig. 9.9b, the long axes of the

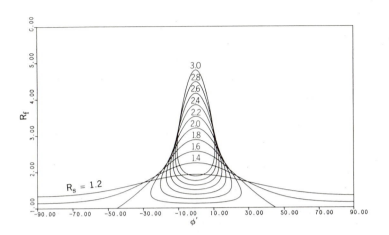

FIGURE 9.10 R_f/ϕ' curves for R_i = 1.6 and a variety of strain ratios.

deformed pebbles have a range of orientations. This is termed the <u>fluctuation</u>
(Cloos, 1947, p. 861), and it can be can obtained from the R_f/ϕ' curve.

This example illustrates one of two distinct cases which arise as the result of
the homogeneous deformation of elliptical particles--namely when the strain ratio is
greater than the initial shape ratio. If the reverse is true, that is, if the
strain ratio is less than the initial shape ratio, quite a different curve results
(Fig. 9.10), though the method of finding R_s and R_i is the same. In this second
case, the fluctuation is F = 180°.

Ramsay (1967, p. 208) has derived an expression which describes how the maximum
fluctuation is related to the strain ratio and initial shape ratio.

$$F = \frac{R_s(R_i^2 - 1)}{[(R_i^2 R_s^2 - 1)(R_s^2 - R_i^2)]^{1/2}}$$

The important feature of this relationship is shown graphically in the inset of Fig.
9.11; for any given initial shape ratio, the fluctuation remains constant at 180°
until

$$R_s = R_i$$

at which point a preferred orientation suddenly appears defined by F = 90° (see inset, Fig. 9.11), that is, the most divergent shapes lie at angles of 45° on either

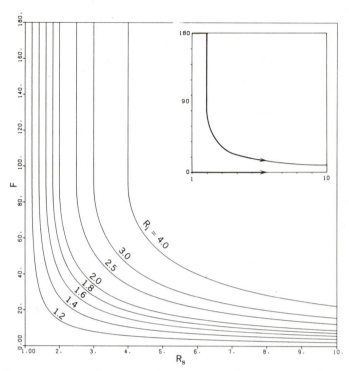

FIGURE 9.11 Graph of the fluctuation (after Ramsay, 1967, p. 208) as a function of the initial pebble shape and the strain ratio.

side of the S_1 direction. In many areas, slaty cleavage appears quite abruptly, and Elliott (1970, p. 2232) has suggested that this cleavage front marks such a sudden onset of preferred orientation.

With its constant ratio initial shapes this example is not very realistic. If a variety of distinct initial shapes are present, the R_f/ϕ' graphs consist of a series of nested curves, one for each R_i (Fig. 9.12).

More generally yet, real data will not plot on such distinct curves, but will appear as a scatter of points, reflecting the continuous variation of initial

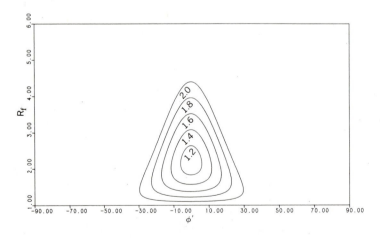

FIGURE 9.12 R_f/ϕ' graph showing the curves for a variety of original shapes strained by $R_s = 2.2$.

shapes. Dunnet (1969) compared such scattered diagram with a series of R_f/ϕ' graphs constructed for a range of values of R_s, as in Fig. 9.12, and visually estimated the best fit.

Lisle (1977, p. 383) has suggested a method by which this fitting process may be improved. If measurements of the pebble shapes and orientations were made in the undeformed state, the resulting R_i/ϕ' diagram would appear as in Fig. 9.13a; with no preferred orientation each subarea of 9° width would be expected to contain 5% of the total number of markers. By finding the deformed equivalents of these $\phi = $ constant lines, we obtain a family of curves which divide the R_f/ϕ' graph into subareas containing the same number of points (Fig. 9.13b).

These curves are calculated from equations derived by Ramsay (1967, Eq. 5-22, p. 206, and Eq. 2-27, p. 209). The result is

$$R_f^2 = \frac{\tan 2\phi(R_s^2 - \tan^2\phi') - 2 R_s \tan \phi'}{\tan 2\phi(1 - R_s^2 \tan^2\phi') - 2 R_s \tan \phi}$$

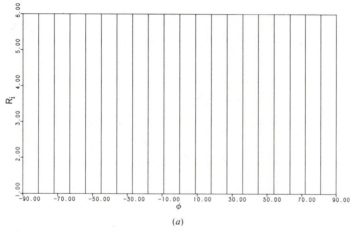

(a)

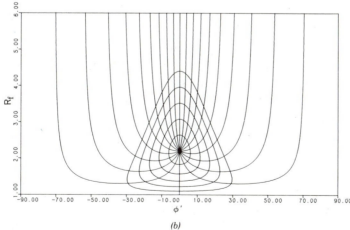

(b)

FIGURE 9.13 Shape vs orientation graphs: (a) before strain--with no preferred orientation, each 9° subarea is expected to contain 9/180 = 5% of the total markers, (b) after strain--curves of constant R_i and ϕ are shown (after Lisle, 1977, p. 384).

except that the $\phi' = 45°$ curve is (Dunnet and Siddans, 1971, p. 314),

$$R_f^2 = \frac{\tan^2\phi - R_s^2}{R_s^2 \tan^2\phi - 1}$$

A set of these curves can then be produced for a range of R_s values. The strain ratio is determined by calculating the Chi-square statistic,

$$\chi^2 = \sum_{i=1}^{n} (O_i - E)^2/E$$

where O_i is the observed number of points in the ith area, and $E = N/n$, where N is the total number of markers and n is the number of subareas bounded by the adjacent ϕ curves. The family of curves giving the lowest value of χ^2 is taken as the best fit. This approach has an additional advantage; the value of χ^2 is also an indication of the fit of the data to the assumed model. Finally, the entire procedure can be programmed for a computer solution (Peach and Lisle, 1979).

In applying this R_f/ϕ technique to naturally deformed materials, there are several additional factors which may limit its use. For example, it assumes that the initial fabric of the elliptical markers was random in the undeformed state. Many naturally occuring sedimentary fabrics do show some degree of preferred orientation. In a study of the simulated deformation of such fabrics, Seymour and Boulter (1979) showed that large errors may results if it is mistakenly assumed that they were originally uniform. Also, if there is a ductility contrast between the elliptical objects and the matrix material, an additional component of rotation will be present which may invalidate the strain analysis (De Paor, 1980).

9.7 INHOMOGENEOUS DEFORMATION

Some important aspects of more general types of deformation can also be illustrated on a card deck. One simply deforms the deck inhomogeneously (Fig. 9.14). The structure modeled here is a shear zone, that is, a tabular body of sheared rock bounded on both sides by undeformed material.

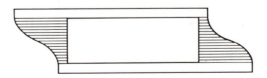

FIGURE 9.14   Shear zone with a card deck; the area within the box is shown in
Fig. 9.15.

The geometric details of such inhomogeneous deformations can be followed more closely if a number of small circles are stamped on the deck; this can be accomplished easily with a slip-on pencil eraser and an ink pad. After shearing the deck, the resulting ellipses show clearly how the strain varies across the deck (Fig. 9.15a). As can be seen, the most intense strain occurs in the center of the zone and dies away in both directions to become zero at the boundaries.

An alternative approach for presenting this same information is to construct two sets of curves, called <u>trajectories</u>, which are everywhere tangent to the principal axes. These are shown in Fig. 9.15b. Note that the magnitudes of the principal stretches vary along these curves, thus only orientational information is given directly. However, the S_1 trajectories converge toward regions of greater strain, so that the patterns do indicate the variations in the values of the principal stretches as well, and it is possible to extract this information from such patterns.

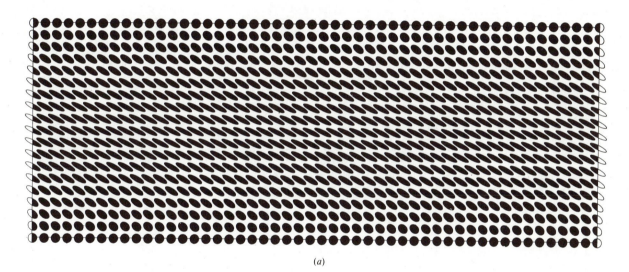

(a)

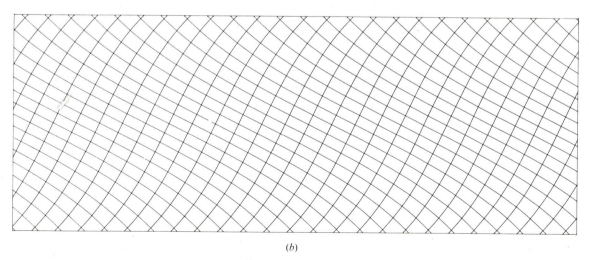

(b)

FIGURE 9.15 Model shear zone: (a) distribution of strain ellipses; (b) strain trajectories.

In real shear zones, the curved S_1 trajectories are commonly marked by a foliation usually referred to as s-surfaces (from schistosity). In coarse-grained rocks,
the shear planes may also be evident, and these are denoted c-surfaces (from
cisaillement = shear); in the model these are the planes of the cards themselves.
When these two intersecting surfaces are both visible, the sense of shear can be
detemined immediately (Simpson and Schmid, 1983, p. 1284).

This correspondence of foliation and S_1 trajectories permits the strain distribution and the displacments to be determined in naturally occuring zones. By measuring the slope of the foliation at a series of points across a zone, the distribution
of θ' which defines the orientation of the strain ellipse to be plotted (Fig.
9.16a). Knowing θ', and using Eq. 9.16 the distribution of γ can then also be
plotted (Fig. 9.16b).

From the shear strain at a number of points, the local angle of shear is also
known. The displacement curve can then be estimated by a graphical integration technique. If a series of short lines parallel to the known local angle of shear are
plotted across the zone, the displacement curve can then be sketch in (Fig. 9.16c).
From this, the total displacment across the zone can then also be found.

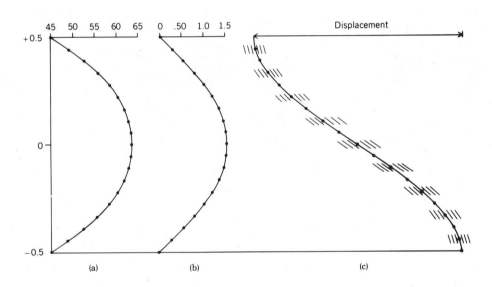

FIGURE 9.16 Analysis of shear zones: (a) distribution of θ'; (b) distribution of γ; (c) displacement by graphic integration of γ.

9.8 MOTION LEADING TO DEFORMATION

The study of deformation is concerned soley with a comparison of the initial
and final configurations. However, the motion, or <u>flow</u>, by which the deformed state

evolves is also of interest, and the card-deck models can be used to illustrate some aspects of the geometrical changes which lead to a state of homogeneous deformation.

An easy and instructive way to appreciate this evolution is to perform the basic circle-to-ellipse experiment by shearing the deck through a series of small increments. Initially, this can be done fairly rapidly, perhaps several times. However, a much more precise picture can be obtained by plotting certain changes in length and orientation as a function of an increasing angle of shear.

A graph showing the change in orientation of the long axis of the ellipse with progressive shear is given in Fig. 9.17a. At small values of ψ the orientation of the strain ellipse is difficult to determine with any accuracy; in particular, it will not be readily apparent but the intial orientation is given by $\theta' = 45°$. Note also that with increasing shear the S_1 direction approaches the shear direction.

A graph of the changing magnitudes of the principal stretches, and especially their ratio, clearly indicates how the strain ellipse evolves in progressive simple shear (Fig. 9.17b).

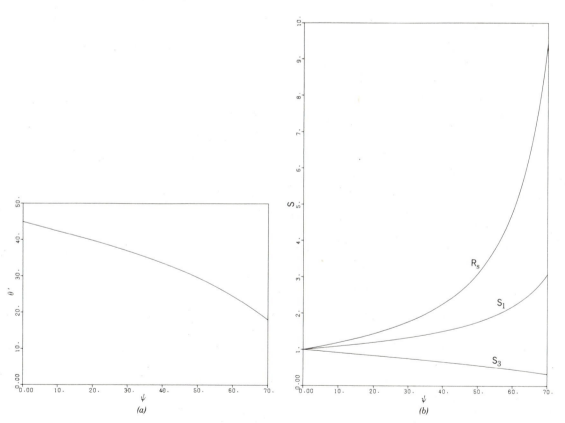

FIGURE 9.17 Parameters of simple shear as functions of ψ: (a) orientation of the S_1 direction; (b) values of S_1, S_3, and R_s.

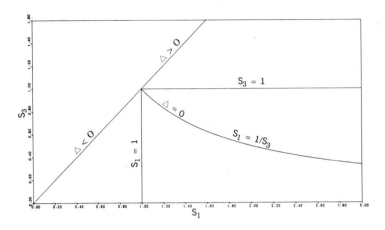

FIGURE 9.18 Graph of strain ellipses; the constant-area strain path is shown. Ellipses involving an area increase plot above this curve, those involving an area decrease plot below.

A plot of the values of S_1 against S_3 is an especially useful aid in visualizing the intermediate stages through which all strain ellipses pass (Fig. 9.18). The curve depiciting this evolution, and by extension the evolution itself, is called the <u>strain path</u>. For constant area ellipses, including simple shear ellipses, the path is defined by the hyperbola,

$$S_1 = 1/S_3$$

A more complete history of the deformation takes the angle of rotation into account with the use of a three-dimensional graph (Fig. 9.19). The curve on this graph may be called the <u>deformation path</u>. This particular plot also makes clear

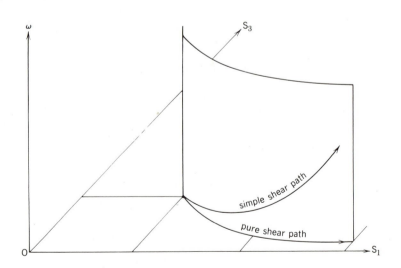

FIGURE 9.19 Graph including the angle of rotation; the surface of constant area is shown (after Ramsay, 1967, p. 96).

that there are many paths for constant area deformation besides simple shear; the constant area path which lies in the plane of S_1 and S_3 defines the special irrotational deformation called pure shear.

Another useful approach is to observe the change in orientation of a line making an initial angle α with the shear direction (Fig. 9.20). An analytical expression for the angle α' the same line makes after deformation can be obtained directly from this illustration; it is

(9.25) $\cot \alpha' = \gamma + \cot \alpha$

This formula is the basis for a graph of the history of a series of lines originally oriented at angles of 10°-170° (Fig. 9.21). As can be seen, all lines rotate toward the shear direction.

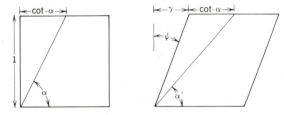

FIGURE 9.20   Change in orientation of a line originally making an angle with the shear direction (after Ramsay, 1967, p. 87).

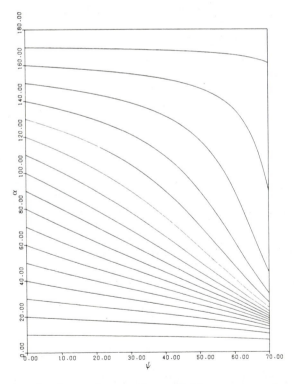

FIGURE 9.21   Graph of the orientation of lines as a function of an increasing angle of shear.

The history of the stretch associated with these same lines is also of interest. From Fig. 9.20, the original and deformed lengths of a line are

$$\ell = 1/\sin\,\alpha \qquad\text{and}\qquad \ell' = 1/\sin\,\alpha'$$

With these in Eq. 9.2 we have an expression for the stretch,

(9.26) $\qquad S = \sin\,\alpha/\sin\,\alpha'$

Graphs of this relationship are shown in Fig. 9.22. There are two distinct types of strain histories. Lines with original orientational angles less than 90° have a his-

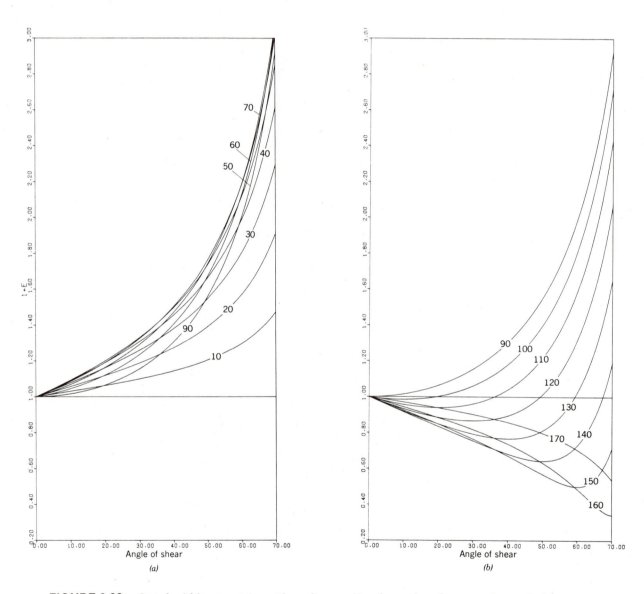

FIGURE 9.22 Graph illustrating the change in length of a series of lines as a function of an increasing angle of shear: (a) for α < 90° all lines have a history of lengthening; (b) for α > 90° the behavior is complicated.

tory of continuous lengthening, though the rates of extension vary with both the values of α and ψ (Fig. 9.22a). On the other hand, lines whose original angles are
greater than 90° have more complex histories (Fig. 9.22b), involving three stages:

1. Initially all such lines shorten.

2. Sooner or later, each line reaches a minimum value of S and thereafter
   lengthens. For lines defined by original angles just slightly greater
   than 90° this minimum is reached after only a small shear, whilen those
   originally closer to 180° require a much greater shear to reach this
   minimum. Careful examination of the geometry of the card deck will
   show that this minimum is reached when $\alpha' = 90°$. The stretch at this
   minimum, found by substituting $\alpha' = 90°$ into Eq. 9.26, is given by sin
   α. From Eq. 9.25, the shear required for this minimum is given by
   $-1/\tan \alpha$. For example, when the line defined by $\alpha = 120°$ has an
   orientation of $\alpha' = 90°$ the stretch is 0.87, and this requires an angle
   of shear $\psi = 30°$.

3. As a line continues to lengthen, it passes the through the point where
   $S = 1$ (or $e = 0$), and thereafter its deformed length is greater than
   its intial length. When $e = 0$, the line coincides with the line of no
   finite longitudinal strain.

The geometric relations of these several types of behavior experienced by lines
during progressive simple shear can be clarified by subdividing the original circle
(Fig. 9.23a) and the ellipse (Fig. 9.23b) into three pairs of sectors, or zones.

Zone 1 contains the S_1-direction and all lines have enjoyed a finite elongation. The boundaries of this zone are the two lines of no finite longitudinal
strain. Lines in Subzone 1a have a continuous history of lengthening. Thin competent bands lying in this sector will be stretched and may develop boudins. Lines in

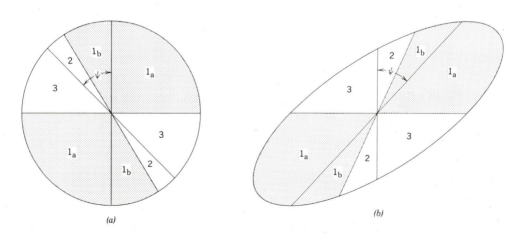

(a)   *(b)*

FIGURE 9.23 Arrangement of zones of elongation and contraction in simple
shear: (a) before deformation; (b) after deformation.

Subzone 1b experience an initial period of shortening, with the result that early formed folds may be stretched, possibly with the development of boudins on the fold limbs. The boundary between these two subzones is the line initially at $\alpha = 90°$.

Zone 2, like Subzone 1b, contains lines with a history of shortening followed by lengthening, but all lines still exhibit a finite shortening. Early folds may begin to be disrupted.

Zone 3 contains the S_3-direction, and all lines within this sector have a continuous history of shortening; ptygmatic folds may be found with this orientations. The boundary between Zone 2 and Zone 3 is the line defined by $\alpha' = 90°$.

TABLE 9.1   Zones of the simple shear strain ellipse.

Zone	Boundary	History	Structure	Net Change
a		Continuous lengthening	Boudins	S > 1
1 — $\alpha = 90°$				
b	— S = 1 —	Early shortening, late lengthening	Folds unfolded or boudinaged	
2	— $\alpha' = 90°$ —			S < 1
3		Continuous shortening	Folds	

It should be emphasized that all these changes, summarized in Table 9.1, occur simultaneously during the simple-shear flow. Other patterns of flow may give identical final strain ellipses, but the histories, and therefore the details of the minor structures, are path dependent; in particular, different rotational histories may give markedly different zonal arrangements. Ramsay (1967, p. 114-120) gives an extended discussion of this matter, together with some excellent photographs of the types of structures which may be associated with such complex behavior.

EXERCISES

1. Perform the circle-to-ellipse card-deck experiment, and make the necessary measurements to determine the values of γ, S_1, S_3, θ, θ', and ω.

2. With the angle ψ used in your experiment, calculate the values of these same parameters. With care your experimental results should be within 5% of these calculated values.

3. Perform a progressive card deck experiment and plot the stages through which simple shear ellipses pass using the type of graph illustrated in Fig. 9.17

4. Perform a progressive card deck experiment and plot the orientation and strain histories of the lines defined by $\alpha = 65°$ and $\alpha = 115°$.

5. Produce a model shear zone by heterogeneously deforming a deck on which a large number of small circles are stamped.

6. Using the example of a naturally occuring shear zone of Fig. X9.1, determine the displacements which are responsible for the structure--assume a unit width.

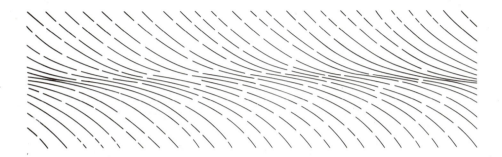

FIGURE X9.1

10
Strain in Rocks

10.1 INTRODUCTION

Performing an ellipse experiment with a card deck clearly demonstrates that the lengths and the orientations of lines and the angles between pairs of lines generally change as the result of deformation. Given suitable geological material the stretch and shear strain may be determined from measurements made on deformed objects whose original shape or dimensions are known. From such measurements it may then be possible to determine something of the shape, size and orientation of the strain ellipse. For example, measurement of the deformed shape of an originally spherical oolite yields the orientation and strain ratio directly. Under certain circumstances this same information can be obtained from diversely oriented, originally ellipsoidal pebbles, as discussed in Chapter 9.

Here some additional examples from which two-dimensional strain information can be obtained are given in order to introduce the basic approach. Many more examples can be found in the excellent book by Ramsay and Huber (1983).

Before we embark on a description of the full analytical method, it will be useful to show that in some situations the shape and orientation of the strain ellipse may be obtained simply and directly using a purely geometrical construction.

10.2 SIMPLE STRAIN ANALYSIS

Fossils often possess planes of symmetry, known angular relationships, or proportions which are constant in individuals of a given species. They are, therefore, common objects of known original shape.

PROBLEM

Given a collection of deformed brachiopods (Fig. 10.1a), construct the strain ellipse.

CONSTRUCTION (after Wellman, 1962)

1. Transfer the hinge line and the symmetry line of each individual deformed fossil to a tracing sheet (Fig. 10.1b).

2. On this tracing draw a line of arbitrary orientation and length; it
   should be at least 10 cm long and not parallel to any fossil line (see
   line AB in Fig. 10.1b).

3. Using the traces of each deformed shell in turn, draw a pair of lines
   parallel to the hinge and another pair parallel to the symmetry line
   through the points A and B giving a parallelogram (see the example
   drawn for Shell No. 8 in Fig. 10.1c)

4. Through all the pairs of fossil points determined in this way, including points A and B, sketch a best fit ellipse, and add the major
   and minor axes. This represents the strain ellipse.

5. Measure the orientation of the principal axes, and their lengths.

ANSWER

Since the size of the constructed ellipse depends entirely on the length
of line AB, the absolute lengths of the semi-axes can have no meaning.

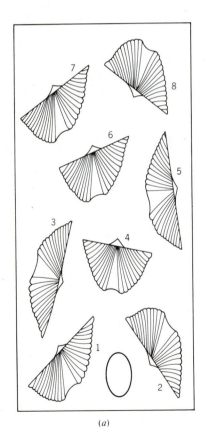

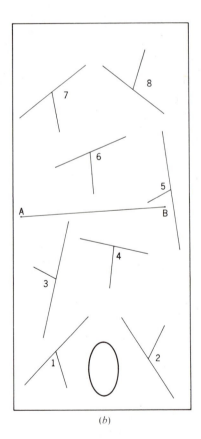

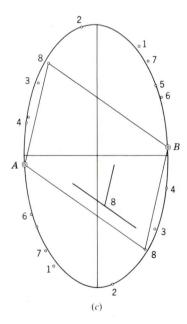

(a)   (b)   (c)

FIGURE 10.1   Simple graphical method for determining the shape and orienta-
              tion of the strain ellipse: (a) slab of brachiopods; (b) traces
              of hinge lines and lines of symmetry; (c) strain ellipse.

However, their ratio is independent of the size of the ellipse, and is found to be $R_s = 1.7$. The S_1-direction makes an angle of $10°$ with the hinge of Shell No. 5. These results can be checked against the small ellipse, which was a circle before deformation.

In order to see why this method works, imagine having made the same construction before deformation. Because each pair of fossil lines was originally perpendicular, rectangles rather than parallelograms would have resulted. Collectively, the corners of all these rectangles would have defined a circle with AB as diameter. It is this circle from which the constructed strain ellipse is derived.

Now reexamine the deformed brachiopods. The deformed shape of each shell is a function of orientation. Strictly, all right angles have been eliminated. However, Shell No. 3 is still nearly symmetrical; this is also the longest and narrowest form. Similarly, Shell No. 4 also retains close to a 90° angle, but it is deformed in the opposite way to a short, broad form. Because the principal axes are the only pair of lines which remain perpendicular, the direction of maximum elongation must nearly coincide with the hinge line of Shell No. 3, and the symmetry line of Shell No. 4, and the direction of minimum elongation is perpendicular to these lines. Thus one can estimate the orientation of the principal axes by inspection.

The chance of finding such a large number of deformed fossils exposed on a single plane is remote. We will, therefore, have to find ways of handling more realistic situations. In order to do this, it will first be necessary to explore the geometry of the strain ellipse in some detail.

In order to describe the changes experienced by lines and angles we need to refer both the initial circle and the derived strain ellipse to a set of coordinate axes. It is most convenient to chose the coordinate origin to be a material point at the center of the original circle, and the two orthogonal axes to be material lines which mark the principal axes in the undeformed state. As we have seen in the card-deck experiments, these two directions will then also be orthogonal in the deformed state, where they will be the principal axes of the strain ellipse. It should be noted that by this choice we have eliminated from consideration the translational and rotational components of the deformation; we do this to concentrate on the geometric properties of the ellipse itself.

10.3 GEOMETRY OF THE STRAIN ELLIPSE

Coordinate points in the undeformed state are labeled (x,y). The equation of the reference circle of unit radius is then

$$x^2 + y^2 = 1$$

Similarly, coordinate points in the deformed state are labeled (x',y'), and strain ellipse has the equation

$$(10.1) \qquad \frac{x'^2}{S_1^2} + \frac{y'^2}{S_2^2} = 1$$

Unlike the ellipse resulting from simple shear, we cannot assume at the outset that we are dealing with the maximum and minimum principal stretches. Hence in this equation of the ellipse, the semi-axes are arbitrarily labeled S_1 and S_2. The results will, however, apply to any principal plane of the strain ellipsoid.

As a result of strain, a radius OP of the reference circle (Fig. 10.2a) becomes the radius OP' of the ellipse, and its new length is the stretch S (Fig. 10.2b). Also, the original right angle between the radius OP and the tangent at P changes, and the measure of this change is the angle of shear ψ. Geometrically, ψ is the angle between the radius of the ellipse OP' and the normal to the tangent at P'.

CHANGE IN LENGTH

In our coordinate systems, a line of unit length along the x-axis becomes the semi-major axis of the strain ellipse of length S_1. A line of length x becomes

$$(10.2) \qquad x' = S_1 x$$

Similarly, a line of unit length along the y-axis becomes the semi-minor axis of length S_2, and a line of length y becomes

$$(10.3) \qquad y' = S_2 y$$

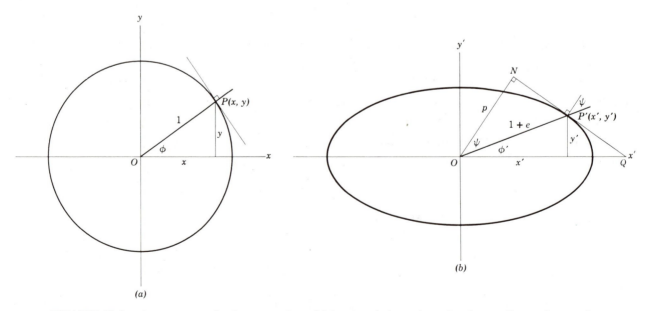

(a)

(b)

FIGURE 10.2 Geometry of the strain ellipse: (a) unit circle referred to the xy axes; (b) strain ellipse referred to the x'y' axes.

From Fig. 10.2a,

$$x = \cos \phi \qquad \text{and} \qquad y = \sin \phi$$

where the angle ϕ is measured from the x-axis in the undeformed state. Using these two expressions in Eqs. 10.2 and 10.3, we obtain

(10.4a) $\qquad x' = S_1 \cos \phi$

(10.4b) $\qquad y' = S_2 \sin \phi$

Also from Fig. 10.2b and the Pythagorean theorem,

$$S^2 = x'^2 + y'^2$$

or with Eqs. 10.4

(10.5) $\qquad S^2 = S_1^2 \cos^2 \phi + S_2^2 \sin^2 \phi$

It is now convenient to introduce a new parameter of change of length. The square of the stretch is the quadratic elongation λ (lambda),

(10.6) $\qquad \lambda = (1+e)^2 = S^2$

The principal quadratic elongations are then

$$\lambda_1 = (1+e_1)^2 = S_1^2 \qquad \text{and} \qquad \lambda_2 = (1+e_2)^2 = S_2^2$$

With these new parameters Eq. 10.5 can be written more compactly as,

(10.7) $\qquad \lambda = \lambda_1 \cos^2 \phi + \lambda_2 \sin^2 \phi$

Thus given the principal stretches, the deformed length of a line in a direction defined by ϕ in the undeformed state can be determined.

However, we are more interested in lines whose orientation is measured in the deformed state, that is, lines whose orientation is given by the angle ϕ'. From Fig. 10.2b

$$\cos \phi' = x'/S$$

Using the expression for x' from Eq. 10.4a, this becomes

$$\cos \phi' = (S_1 \cos \phi)/S$$

or, after rearranging,

(10.8) $\qquad \cos \phi = (S \cos \phi')/S_1$

Similarly, with Eq. 10.4b

(10.9) $\sin \phi = (S \sin \phi')/S_2$

Using Eqs. 10.8 and 10.9 in the identity $\sin^2\phi + \cos^2\phi = 1$, and dividing through by S^2, gives

(10.10) $\dfrac{1}{S^2} = \dfrac{\cos^2\phi'}{S_1^2} + \dfrac{\sin^2\phi'}{S_2^2}$

Again for convenience we define yet another parameter of change of length which is the reciprocal of the square of the stretch, called the reciprocal quadratic elongation,

(10.11) $\lambda' = 1/(1+e)^2 = 1/S^2$

The principal reciprocal quadratic elongations are then

$\lambda_1' = 1/S_1^2 \qquad \text{and} \qquad \lambda_2' = 1/S_2^2$

With these Eq. 10.10 becomes

(10.12) $\lambda' = \lambda_1'\cos^2\phi' + \lambda_2'\sin^2\phi'$

and we can now find the longitudinal strain associated with any line whose orientation is given in the deformed state by the angle ϕ'.

CHANGE IN ORIENTATION

In addition to changing length, a line generally also changes orientation. From Fig. 10.2a,

(10.13) $\tan \phi = y/x$

and similarly from Fig. 10.2b,

(10.14) $\tan \phi' = y'/x'$

With Eqs. 10.2 and 10.3, this latter equation becomes

(10.15) $\tan \phi' = S_2 y/S_1 x$

or, with Eq. 10.13, and using the definition of R_s, we have the useful result,

(10.16) $\tan \phi = R_s \tan \phi'$

which is sometimes called Wettstein's equation. Note that the change in orientation depends only on the ratio of the principal stretches, not on their absolute values.

CHANGE OF AN ANGLE

Every radius of the ellipse also has an angle of shear associated with it, and we now wish to obtain an expression for this angle. First, we need the equation for the line tangent to an ellipse. This is obtained by replacing x'^2 and y'^2 in Eq. 10.1 with $x'x*$ and $y'y*$, where $(x*, y*)$ is the point of tangency. We then have

$$\frac{x'x*}{S_1^2} + \frac{y'y*}{S_2^2} = 1$$

The coordinates of the point on the ellipse are given in terms of the angle ϕ by Eq. 10.4. By substitution we then obtain

(10.17)
$$\frac{\cos \phi}{S_1} x' + \frac{\sin \phi}{S_2} y' = 1$$

Next, we need the length of the perpendicular line segment from the origin to this tangent (p = ON in Fig. 10.2b). First, we will express this length in terms of the principal stretches and the angle ϕ. To simplify the manipulations, it will be useful to abbreviate the equation of the tangent line as

$$ax' + by' = 1$$

where a and b represent the corresponding coefficients of Eq. 10.17. This equation in slope-intercept form is

$$y' = -(a/b)x' + (1/b)$$

where $-(a/b)$ is the slope of the tangent line, and $(1/b)$ is the y'-intercept. The x'-intercept is then $(1/a)$; in Fig. 10.2b $(1/a) = OQ$. Defining $\alpha = \psi + \phi'$, then

(10.18) $\cos \alpha = p/(1/a) = ap$ or $p = (\cos \alpha)/a$

Because the slope of the tangent line is $-(a/b)$, the slope of its normal ON is the negative reciprocal, or

$$\tan \alpha = b/a$$

In a right triangle with one angle α and sides a and b, the hypotenuse is equal to $(a^2 + b^2)^{1/2}$. Then for this same triangle,

(10.19) $\cos \alpha = a/(a^2 + b^2)^{1/2}$

Using this in Eq. 10.18 gives, after squaring and rearranging,

$$1/p^2 = a^2 + b^2$$

Reverting to the full expression for a and b, we then have the equation for p,

$$\frac{1}{p^2} = \frac{\cos^2\phi}{S_1^2} + \frac{\sin^2\phi}{S_2^2}$$

With the principal reciprocal quadratic elongations, this becomes

(10.20) $1/p^2 = (\cos^2\phi)/\lambda_1 + (\sin^2\phi)/\lambda_2$

Next, we express the length p in terms of the angle of shear. Again from Fig. 10.2b,

$$\sec \psi = (S)/p$$

Squaring this, substituting the result into the identity $\tan^2\psi = \sec^2\psi - 1$, and using the definitions $\gamma = \tan \psi$ and $\lambda = S^2$, we have

$$\gamma^2 = (\lambda/p^2) - 1$$

Now, using the expression for $1/p^2$ from Eq. 10.20, and the expression for λ from Eq. 10.7, we obtain

(10.21) $\gamma^2 = [\lambda_1\cos^2\phi + \lambda_2\sin^2\phi][(\cos^2\phi)/\lambda_1 + (\sin^2\phi)/\lambda_2] - 1$

Expanded, this is a fourth degree equation of seemingly complex form. However, if the identity

$$1 = (\sin^2\phi + \cos^2\phi)^2$$

is substituted for the final 1, on expanding, combining terms and taking the square root a considerable simplification results, leaving

(10.22) $\gamma = [(\lambda_1 - \lambda_2)/(\lambda_1\lambda_2)^{1/2}]\sin \phi \cos \phi$

As before, the need is for an expression in terms of the angle ϕ'. This can be obtained by substituting Eqs. 10.8 and 10.9; after rearranging we have the result,

$$\gamma/\lambda = [(1/\lambda_1) - (1/\lambda_2)]\sin \phi'\cos \phi'$$

Then using reciprocal quadratic elongations, and defining a new but nameless parameter of shear strain,

(10.23) $\gamma' = \gamma/\lambda$

we finally have

$$(10.24) \qquad \gamma' = (\lambda_1' - \lambda_2')\sin \phi'\cos \phi'$$

and with this expression we can easily determine the shear strain associated with any line whose orientation is given by the angle ϕ'.

There is an alternative approach to expressing the relationship between the angles ψ and ϕ' and the principal stretches which will be useful later. Geometrically, the coefficients of the general equation of a line define the slope of its normal. Hence from Eq. 10.17 and Fig. 10.2b,

$$\tan(\psi+\phi') = \frac{\sin \phi/S_2}{\cos \phi/S_1} = R_s \tan \phi$$

With Eq. 10.16 and rearranging, we then obtain,

$$(10.25) \qquad R_s^2 = \frac{\tan(\psi+\phi')}{\tan \phi'}$$

Note that whatever the sign of measured angle of shear, it must be treated as a positive quantity when used in this equation.

This result can be used to obtain a useful expression for γ in terms of strain ratio and orientational angle. Introducing the identity for the tangent of the sum of two angles,

$$R_s^2 = \frac{\tan \phi' + \tan \psi}{\tan \phi'(1 - \tan \phi'\tan \psi)}$$

Cross multiplying, collecting terms and using $\gamma = \tan \psi$, this becomes

$$(10.26) \qquad \gamma = \frac{(R_s^2 - 1)\tan \phi'}{1 + R_s^2 \tan \phi'}$$

10.4 MOHR CIRCLE FOR FINITE STRAIN

The effort of the previous section, and especially the introduction of the parameters λ' and γ', was aimed at obtaining the expressions for the change in length and the change of a right angle of Eqs. 10.12 and 10.24:

$$\lambda' = \lambda_1'\cos^2\phi' + \lambda_2'\sin^2\phi'$$

$$\gamma' = (\lambda_1' - \lambda_2')\sin \phi'\cos \phi'$$

These are formally identical to Eqs. 7.13 for the normal and shearing tractions acting on a plane. And just as in that case, we may convert these two equations to a much more useful form by substituting the double angle identities of Eqs. 7.14, to give,

$$(10.27a) \qquad \lambda' = \frac{\lambda_1' + \lambda_2'}{2} - \frac{\lambda_2' - \lambda_1'}{2} \cos 2\phi'$$

$$(10.27b) \qquad \gamma' = \frac{\lambda_2' - \lambda_1'}{2} \sin 2\phi'$$

These describe the <u>Mohr circle</u> for finite strain, and like its counterpart for stress, this property of Eqs. 10.27 allows a wide variety of problems involving the parameters λ' and γ' to be solved graphically (Fig. 10.3a). The main features of this construction is a circle on the horizontal $+\lambda'$-axis, whose center and radius are given by

$$c = (\lambda_1' + \lambda_2')/2 \qquad \text{and} \qquad r = (\lambda_2' - \lambda_1')/2$$

This circle is the locus of all permissible values of λ' and γ' for a given state of homogeneous strain.

There are several auxiliary constructions which can be performed on this diagram which give additional useful information. If a line from the origin to a point $P(\lambda', \gamma')$ is drawn, the angle of shear associated with this particular direction appears directly as the slope angle, thus bypassing the mathematically convenient, but otherwise obscure parameter γ'. The construction of this line may be used in sever-

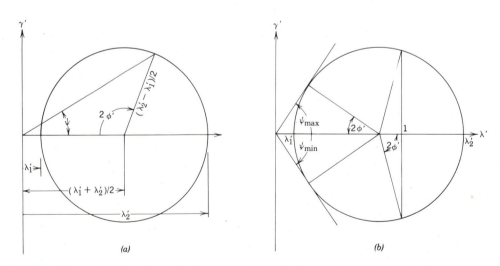

FIGURE 10.3 Mohr circle for finite strain: (a) basic geometrical properties; (b) maximum angle of shear and the lines of no finite longitudinal strain.

al different ways. For example, the algebraic maximum and minimum values of the angle of shear and their orientations may be found by drawing lines passing through the origin which are tangent to the circle (Fig. 10.3b).

Also, the orientation of the lines of no finite longitudinal strain can be found from the points where the vertical line $\lambda' = 1$ intersects the circle (Fig. 10.3b).

Obtaining the Mohr circle equations is sufficient justification for introducing the two finite-strain parameters λ' and γ'. However, it may be useful to give an additional geometrical interpretation. The equation of a strain ellipse with its axes coinciding with the coordinate axes is

$$\lambda'_1 x'^2 + \lambda'_2 y'^2 = 1$$

and we see that the two principal reciprocal quadratic elongations are simply the coefficients of the x' and y' terms. These are measures of the lengths of the two semi-axes; this equation should be compared with Eq. 10.1. In the Mohr circle, these two coefficients are found at opposite ends of a diameter lying along the λ'-axis.

For any other diameter of the Mohr circle, representing two mutually perpendicular directions, there are four measures--two for the change in length and two for the change of the original right angle associated with each of these two directions. As in the case of the four stress components in two dimension (Eq. 7.6), we may also write these four components as a matrix,

$$\begin{bmatrix} \lambda'_{xx} & \gamma'_{xy} \\ \gamma'_{yx} & \lambda'_{yy} \end{bmatrix}$$

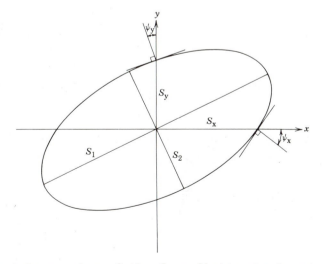

FIGURE 10.4 Geometric meaning of the four finite strain parameters associated with a diameter of the Mohr circle.

As before, the two parameters of shear strain are equal.

With these four components we may then write the general equation of a strain ellipse centered at the origin as

$$\lambda'_{xx} x'^2 + 2\gamma'_{xy} x'y' + \lambda'_{yy} y'^2 = 1$$

and once again we see a direct correspondence between strain parameters and coefficients. Each of these coefficients also has a geometric meaning. This can be most easily illustrated with a specific example. The coefficients in the equation for the ellipse of Fig. 10.4 are

$$\begin{bmatrix} \lambda'_{xx} & \gamma'_{xy} \\ \gamma'_{yx} & \lambda'_{yy} \end{bmatrix} = \begin{bmatrix} 0.76791 & 0.57435 \\ 0.57435 & 1.73209 \end{bmatrix}$$

The stretches in the two coordinate directions are then

$$S_x = 1/\sqrt{\lambda'_{xx}} = 1.14116$$

$$S_y = 1/\sqrt{\lambda'_{yy}} = 0.75983$$

The shear strains, together with the corresponding angles of shear in these same two directions are

$$\gamma_x = \gamma'_{xy}/\lambda'_{xx} = 0.74794, \quad (\psi_x = 36.8°)$$

$$\gamma_y = \gamma'_{yx}/\lambda'_{yy} = 0.33159, \quad (\psi_y = 18.3°)$$

10.5 SOLVING PROBLEMS WITH THE MOHR CIRCLE

The Mohr circle construction is a very useful way of obtaining solutions to problems involving the strain parameters. The simplest type of problem which may be solved with the aid of a Mohr circle is the determination of the parameters of longitudinal and shear strain associated with a specific direction in a given strain ellipse.

Just as in the case of the Mohr construction for stress, we need a sign convention for the angle of shear. The sense of shearing is taken positive if it is measured in an anticlockwise direction (this is the same as the sense of shearing traction, so it should be easily remembered).

PROBLEM

Given a strain ellipse (Fig. 10.5a) whose principal stretches are $S_1 = \sqrt{2}$

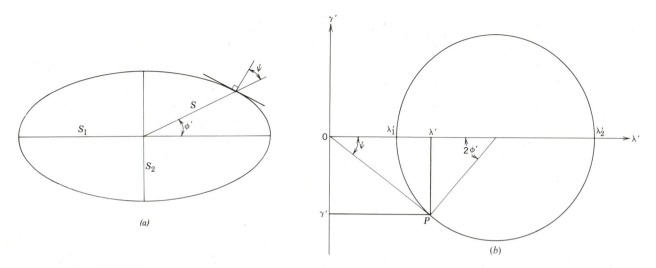

FIGURE 10.5 Strain parameters associated with a particular direction: (a) line of interest in the strain ellipse; (b) strain parameters associated with this line on the Mohr circle.

and $S_2 = 1/\sqrt{2}$, what is the stretch and shear strain associated with a line making an angle of $\phi' = +25°$ with the long axis.

CONSTRUCTION (Fig. 10.5b)

1. Convert these principal stretches to the corresponding values of the principal reciprocal quadratic elongations:

$$\lambda_1' = 1/S_1^2 = 0.5 \qquad \text{and} \qquad \lambda_2' = 1/S_2^2 = 2.0$$

2. On a pair of orthogonal axes locate the center of the circle at the horizontal λ'-axis at $c = 1.25$ units to the right of the origin with a radius of $r = 0.75$ units.

3. A radius of this circle making an angle of $2\phi' = 50°$, measured anticlockwise from λ_1', locates a point P on the circumference whose coordinates are the required values of λ' and γ'.

ANSWER

Scaling off these values gives $\lambda' = 0.77$ and $\gamma' = 0.57$; within the accuracy of the graphical construction these are the same values found by using Eqs. 10.26. These results may be converted to more recognizable form:

$$S = 1/\sqrt{\lambda'} = 1.14 \qquad \text{and} \qquad \gamma = \gamma'/\lambda' = 0.74 \ (\psi = -36°)$$

As with any graphic work, the accuracy of the Mohr circle construction is a function of scale. The strain parameters can be plotted and read to the second decimal point if care is taken and the circle is drawn large enough; a circle with a radius of about 10 cm is generally satisfactory

The real power of the Mohr circle construction, however, lies not with the solution of this relatively simple problem, but with a much more important class of problems--finding the principal stretches and their orientation from measurements made on deformed rocks. The basic approach is to find enough information to be able to construct the circle. Once drawn, the principal stretches and their orientation are then easily found.

PROBLEM

Suppose the stretch and shear strain have been determined for each of two directions a and b on an exposed plane.

$$S_a = 1.35, \qquad \psi_a = +26°$$

$$S_b = 1.18, \qquad \psi_b = -36°$$

Find the magnitudes of the principal stretches and their orientations.

CONSTRUCTION (Fig. 10.6)

1. Calculate the strain parameters $\lambda' = 1/S^2$ and $\gamma' = \lambda' \tan \psi$ for each line:

$$A = (\lambda'_a, \gamma'_a) = (0.549, +0.268)$$

$$B = (\lambda'_b, \gamma'_b) = (0.718, -0.522)$$

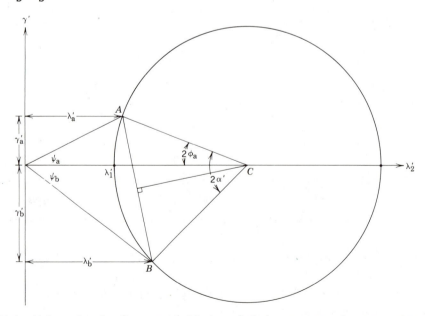

FIGURE 10.6   Mohr circle from two lines with known strain parameters.

2. Plot A and B as points on a set of axes using a convenient scale in either of two ways:

   a. at the intersection of the vertical line λ' = constant and horizontal line γ' = constant.

   b. at the intersection of the vertical line λ' = constant and the line passing through the origin with slope angle ψ.

3. Because A and B must lie on the circumference of a circle, the line AB is a chord of that circle, and its perpendicular bisector intersects the λ'-axis to locate point C. With this as center, and a radius of AC=BC, complete the circle.

4. The two points where this circle cuts the λ'-axis represent the principal quadratic elongations λ_1' and λ_2'. Their magnitudes are found by scaling off the distances from the origin.

5. The orientation of the λ_1' direction can be found from either of the deformed lines. For example, the angle between λ_1' and the radius AC is equal to $2\phi_a'$. The orientation of the long axis of the strain ellipse is found by measuring off from deformed line a an angle equal to half the measured angle and in the opposite sense. Equivalently, point B could be used to locate the long axis relative to line b.

ANSWER

The principal quadratic elongations are λ_1' = 0.5 and λ_2' = 2.0; λ_1' lies between A and B and makes an angle of $2\phi_a'$ = 24° with A. Hence the long axis of the strain ellipse lies between the two lines and makes an angle of ϕ_a' = 12° with line a.

In practice, it would unusual to be able to determine both strain parameters for each of two directions, and thus this graphical method must be adapted to still more realistic situations.

10.6 STRAIN ELLIPSE FROM MEASURED EXTENSIONS

If the extensions associated with two lines and the angular relationships of these two lines to the principal directions are known, the magnitudes of the principal stretches can be found. By way of introduction, we adopt the simplest method of measuring the extensions (Ramsay, 1967, p. 248); other more involved methods yield better results (Hossain, 1979; Ferguson, 1981; Ferguson and Lloyd, 1984).

PROBLEM

Two stretched belemnites are exposed on the plane of slaty cleavage (Fig. 10.7a). A prominant lineation on this plane is identified as the long axis of the strain ellipse. Find the principal stretches.

APPROACH

It is useful to make a sketch the Mohr circle to show what is known and what is needed (Fig. 10.7b). As can be seen, we need to construct the isoceles triangle which has the chord AB as its base, and $2\alpha'$ as it apex angle. For future reference, note that

$$2\alpha' = 2(\phi_a' + \phi_b')$$

hence the two base angles are equal to $(180° - 2\alpha')/2$.

CONSTRUCTION

1. For a deformed belemnite both the original length (the sum of the now separated segments) and the final length are known. By measurement we obtain:

$$S_a = 2.1 \quad (\lambda_a' = 0.227), \quad \phi_a' = 15°$$

$$S_b = 1.8 \quad (\lambda_b' = 0.309), \quad \phi_b' = 30°$$

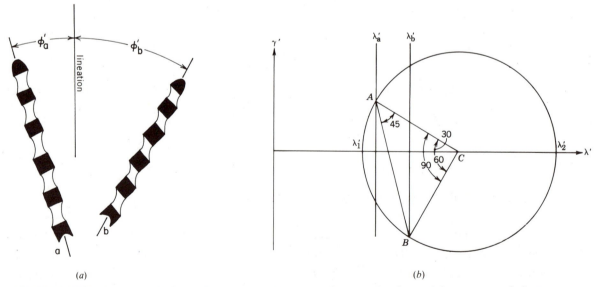

(a)   (b)

FIGURE 10.7 Deformed fossils exposed on a plane of slaty cleavage: (a) two stretched belemnites; (b) sketch of the Mohr circle.

2. Draw the vertical γ'-axis, and add two parallel lines to the right at scale distances equal to the values of λ_a' and λ_b' (Fig. 10.8a).

3. Arbitrarily locate point A on the λ_a' line, and through it construct a line making an angle of $2\phi_a' = 30°$ with the yet to be drawn horizontal axis.

4. The vertex angle of the isoceles triangle ABC (cf. Fig. 10.6b) is equal to $(2\phi_a'+2\phi_b') = 90°$. Each base angles is then 45°; construct a second line at A making this angle with the radius line AC. This locates point B on the λ_b' line.

5. Construct a line at B making this same base angle with AB; the intersection of this line with the first line through A locates the center C.

6. Complete the circle using this center and AC=BC as radius (Fig. 10.8b).

7. Draw in the horizontal λ'-axis through C. The points where the circle cuts this axis represent the values of the principal reciprocal quadratic elongations.

ANSWER

$$\lambda_1' = 0.197 \quad (S_1 = 2.25)$$

$$\lambda_2' = 0.644 \quad (S_2 = 1.25)$$

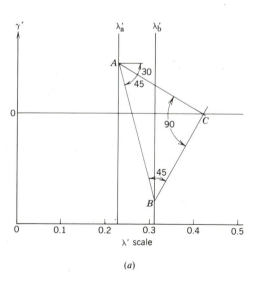

(a)

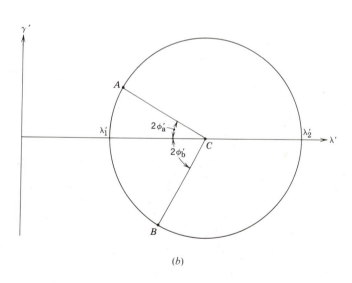

(b)

FIGURE 10.8 Solution of the belemnite problem: (a) center and radius of the circle; (b) the circle and the principal reciprocal quadratic elongations.

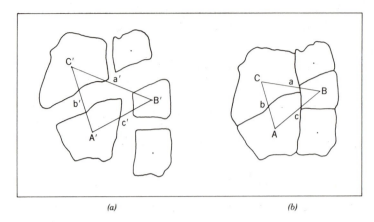

FIGURE 10.9 Chocolate tablet boudinage of a pyrite layer in slate (after Casey, Dietrich and Ramsay, 1983, p. 218): (a) as found; (b) as reconstructed.

Clearly, the strain recorded by these stretched belemnites is not homogeneous. Their use as strain markers is justified if they are small compared to the volume of homogeneously deformed rock which contains them. We have met this dependency on scale before when we drew relatively large forms on card decks.

If we are not be able to identify the principal directions we then need to determine stretches in three independent directions in a plane in order to construct a Mohr circle.

PROBLEM

A thin plate of pyrite has been broken into a number of pieces and these pieces have been pulled apart in all directions in a plane (Fig. 10.9a). Determine the magnitude of the principal stretches and their orientation.

APPROACH

We recover the original configuration of the layer by reassembling the broken pieces (Fig. 10.9b). Then using the centers of gravity of three adjacent pieces we have the lengths of three original lines in the undeformed state. Similarly, we have the lengths of these same lengths after deformation (Fig. 10.9a). Measuring these original and final lengths, we then calculate the three stretches and the corresponding reciprocal quadratic elongations:

$$S_a = 1.493, \quad \lambda'_a = 0.447$$

$$S_b = 1.426, \quad \lambda'_b = 0.492$$

$$S_c = 1.167, \quad \lambda'_c = 0.734$$

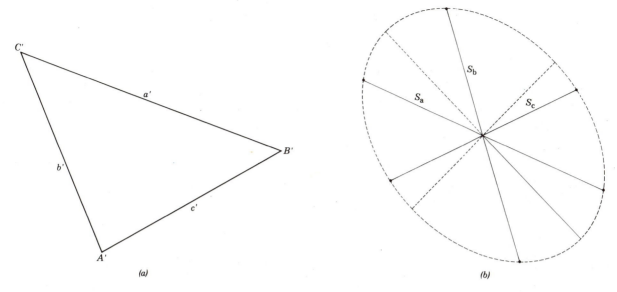

FIGURE 10.10 An estimate of the ellipse obtained directly from the measured stretches: (a) the deformed triangle; (b) the three stretches in their proper orientation.

It is useful to obtain an estimate of the shape and orientation of the strain ellipse before starting the full Mohr construction. First, draw three lines parallel to the sides of the deformed triangle (Fig. 10.10a) so that they emanate from a single point (Fig. 10.10b). Then along each of these three directions plot the lengths of the measured stretches. Each of these segments is a radius, or doubled a diameter of the strain ellipses, which can be sketched with a fair degree of accuracy. From this ellipse we can see that the S_1-direction lies between lines a and b, and that it is a bit closer to line a. From this figure we can also obtain the angles between the three lines; they are

$$a':b' = 49° \qquad b':c' = 82° \qquad c':a' = 49°$$

CONSTRUCTION

1. In the manner of Fig. 10.8a, draw the three vertical lines at scale distances equal to the values of λ_a', λ_b', and λ_c' (Fig. 10.11a).

2. From point B arbitrarily located on the intermediate λ_b' line construct line AB at an angle of $a':b' = 49°$ and line BC at an angle of $b':c' = 82°$ to the line $\lambda_b' = $ constant.

3. This locates three points A, B and C which lie on the circumference of the Mohr circle. The perpendicular bisectors of each of these chords intersect to locate the center, and the horizontal λ'-axis is then added. After completing the circle, read off the values of λ_1' and λ_2'.

4. The circle cuts each of the vertical λ' lines in two places, giving a
   total of six points. Only three satisfy the orientation of the three
   deformed lines, and the easiest way to determine which is to compare
   Figs. 10.10b and 10.11a. On the preliminary ellipse, line a lies anticlockwise from the long axis, and therefore on the Mohr circle we can
   identify point A as also being anticlockwise from λ_1 (Fig. 10.11b).
   The two other points can be similarly identified. Note also that the
   angles between the radii to each of these three points are twice the
   angles measured in Fig. 10.10b.

5. The orientation of the principal axes can now be determined by measuring the angle from any one of the three valid points to λ_1'.

ANSWER

The values of the principal quadratic elongations and the corresponding
principal stretches are

$$\lambda_1' = 0.404; \; S_1 = 1.57$$

$$\lambda_2' = 0.782; \; S_2 = 1.13$$

The angle between λ_1' and A is $2\phi' = 40°$, thus the long axis of the strain
ellipse makes an angle of $\phi' = 20°$ from side a' measured anticlockwise.

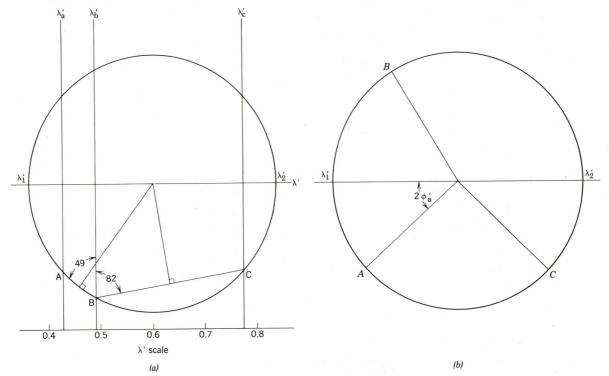

FIGURE 10.11 Solution of the chocolate tablet boudinage problem: (a) construction of the Mohr circle; (b) orientations of the principal
directions.

10.7 STRAIN ELLIPSE FROM MEASURED ANGLES

Features from which angular changes of lines can be determined are more common than lines of known original length. For example, a deformed, originally bilaterally symmetrical fossil gives an angle of shear directly.

PROBLEM

A single deformed trilobite is exposed on a plane of slaty cleavage. A lineation on this plane marks the long axis of the strain ellipse (Fig. 10.12a). Find the strain ratio.

CONSTRUCTION

1. Two pieces of information are needed: the angle of shear associated with the line of original symmetry of the trilobite ($\psi = 35°$), and the angle this line makes with the long axis of the strain ellipse ($\phi' = 19°$).

2. On a set of $\lambda'\gamma'$-axes, draw a line through the origin making an angle equal to the measured value of $\psi = +35°$ (Fig. 10.12b).

3. At a convenient distance along the λ'-axis arbitrarily locate the center C of the Mohr circle. From this center draw a line making an angle of $2\phi' = 38°$ with the λ'-axis to intersect the first line at point P.

4. Draw in the circle with CP as radius, and measure the distances from the origin to each of the two intercepts.

ANSWER

Because the center of the circle is arbitrarily located, we can not uniquely determine the two principal reciprocal quadratic elongations; the scale

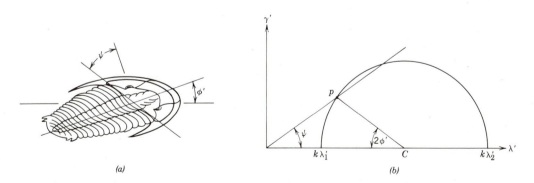

(a) *(b)*

FIGURE 10.12 Deformed trilobite: (a) strained form in relation to the stretching lineation; (b) construction of the Mohr circle.

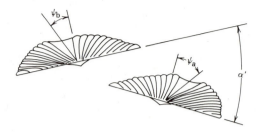

FIGURE 10.13 Two deformed brachiopods in a plane.

factor k is unknown. We can, however, calculate the strain ratio from

$$R_s^2 = k\lambda_2' / k\lambda_1'$$

where k is the unknown scale factor. Measurement of the two intercepts yields R_s = 2.0. We can check this result by using Eq. 10.25. Note that with only the strain ration any area change remains unknown.

If the principal axes can not be identified, then two angles of shear are needed to determine the strain ratio.

PROBLEM

From the two deformed brachiopods of Fig. 10.13, determine the ratio of the principal stretches and their orientation.

CONSTRUCTION

1. Measure the two angles of shear associated with each hinge line, and the angle between the two hinge lines, giving:

$$\lambda_a = +35°, \qquad \lambda_b = -25°, \qquad \alpha' = 29°$$

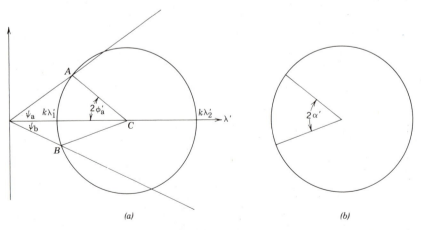

(a)   *(b)*

FIGURE 10.14 Solution of the brachiopod problem: (a) coordinate axes and two angles of shear; (b) separate circle with two radii.

2. On a pair of coordinate axes draw lines which make angles of ψ_a and ψ_b with the λ'-axis which pass through the origin (Fig. 10.14a).

3. On a separate sheet of tracing paper, draw a circular arc of convenient radius and with two radii making an angle of $2\alpha'=58°$ (Fig. 10.14b).

4. Superimpose this sheet on the coordinate axes, and by moving the center along the λ'-axis and at the same time rotating the sheet, find the position where the intersections of the arc with the two sloping ψ lines lie at the ends of the two radii, thus locating the center of the circle. Draw in this circle and label the two points A and B (Fig. 10.14a).

5. Measure the two intercepts $k\lambda_1'$ and $k\lambda_2'$ and calculate R_s. Also measure the angle between the λ_2' point and either A or B; for example, using A gives the angle $2\phi_a'$. The S_1-direction is found by measuring ϕ_a' from the hinge of shell A toward shell B.

ANSWER

The strain ratio $R_s = 2.0$, and the λ_1-direction makes an angle of $\phi'=19°$ with the hinge of shell A.

This method will also work when both the measured angles of shear have the same sign. It will be found, however, that there are two geometric ways of fitting the circle to the raw data. Of several ways of resolving this ambiguity, the simplest is to convert one of the angles to the opposite sign, and proceed as above. This is done by using as the reference line the second side of the angle. For example, the brachiopod with $\psi = +25°$ in Fig. 10.13, becomes $\psi = -25°$, when referred to the line of symmetry, rather than the hinge line. The angle used to define the relative orientation is now measured from this symmetry line.

This method is not restricted to original right angles. The following problem illustrates how other known angles may be used.

PROBLEM

From the deformed glass shard typically found in welded tuff, determine the strain associated with their deformation (Fig. 10.15a). Assume that the shard originally formed at the junction of three bubbles, and thus had angles of 120° separating the limbs.

CONSTRUCTION

1. Reassemble the prongs of the deformed shard into an equivalent scalene triangle with sides a', b' and c' (Fig. 10.15b).

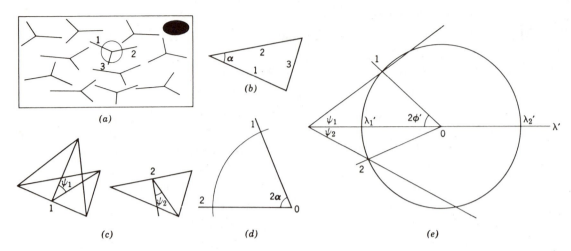

FIGURE 10.15 Strain from deformed glass shards in welded tuff: (a) simulated shards (after Ragan and Sheridan, 1972); (b) equivalent scalene triangle; (c) constructed angles of shear; (d) relative orientation of shard limbs using angle α'; (e) Mohr circle.

2. Originally this triangle was equilateral, hence by comparing shapes before and after deformation, the angle of shear associated with any side may be determined in exactly the same manner as in Fig. 10.11--see Fig. 10.15c. The results are

$$\psi_a = -35°, \qquad \psi_b = +25°, \qquad \alpha' = 23°$$

3. Proceeding just as in the previous example, construct lines with slopes equal to the two angles of shear, and on a tracing sheet draw an arc with two radii at an angle of $2\alpha'$. Superimposing this tracing on the coordinate axes, find the position of fit (Fig. 10.15d). Measure the two intercepts and one of the orientational angles $2\phi'$.

ANSWER

The strain ratio is found to be $R_s = 1.93$, and S_1 makes an angle of $\phi'=11°$ measured from A toward B. As will be seen in Fig. 10.15a, the trace of the foliation, as marked by the preferential alignment of one limb of each shard, is approximately parallel to the long axis of the strain ellipse. Because this foliation is approximately horizontal over large distances in welded ash-flow sheets, the deformation must be of a compactional nature with all changes occurring vertically. Hence, $S_1=1$ and $S_2=1/R_s$, and we can now compute the area (volume) change using Eq. 9.10. The results are that $\Delta = 0.48$--there has been a volume reduction of almost one half.

10.8 ROTATION

If the pre-deformation orientation of a line or plane is known, the rotational component of the deformation can be determined by a two step calculation. With the angle ϕ' between the long axis of the strain ellipse and the line or trace of the plane, together with the strain ratio, the original angle ϕ can be found with Eq. 10.16. The difference between the inclination of this line in the unstrained state and its pre-deformational inclination is then the angle of rotation. One situation where this might be possible is if sedimentary bedding is recognizable, and if it can be confidently assumed to have been originally horizontal, and if the strain is really two dimensional (Fig. 10.16).

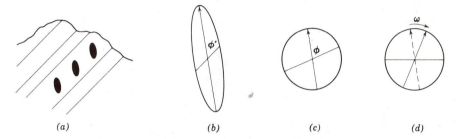

    (a)   (b)   (c)   (d)

FIGURE 10.16 Rotational component (after Ramsay, 1969, p. 54): (a) deformed sedimentary beds; (b) trace of bedding in relation to strain ellipse; (c) strain removed; (d) rotation to horizontality.

EXERCISES

1. Reproduce the fully labeled Mohr circle of Fig. 10.3a from memory. Then derive Eqs. 10.26 from this figure. Also show that the angle labeled ψ is indeed the angle of shear associated with the direction represented by point P.

2. Using the chocolate tablet structure of Fig. X10.1, determine orientation and magnitude of the principal stretches for triangle ABD and BCE. Note that the results will not be exactly the same as in the example.

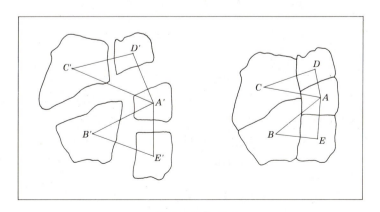

FIGURE X10.1

3. Using the deformed brachiopods of Fig. X10.2, determine the orientation and shape of the strain ellipse using Wellman's method. Check your result by constructing a Mohr's circle diagram using two of the shells.

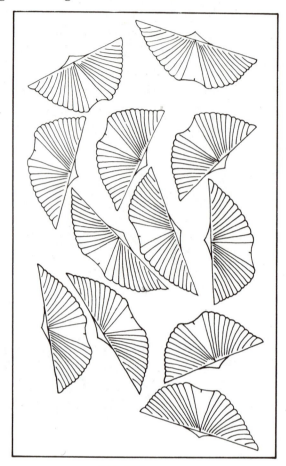

FIGURE X10.2

4. Analyse the strain in the simulated welded tuff of Fig. X10.3. Assuming the deformation to be wholly compactional, calculate the volume change.

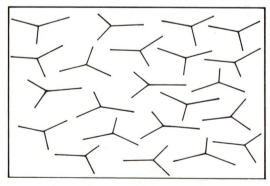

FIGURE X10.3

11
Description and Classification of Folds

11.1 INTRODUCTION

A <u>fold</u> is a distortion of a volume of rock material that manifests itself as a bend or nest of bends in linear or planar elements within the material (Hansen, 1971, p. 8). Most folds involve elements that originally defined a plane. Bedding is the common example. This is an important case because the fold then represents an important indicator of the nature of the deformation; in particular, its geometric features can be correlated with certain aspects of distortion and rotation. However, folds may also develop from originally curved elements, and the problem of relating the features of such folds to the deformation is much more severe.

<u>Folding</u> occurs when pre-existing elements are transformed into new curviplanar or curvilinear configurations, whatever their original condition. Thus folding is just an inhomogeneous deformation which acts on a body of material originally containing linear or planar elements. However, it is worth emphasizing that a deformation which produces a fold in one situation may not in another. Planar or linear elements may be entirely absent from the rock mass, and therefore there is nothing to mark folds. It is also conceivable that initially curved elements might become planar or linear, or that the elements may be so oriented as to remain planar or linear (Ramsay, 1967, p. 473).

In the following sections, a number of relatively simple geometric properties of folded surfaces are explored; the methods and terminology follow closely Fleuty (1964) and Ramsay (1967, Ch. 7). For further details these two works, together with the books by Turner and Weiss (1963), and by Hansen (1971) should be consulted.

11.2 DESCRIPTION OF SINGLE SURFACES

Single curviplanar surfaces may have a wide variety of forms, ranging from comparatively simple, such as shown in Fig. 11.1, to exceedingly complex. The geometry of even a relatively simple curved surface may be quite difficult to describe in detail. There are mathematical methods for describing such surfaces but these have found little practical use, and alternative methods are required; one such method is to use structure contours (see Chapter 18).

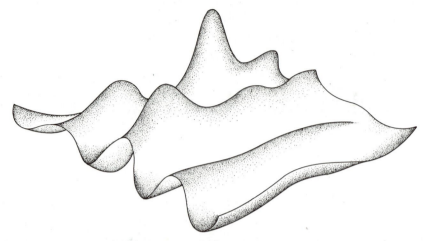

FIGURE 11.1 A single curviplanar surface; the fold in the front is approximately cylindrical.

Fortunately, it is meaningful to restrict our initial consideration to a much
simpler class of surfaces. Many naturally folded surfaces have shapes which closely
approach the form of cylinders, or are made up of approximately cylindrical parts.
A cylindrical surface is defined as generated by a line moved parallel to itself in
space. The orientation of the generating line is a directional property of the entire surface, and has no particular location. It is analogous to a crystallographic
axis, and is called the fold axis. One of the most important geometric features of
cylindrical surfaces is that their shape can be fully represented in a cross-section
drawn perpendicular to the axial direction. This section is called the profile of
the cylindrically folded surface, or simply the fold profile.

Each trace of a folded surface on the profile plane is a curve, and each curve
has several geometric features which serve to identify certain points on it. The
crest, or high point and the trough or low point on the curves are two such features. In three-dimensions, each of these points is the intersection of a line on
the profile plane. These are the crest line and the trough line, and they are paral-

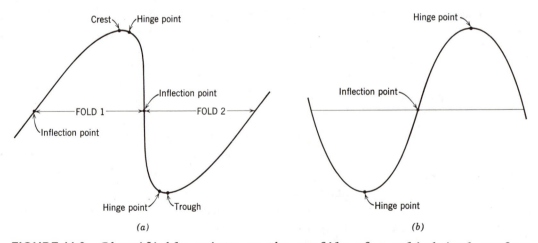

FIGURE 11.2 Identifiable points on the profile of a cylindrical surface.

lel to the axial direction. The location of both of these is dependent on the orientation of the surface relative to horizontal. On the other hand, the point of maximum curvature, or <u>hinge</u> point, and the point where the curve changes from concave to convex, or <u>inflection</u> point, are independent of any reference frame, and are, therefore, spatially <u>invariant</u>; such features serve to describe the geometry of folds more fundamentally (Fig. 11.2).

Often single hinge and inflection points alternate. In three dimensions such points lie on <u>hinge lines</u> and <u>inflection lines</u>, and it is convenient to consider a single fold as the portion of the curved surface between the two inflection lines. If a portion of profile curve has the form of a circular arc, the surface does not have a specific hinge point or hinge line; in such instances it is arbitrarily identified as the bisector of the circular segment. Similarly, there will be no inflection point if the transition from concave to convex involves a straight segment; the inflection point is then taken to be the midpoint of this straight section.

The terms <u>hinge</u> <u>zone</u> and <u>fold</u> <u>limb</u>, and the distinction between them, have been precisely defined in a form which is useful for some purposes (Ramsay, 1967, p. 345). Here, however, a more general meaning is adopted which follows conventional usage (Dennis, 1967, p. 88, 102). The hinge zone is considered to be that portion of the curved surface adjacent to the hinge line, and the fold limb to be that part of the surface adjacent to the inflection line. The proportion of the entire curved surface which may be considered to be hinge zone and limb may vary. The extreme examples are if the hinge zone is reduced to a line, and if the limb is represented by the inflection line. These two cases are illustrated in Fig. 11.3.

An additional descriptive element is the <u>interlimb</u> angle, which is defined as the minimum angle between the limbs as measured in the profile plane, or, alternatively, between the lines tangent to the curve at the inflection points (Fig. 11.4). This angle describes the "tightness" of the fold. For general purposes, however, it is often sufficient to categorize the angular relationship between fold limbs with descriptive adjectives. The terms gentle, open, close, tight, isoclinal and mushroom are commonly used. Fleuty (1964, p. 470) has suggested that these terms be restricted to specific ranges of interlimb angles (see Table 11.1).

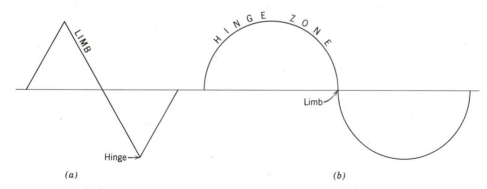

FIGURE 11.3 Differing proportions of a fold made up of hinge zone and limb.

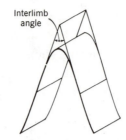

FIGURE 11.4 The interlimb angle.

TABLE 11.1 Terms for the tightness of a fold (after Fleuty, 1964, p. 470)

Interlimb angle	Description of fold
180°–120°	Gentle
120°–70°	Open
70°–30°	Close
30°–0°	Tight
0°	Isoclinal
Negative angles	Mushroom

Symmetry is another invariant feature of cylindrically folded surfaces. Considering only the shape of the surface, every cylindrical fold has at least one plane of symmetry which is perpendicular to the axis. If, in addition, the plane passing through the hinge line and bisecting the interlimb angle is also a plane of symmetry, the fold shape in profile is said to be symmetric, and the fold as a whole possesses orthorhombic symmetry. A series of linked folds are symmetrical if each member is symmetric, and if the pattern is strictly periodic. A consequence is that the two enveloping surfaces are planar and parallel, and the surface containing all the inflection points, or median surface, is mid-way between the two enveloping surfaces. These features of symmetric folds makes it easy to describe the dimensions of the folds in terms of amplitude and wavelength (see Fig. 11.5a). Fold shape is related, in part, to the ratio of these two measures.

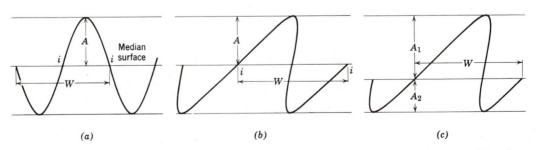

FIGURE 11.5 Dimensions folds; W = wave length, and A = amplitude.

If the trace of the hinge surface is not a plane of symmetry the fold is said to be asymmetric. With only a single remaining plane of symmetry, the fold as a whole has only monoclinic symmetry. For asymmetric folds it is useful to describe the sense of the asymmetry. This is commonly done in terms of vergence, or the direction in which the antiformal hinges have been displaced relative to the synformal hinges. For the two profile curves of Fig. 11.5b,c the vergence is to the right. If the fold axis is approximately horizontal, the vergence is described unambiguously by giving its azimuth (Bell, 1983). For folds with plunging axes, a useful alternative describes the sense of asymmetry as clockwise or anticlockwise when viewed in a down-plunge direction; the profiles of Fig. 11.5b,c show clockwise asymmetry.

The description of the dimensions of asymmetric folds becomes increasingly involved as the degree of asymmetry increases (Fig. 11.5b,c); more complete schemes have been suggested by Fleuty (1964, p. 467), by Ramsay (1967, p. 351), and by Hansen (1971, p. 9).

11.3 RELATIONSHIPS BETWEEN ADJACENT SURFACES

Since folds almost invariably involve more than one surface, additional terms and methods are needed to establish the spatial and geometric relationship between the adjacent curviplanar surfaces making up the fold. The locus of all hinge lines is one important feature, especially from the point of view of field mapping. This surface is often referred to as the axial plane or the axial surface, but it is not directly related to the axis; indeed noncylindrical folds may possess such a surface without having an axis. It is more appropriately called the hinge surface (Fig. 11.6). Preferably axial plane should be reserved for the plane parallel to the

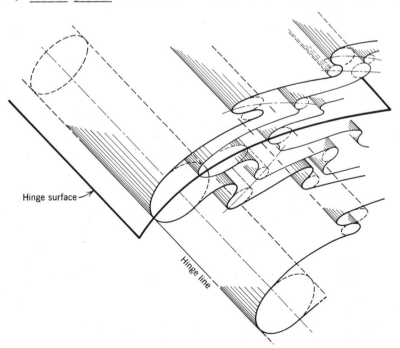

FIGURE 11.6 Hinge surface of cylindrical folds (after Wilson, 1982, p. 15).

hinge surface throughout the entire cylindrical fold, as in the phrase <u>axial</u> <u>plane</u> cleavage. This distinction is not commonly made, and if you too chose not to adopt it, make sure that you understand that the phrase <u>axial</u> <u>plane</u> has two quite different meanings.

In addition to the hinge surface, there is also an <u>inflection</u> surface, which is the locus of inflection lines on successive surfaces, and similarly, there are crestal and trough surfaces.

The geometrical relationship between any two adjacent curved surfaces depends on their relative curvature, and on the distance between them. The simplest and most sensitive way of defining this relationship is to construct lines of equal dip, or apparent dip, on the profile plane between the two surfaces. These lines are called <u>isogons</u> (Ickes, 1923; Elliott, 1965, 1968; Ramsay, 1967, p. 363). Not only can the resulting patterns aid in distinguishing accurately between different fold forms, but the use of dip isogons also leads to a classification of fold geometry which is simple to apply and easy to remember.

CONSTRUCTION OF ISOGONS

1. Obtain a profile of the folds to be analyzed. The most direct and accurate method is to trace the various curves from a photograph taken along the fold axis. If such a view is not possible, either because of lack of exposure or large size, the profile view may be constructed from a carefully drawn map (see Chapter 14).

2. On each trace of two adjacent surfaces, construct a series dip or apparent dip lines tangent to the curves. A 10° interval is often convenient, but the spacing should be dictated by the actual fold form and the amount of detail needed in each situation.

3. Connect points of equal dip or apparent dip on the two adjacent curves with a straight line--these are the isogons.

The pairs of parallel tangents can be constructed quite easily with the aid of a protractor and a triangle, as shown in Fig. 11.7. Marjoribanks (1974) has described an instrument which is useful if large numbers of folds are to be analyzed.

The placement of the individual isogons depends on the orientation of the fold. The pattern of isogons is, however, an invariant feature of the fold shape. Therefore, it is usually convenient to reoriented the fold so that the hinge point coincides with the $\alpha = 0$ tangent line (Fig. 11.8).

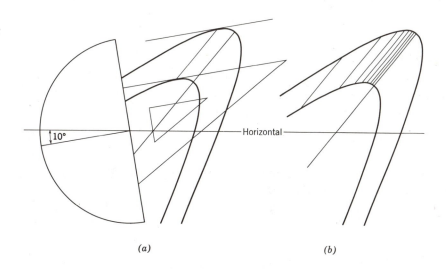

(a)   (b)

FIGURE 11.7   Construction of dip isogons.

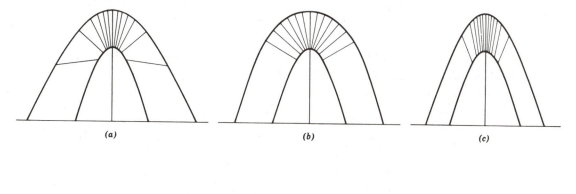

(a)   (b)   (c)

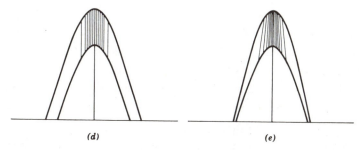

(d)   (e)

FIGURE 11.8   Isogon classification: (a) strongly convergent (1A), (b) parallel (1B), (c) weakly convergent (1C), (d) similar (2), (e) divergent (3) (after Ramsay, 1967, p. 365).

11.4 ISOGON CLASSIFICATION

Generally, the isogons of a particular layer will not be parallel, and the degree of departure from parallelism, together with the direction of convergence or divergence forms the basis of the classification (Ramsay, 1967, p. 365). For consistency, the inner arc of the fold is taken as the reference point for statements of the direction of isogon convergence. Five isogon patterns are recognizable, including three general and two special cases:

1A. <u>Folds</u> <u>with</u> <u>strongly</u> <u>convergent</u> <u>isogons</u>: the curvature of the outer surface is less than that of the inner, and the smallest distance between the two surfaces occurs along the trace of the hinge surface (Fig. 11.8a)

1B. <u>Parallel</u> <u>folds</u>: the inner surface has a greater curvature than the outer one, but their relationship is such that each isogon is perpendicular to the tangents (Fig. 11.8b). The name is derived from the fact that the distance between the two curves is everywhere constant; this measure is called the orthogonal thickness of the layer.

1C. <u>Folds</u> <u>with</u> <u>weakly</u> <u>convergent</u> <u>isogons</u>: the curvature of the inner surface is still greater, but the spacing between the two curves is greatest at the hinge (Fig. 11.8c).

2. <u>Similar</u> <u>folds</u>: both curves are identical and the isogons are parallel. In this special case, the distance between the two curves as measured along the isogons is constant, a measure known as axial-plane thickness (Fig. 11.8d).

3. <u>Folds</u> <u>with</u> <u>divergent</u> <u>isogons</u>: the curvature of the inner arc is less than that of the outer arc (Fig. 11.8e)

In addition to classifying a fold, it is also of interest to determine the particular pattern of the isogons within a class of a specific single layer. This can be done in several ways. Ramsay (1967, p. 359f) plotted the normalized orthogonal and axial-plane thickness as a function of the dip angle α. Because it can be obtained directly from the isogon construction, a simpler, more direct approach is to record the variation in the isogon angle ϕ as a function of α (Hudleston, 1973, p. 7). This is the angle each isogon makes with the normal to the tangent (Fig. 11.9a); it is taken as positive if measured in a clockwise direction from the normal and as negative if measured in a anticlockwise direction. A sign is also applied to the dip angle α to distinguish the two limbs; it is positive on the right limbs of antiforms or left limbs of synforms.

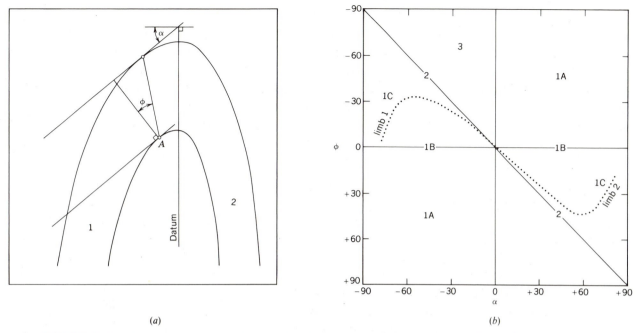

(a)   (b)

FIGURE 11.9 Isogon pattern in a single layer: (a) construction of the isogon
angle φ, (b) graph of φ vs α (after Hudleston, 1973, p. 6-7)

The variation of φ with α is represented by a curve on a graph (Fig. 11.9b).
Such a plot is important because the variation in shape gives an important clue to
mechanics of folding.

This same type of graph can be used to depict the variation in the orientation
of cleavage around a fold (Treagus, 1982). The angle β which the trace of the cleavage makes with the normal to the tangent lines is plotted as a function of the dip.

11.5 FOLD ORIENTATION

The orientation of a fold is completely defined by the attitude of the hinge
and hinge surface, and a statement of the direction of closure. There are three descriptive terms which describe the direction of closure. An <u>antiform</u> is a fold
which closes upward, and a <u>synform</u> is a fold which closes downward. A fold which
closes sideways is called a <u>neutral</u> fold; strictly, a neutral fold is one whose
hinge line pitches at an angle of 80°-90° on the hinge surface. The terms anticline
and syncline are reserved for folds with older and younger rocks, respectively, in
their cores. Many anticlines are also antiforms, and all anticlines start their existence as antiforms. It is possible, however, for an anticline to be turned completely over so that it closes downward; such a fold would be described as an anticline in synformal position, or, simply, a synformal anticline.

The angles of dip and plunge fix the attitude of the hinge surface and hinge
line with respect to horizontal, and these two angles are also the basis for a de-

scriptive nomenclature. In an effort to standardize usage and to increase precision, Fleuty (1964) suggested precise limits to a series of traditional terms for both dip and plunge (see Table 11.2). These terms can then be combined to describe fold attitude, for example, a steeply-dipping, gently-plunging fold. Note, however, that because the hinge line is confined to the hinge surface, some combinations of terms are invalid--a gently-inclined, steeply-plunging fold is an impossibility.

TABLE 11.2 Terms describing fold attitude (after Fleuty, 1964, p.483, 486)

Angle	Term	Dip of hinge surface	Plunge of hinge line
0°	Horizontal	Recumbent fold	Horizontal fold
1°–10°	Subhorizontal		
10°–30°	Gentle	Gently inclined fold	Gently plunging fold
30°–60°	Moderate	Moderately inclined fold	Moderately plunging fold
60°–80°	Steep	Steeply inclined fold	Steeply plunging fold
80°–89°	Subvertical	Upright fold	Vertical fold
90°	Vertical		

Neutral folds need additional attention. Recumbent applies to horizontal neutral folds; both hinge line and hinge surface are horizontal. With vertical folds the hinge line and hinge surface are vertical. Both of these terms are included in Table 11.2. A fold intermediate between these two extremes is reclined; this term describes a fold with a hinge surface which dips at angles of 10° to approximately 80°, and a hinge line which has a pitch of more than 80° in this plane. Since the dip can be greater than 80° and the plunge still less than the 80° required for designation as vertical (for example, dip = 82°, plunge = 78°), it is not practical to place a precise upper limit on the dip of the hinge surface. This minor discrepancy is the result of using the plunge angle in the description of some folds, and the pitch for others.

The basic failing of this approach to describing fold orientation is the result of referring the attitude of the hinge line to a vertical plane through the use of the angle of plunge, even though this plane generally bears no relation to fold geometry. A classification base solely on the pitch and dip could be constructed to avoid this, but would itself have drawbacks, the gravest of which is that pitch is often difficult to measure in the field. A simpler and still more precise classification can be achieved by combining both the plunge and pitch based schemes, through the use of a special triangular diagram involving all three variables (Rickard, 1971; see Fig. 11.10). The dip and pitch pair of a particular fold orientation is represented as a point at the intersection of the straight lines whose scales are found along the base and left side of the diagram (Fig. 11.11a). For plunge, the scale on the right and the curves are used. Each point on this graph uniquely repre-

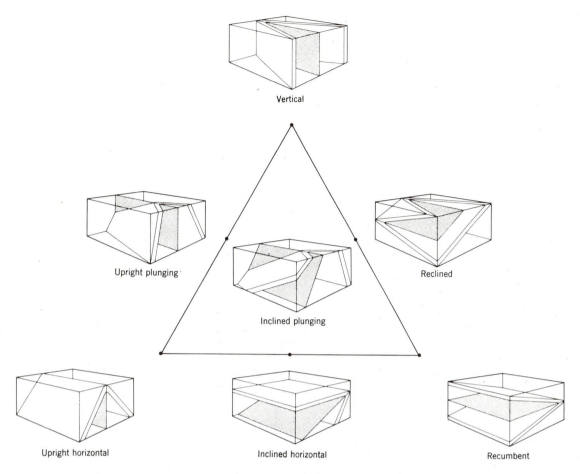

FIGURE 11.10 Graph of all possible fold attitudes.

sents a fold attitude, and all possible attitudes can be represented. It is also useful to include on the diagram the fields representing various catagories of orientation. Only the limiting orientations with respect to horizontal and vertical are needed, and these are shown on Fig. 11.11b. In the few small areas where overlaps occur, the terms describing the folds are optional.

Instead of further subdividing the large field of inclined folds, Rickard proposed that all folds be allocated a two number attitude index according to their position on the diagram. This index consists of the angles of dip D and plunge P. The precise attitude of any fold can then be expresses by a simple descriptive name, followed by the index in parentheses. The following examples illustrate the method, and each is also plotted on Fig. 11.10a:

a. Upright fold ($D_{85}P_{20}$): according to the terminology of Table 11.2 this is fully described as an upright, gently-plunging fold.

b. Inclined fold ($D_{70}P_{45}$): a steeply-inclined, moderately-plunging fold.

c. Reclined fold ($D_{56}P_{55}$): a moderately-inclined reclined fold.

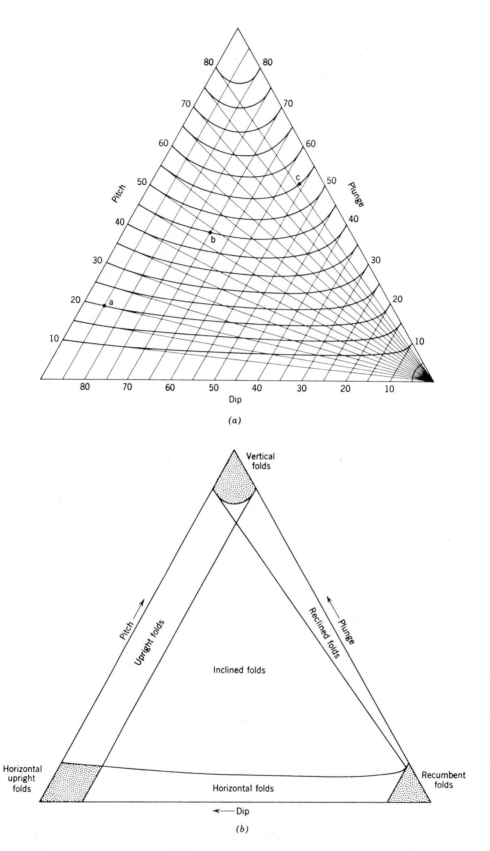

FIGURE 11.11 Fold attitude: (a) triangular grid for plotting fold attitude, (b) use of grid in classifying fold attitude (after Rickard, 1971).

This diagram could also be used to bring out additional details about the folds of an area. For example, if the folds progressively changed orientation in some geographic direction, or, if an element of fold geometry, such as the interlimb angle, changed with attitude, these variations could be brought by a series of points, perhaps with a curve drawn through them.

11.6 ASSOCIATED STRUCTURES

Two geologic structures are commonly found to be associated with folds: cleavage, and minor folds.

Where fold and cleavage develop synchronously, the usual case is that the cleavage closely approximates the orientation of the hinge surface. The qualification "approximate" is needed because this cleavage commonly displays a fan-shaped pattern, and also because it often changes direction abruptly when passing from a layer of one lithology into another.

As a consequence of this axial plane character of the cleavage, there are two additional relationships which are of great importance in the field study of folds. The first is that the line of intersection of the folded layers and the cleavage is parallel to the hinge line, and therefore its orientation gives the axial direction of the fold.

The second is that the angular relationship between the bedding and the cleavage as seen in the profile plane allows the antiformal and synformal hinges to be located from a single exposure. In Fig. 11.12a, a sandstone bed dipping due 40° west overlies a slate with cleavage dipping 80° due west. Where are the folds? If the hinge surface is parallel the cleavage, then an antiformal hinge must lie to the east (Fig. 11.12b), and a synformal hinge to the west. Of course nothing can be said about the size of these folds from the single exposure.

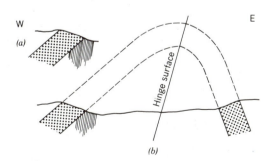

FIGURE 11.12 Cleavage-bedding relationships at a single exposure to locate antiforms and synforms.

Cleavage cutting bedding can be used in still another way. In terms of working out the structure of an area, it is of utmost importance to identify anticlines and synclines. This can be done in a number of ways. If the stratigraphy is known, it is a simple matter to identify the relative age of the rocks in the core of the fold. If they older, then the fold is an anticline, and if they are not it is a syncline. Another way is through the use of sedimentary structures. As deposited sedimentary rocks are said to <u>face</u> upward, and in any subsequent attitude they continues to face toward the side that was originally upward and younger. This direction can be identified from a study of a variety of sedimentary structures, including cross bedding, graded bedding and ripple marks.

It is useful to extend this concept of facing to the folds themselves. In the case of a normal, upright fold the <u>structural facing</u> is upward, and this can be confirmed immediately by determining the direction in which the sedimentary beds of the fold face. Where the folds have overturned limbs, some bed will face downward, and some upward, and especially where exposures are sparse, the direction of structural facing may be obscure. However, this direction can be determined unambiguously by an examination of the beds at the hinges of the folds (Cummins and Shackleton, 1955; Shackleton, 1958).

The direction of structural facing can also be determined at a single outcrop by the application of a simple principle. While the direction of the facing of an individual bed is greatly affected by its local attitude, the component of the facing direction projected onto the axial plane cleavage has a constant direction regardless of the orientation of the bedding (Borrodaile, 1976). The situation is illustrated in Fig. 11.13; while the beds of a series of folds have considerable variation in the direction of younging, the component on the cleavage plane is consistently oriented. The identification of the direction of structural facing of folds through the use of sedimentary structures has come to be called Shackleton's Rule.

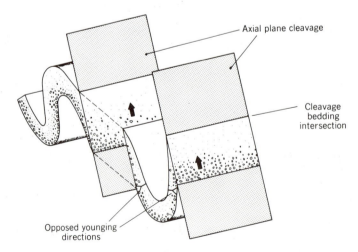

FIGURE 11.13 Opposed younging directions on the flanks of a fold have a consistently oriented younging direction (structural facing direction) on the cleavage surfaces (from Borrodaile, 1976).

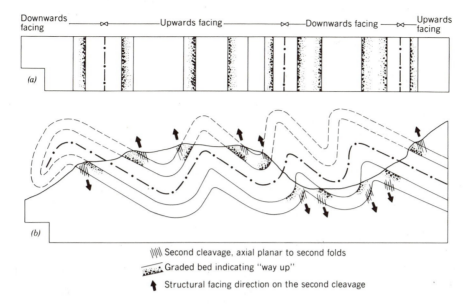

(a)

(b)

|||| Second cleavage, axial planar to second folds

Graded bed indicating "way up"

↑ Structural facing direction on the second cleavage

FIGURE 11.14 Hypothetical area with two phases of folding: (a) map showing the local younging directions; (b) cross section showing the use of these directions to locate the hinge surface of the first fold (from Borrodaile, 1976).

A profile through a hypothetical area with two phases of folding illustrates the practical importance of this rule. A large, westward closing recumbent anticline has been refolded by smaller, nearly upright antiforms and synforms. The younging directions plotted on the map do not readily indicate the location of the now folded hinge surface of the first fold (Fig. 11.14a). When projected onto the axial plane cleavage of the second folds, however, the directions of the structural facings immediately become apparent, and falls into two groups (Fig. 11.14b). The second phase folds above the first hinge surface face upward, and those below it face downward. In this way, the location of the trace of the hinge surface is identified by the reversals in the facing directions.

This example will also make clear why cleavage–bedding relations alone are insufficient to determine the direction of structural facing, and therefore to allow the folds to be identified as anticlines and synclines (cf. Billings, 1973, p. 400f).

The presence of minor folds developed in thin beds may also be used as an aid in working out the structure of an area. It has been observed that these smaller folds often share axes and axial planes with the main fold, a rule of thumb which has become known as Pumpelly's Rule (Pumpelly, Wolfe and Dale, 1894, p. 158). Such folds often show a strong asymmetry with a vergence which is consistently toward the hinges of the antiforms. This is useful in identifying large folds when exposures are poor. A short hand terminology has developed to emphasize these relationship. The strongly asymmetric minor fold on one limb are denoted Z folds and those on the

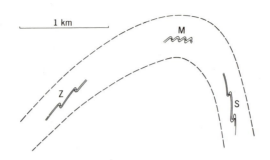

FIGURE 11.15 Z, S and M minor folds.

other are S folds, while the more nearly symmetric folds in the hinge zone are M (or W) folds (see Fig. 11.15).

Such minor folds are commonly referred to as "drag" folds. This is inappropriate on at least two grounds. It introduces genetic connotations into what should be a descriptive terminology. Further, the name implies that such folds formed in response to the slipping of the layers past one another producing something like simple shear in the layer containing the small folds by drag. As we have already seen, the shear direction in simple shear is a direction no finite elongation, and yet it is clear that the small folds have indeed shortened in this direction. This makes the concept of drag as a mechanism of folding questionable, though, of course, the shape of small folds once formed by layer parallel shortening would change shape due to the shear. The most reasonable explanation of Pumpelly's Rule is that the minor folds are small because the layers in which they form are thin, and that they share axes and axial planes with the main folds because the same pattern of deformation is responsible for both.

As with all such rules there are exceptions. The cleavage associated with a fold may not closely parallel the hinge surface. This may be the result of superposition of two differently oriented deformational episodes, or of a more complex single deformation involving a relatively large rotational component. The term transected fold describes this more complex situation (Powell, 1973, p. 1045; Borrodaile, 1978). Minor folds may show similar departures from the orientation of the main fold. When the departures of the cleavage and minor folds are taken into account, even cylindrical folds will have triclinic symmetry. From a practical point of view the existence of such low symmetry folds means that rules given above must be tested before great reliance is placed upon the results that are derived from them.

11.7 NONCYLINDRICAL FOLDS

On the basis of the form of the hinge line and the hinge surface Turner and Weiss (1963, p. 106f; see Fig. 11.15) developed a descriptive terminology and a classification which includes noncylindrical folds. In general, the hinge line may be rectilinear, curvilinear and lying in a plane, or curvilinear but not lying in a plane. Cylindrical folds can have only rectilinear hinge lines, otherwise they are noncylindrical. In a similar way, the hinge surface may be planar, cylindrically curviplanar, or more generally curviplanar. If the hinge surface is planar, the fold is described at being a _plane_ fold; if not, it it a _nonplane_ fold. Subject to the above condition, cylindrical folds may only have a planar or cylindrical hinge surface; all other folds are noncylindrical. As a special case, the hinge line degenerates to a point, and the hinge surface to a line in domical folds; such folds are not included in this classification.

A more quantitative scheme has been suggested by Williams and Chapman (1979) which should be useful for some purposes. Their description incorporates two angles: the interlimb angle and the angle subtended by the hinge in the hinge surface. A broad range of noncylindrical geometries may then be represented on a triangular diagram.

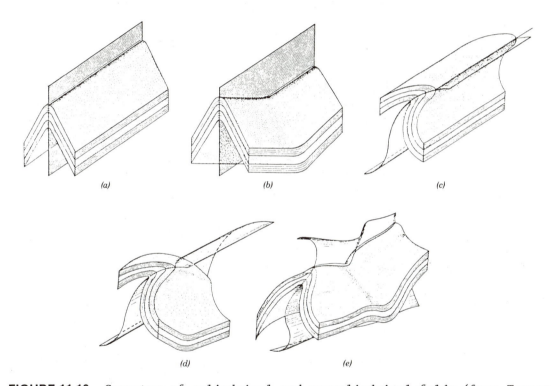

(a)   *(b)*   *(c)*

(d)   *(e)*

FIGURE 11.16 Geometry of cylindrical and noncylindrical folds (from Turner and Weiss, 1963, p. 110): (a) cylindrical plane fold; (b) non-cylindrical plane fold; (c) nonplane cylindrical fold; (c) non-plane noncylindrical with cylindrical hinge surface; (e) non-plane noncylindrical fold with noncylindrical hinge surface.

EXERCISES

1. Using the profile depicted in Fig. Xll.l, construct the following features: the traces of the crest and trough surfaces, the traces of the hinge and inflection

surfaces, dip isogons at 20° intervals throughout the structure.   For one
particular layer, construct a graph of the isogon angle.

FIGURE X11.1

2. Plot the following data on Rickard's triangular diagram, and name the orientation
   of each fold: D = 85°, P = 60°; D = 80°, R = 80°; D = 83°, P = 85°; D = 20°, R =
   25°.

12
Parallel Folds

12.1 INTRODUCTION

A simple type of fold can be modeled with a card deck (Fig. 12.1a). The total deformation is accomplished by the bending of the individual cards, and slippage at the surface of each card is an inevitable consequence of this mechanism. The importance of this slip can be readily appreciated by trying to bend a deck firmly clamped at each end. By preventing slip from occuring, the layered structure of the deck is effectively eliminated, and the result is nearly as rigid as a block of wood. Clear evidence of this <u>bedding-plane</u> slip is found in naturally occuring folds where veins or dikes are offset across bedding planes, and by slickensided and striated bedding surfaces.

Also as a result of the bending, the surface of each layer is stretched along its outer arc and contracted along its inner arc. This, in turn, results in a thinning of the material adjacent to the outer arc and a complementary thickening adjacent to the inner arc of a single layer. These two regions are separated by a surface of no longitudinal strain, called the <u>finite neutral surface</u> (Fig. 12.1b). At the initiation of bending the neutral surface lies along the center line of the layer, but thereafter it migrates. While the changes above and below this surface tend to cancel, there will be, in general, a net change in the thickness. The magnitude of the effect is proportional to the distance from the neutral surface, hence to the thickness of the layer. If the layers are as thin as the cards the effect is slight, and the change in thickness is negligible. Therefore, the total thickness of the pack remains effectively constant, and this can be confirmed by measurements made on the bent card deck. With orthogonal thickness constant throughout, the model fold will be <u>parallel</u>. To the extent that the mechanical properties of a layered rock mass undergoing folding approach those of the card deck, the natural folds will also have parallel form. The important factor in making the comparison between model and real folds is the relationship between layer thickness and

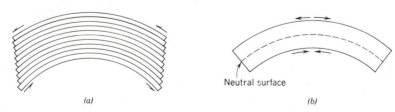

(a)   (b)

Neutral surface

FIGURE 12.1 Models of folds: (a) buckled card deck illustrating flexural slip; (b) buckled single layer showing the finite neutral surface.

radius of curvature.

The physical nature of the card deck imposes a further condition on the shape of this simple model fold--it has a rectilinear hinge line, and is therefore cylindrical. Once the cards are folded, a second oblique bending utilizing the bedding-plane slip mechanism is not possible. This is the same property which gives corrugated sheet metal it rigidity in the direction of the corrugations. If the hinge line is curved, it must be produced by some additional mechanism involving distortions within the plane of the layering. It should be noted here that with a more involved kind of bending, it is possible to produce conical folds with a card deck. Although such noncylindrical folds may be important where cylindrical parallel folds die out, they will not be treated further here (see Wilson, 1967).

The bedding-plane slip mechanism is closely related to simple shear, except that the shear plane is curved and the angle of shear varies in the shear direction. The angle of shear at any point in the layer is closely related to the fold shape, and its value can be easily calculated (Ramsay, 1967, p. 392). Suppose a portion of a folded layer of constant orthogonal thickness is made up of two circular arcs, each defined by a radius of curvature r and an arc which is measured by the angle θ each segment subtends.

First, we need to find an expression for the angle of dip α in terms of the fold shape at any point in the folded layer (Fig. 12.2a). Because the horizontal and inclined lines which define the dip angle at point X are perpendicular to the two corresponding radii which define the angle θ, the dip at this point is,

(12.1) $\qquad \alpha_1 = \theta_1$

Similarly, along the radius passing through point Y on the inner arc,

(12.2) $\qquad \alpha_2 = \theta_1 + \theta_2$

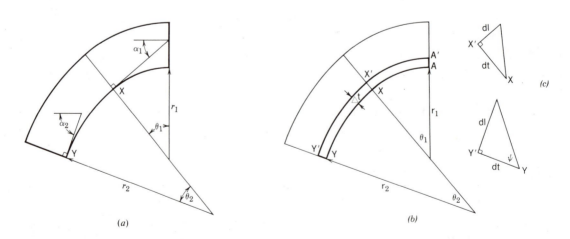

FIGURE 12.2 Flexural slip fold: (a) relationship between θ and α; (b) detail of a thin layer used to derive the shear strain (after Ramsay, 1967, p. 392).

Now consider a thin layer within the packet of layers of thickness Δt (Fig. 12.2b). The length AX along the inner arc is

$$AX = \theta_1 r_1 \qquad (\theta \text{ in radians})$$

and the length of A'X' along the outer arc is

$$A'X' = \theta_1(r_1 + \Delta t)$$

The difference in length of these two arcs is then

(12.3)
$$\Delta \ell = A'X' - AX$$

$$= \theta_1(r_1 + \Delta t) - \theta_1 r_1 = \theta_1 \Delta t$$

Similarly, the difference in length of the two arcs at Y is,

$$\Delta \ell = AY - A'Y'$$

$$= \theta_1 \Delta r + \theta_2 \Delta t = (\theta_1 + \theta_2)\Delta t$$

In the limit, as the layer becomes very thin, we obtain expressions for the shear strain at point X (Fig. 12.2c)

$$\tan \psi = d\ell / \Delta t$$

With Eq. 12.3, this then becomes,

$$\tan \psi = \theta_1$$

Similarly, at point Y

$$\tan \psi = (\theta_1 + \theta_2)$$

Then using Eqs. 12.1 and 12.2, we have the general result for the shear strain at any point within the folded layer.

(12.4)
$$\gamma = \alpha \qquad (\alpha \text{ in radians})$$

This relationship has been solved for values of dip ranging from 0°-90°, and the results presented graphically in Fig. 12.3. With this result, together with Eqs. 5.7 and 5.14, the state of strain at any point in the folded layer can be found.

12.2 PARALLEL FOLDS IN CROSS SECTION

The property of constant orthogonal thickness implies that a line perpendicular to the bounding surface of one layer is also perpendicular to the bounding surfaces of layers above and below. This, and the fact that any curve may be approximated by a series of tangential circular arcs, forms the geometrical basis for reconstructing

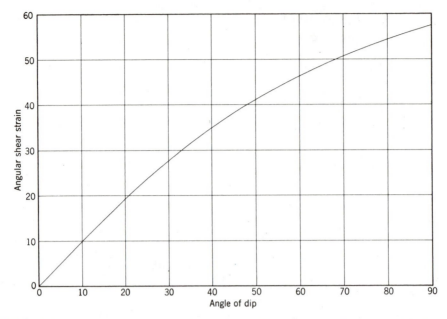

FIGURE 12.3 Graph of ψ as a function of α (after Ramsay, 1967, p. 393).

a series of horizontal parallel folds in vertical cross section. The technique was originally described by Busk (1929), and it is now commonly called "Busking". This construction depends on the elementary proposition that the centers of two tangent circles lie on the straight line which passes through their point of contact and which is perpendicular to the common tangent. In Fig. 12.4, two circle with centers at O_1 and O_2 touch at point A. Line segments O_1A and O_2A are both perpendicular to the tangent AB, and therefore must lie on the same straight line.

The simplest application of this principle is to the problem of constructing the curved traces of the bounding surfaces of a layer between two successive dip lines, as plotted on the profile plane.

PROBLEM

Given angles of dip at two points, draw the curves which pass through each of the points which represent the traces of the layer.

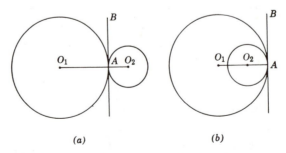

(a)   (b)

FIGURE 12.4 Tangent circles--the geometric basis for reconstructing parallel folds (after Busk, 1929, p. 13).

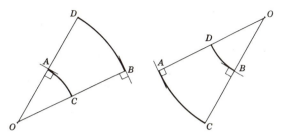

FIGURE 12.5 Tangent arcs though two adjacent dips (after Busk, 1929, p. 14).

CONSTRUCTION (Fig. 12.5)

1. Plot the two dip lines at their proper places on a cross section, and label them A and B. At each point, construct lines OA and OB normal to the respective dip lines to intersect at point O.

2. With OA and OB as radii, and point O as center, draw arcs AC and BD. Note that the thickness of the stratum between A and B is represented by the line segments AD = BC.

This construction is easily extended to the case of three or more dip readings (Fig. 12.6). If A, B and C are the dips at three adjacent locations, first construct the normals to the first pair of dip lines A and B, thus locating center O_1. Then repeat for the second pair B and C, locating center O_2. With appropriate radii, the tangent arc may then be carried across the fold from one dip normal to the next.

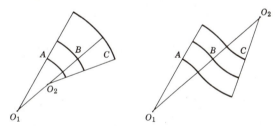

FIGURE 12.6 Tangent arcs through three adjacent dips (after Busk, 1929, p. 16).

If the measured dips at two adjacent points A and B are equal the normals at A and B will be parallel, and the required "arc" will be a straight line (Fig. 12.7). If the two adjacent dips differ only slightly, the normals will intersect at some distance above or below the section, and the required radius may exceed the expansion of even a beam compass. This may be controlled by working at a smaller scale, or the large radius arc may be approximated.

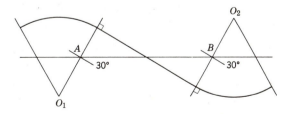

FIGURE 12.7 Reconstruction where two adjacent dips are equal.

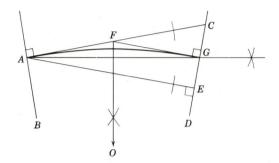

FIGURE 12.8 Reconstruction where the difference between two adjacent dip angles is small.

PROBLEM

Given the dip normals AB and CD, not far from parallel, draw a circular arc through point A (Fig. 12.8).

CONSTRUCTION (after Busk, 1929, p. 21; see Fig. 12.8)

1. Draw AC perpendicular to AB, and AE perpendicular to CD.

2. Construct the bisector of the angle CAE to meet line CD at point G.

3. Construct line GF to be perpendicular to CD.

4. Point G is the intersection of the required arc with CD, and the arc may then be sketched in from A to G. The lines AF and FG, which are perpendicular to their respective normals, are also tangents to the arc, and they may be used in positioning the arc.

The validity of this construction lies in the fact that triangles AOF and GOF are equal, and therefore OA = OG.

The reconstructed fold must, of course, be in accordance with the evidence on the ground. If a recognizable horizon appears repeated along the line of traverse, there may be a discrepancy between its actual and reconstructed location. If one can assume that the error is not due to other factors, such as thinning or faulting, an intermediate dip may be interpolated to adjust the actual and predicted position of the marker horizon. This dip may be interpolated anywhere along the traverse, but in the absence of any further information it is probably best positioned where the control is poorest, that is, where the difference between adjacent dip measurements is large, or the distance between measurements is great, or both.

PROBLEM

Given two adjacent dip readings, construct a curve to pass through both measurement points (Fig. 12.9).

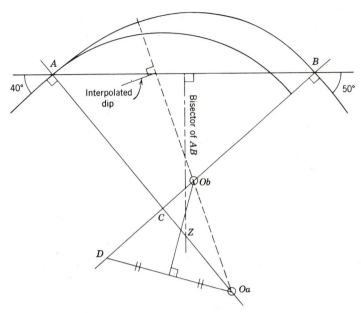

FIGURE 12.9 Interpolation between two measured dips (after Higgins, 1962).

CONSTRUCTION (after Higgins, 1962)

1. Label the lesser dip A, and the greater dip B. Construct normals to each dip line intersecting at C and extending well beyond. Ordinarily C would be used as the center for drawing the circular arc between these two normals.

2. Construct the perpendicular bisector of line segment AB to intersect the normal AC at point Z.

3. Arbitrarily locate center O_a on the normal AC as far beyond Z as is convenient.

4. Find point D on the normal BC, so that BD = AO_a.

5. Find center O_b at the intersection of the perpendicular bisector of line DO_a with BD.

6. With O_a and O_b as centers draw two tangent arcs that define the require curve. The line O_aO_b is normal to the interpolated dip.

With these techniques, the form of folded strata may be reconstructed from map data, or from a field traverse made expressly for the purpose. In either case, the line of section should be as nearly perpendicular to the strike direction of the dipping beds as possible. In preparing the cross section, it is conventional to orient the line so that its eastern or northern end is on the right hand side.

Even in simple, regular folds, it is rarely possible to locate a section line which is exactly normal to all measured strikes, in which case apparent dips in the direction of the section line must be computed before the dip lines are plotted on the cross section. In addition to the correct map locations, the correct elevations of the data points must, of course, also be used in plotting the dip angles.

The number of readings exactly on the chosen section line is never enough. Other measurements may be used by projecting them short distances to the line of section. Generally, they are projected parallel the the local line of strike, but this may require adjustment if the angle between the section line and the dip direction is large, for otherwise projecting strike lines may cross, and the dips will then be reversed on the section line from their relative positions in the field. The need to convert to apparent dips and the use of obliquely projected attitudes means that the folds are not strictly cylindrical, or that the folds plunge, or both, and this introduces errors and uncertainties which must be kept in mind when interpreting the form of reconstructed folds.

PROBLEM

Given the following dip angles, already corrected or projected to an east-west line as required, reconstruct the form of the folds.

A = 20°E   F = 25°E
B = 10°W   G = 75°E
C = 45°W   H = 50°E
D = 10°W   I = 20°E
E = horizontal

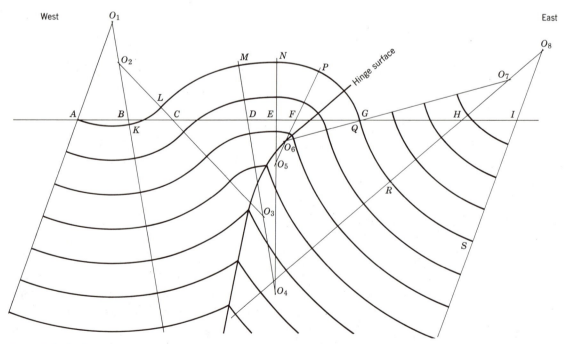

FIGURE 12.10 Full reconstruction of parallel folds by tangent arcs, showing the trace of the hinge surface (after Busk, 1929, p. 19).

CONSTRUCTION (after Busk, 1929, p. 19; see Fig. 12.10)

1. Construct normals to each of the dip lines. These intersect successively at points O_1 through O_8.

2. With O_1 as center and O_1A as radius, draw an arc to the next normal O_2B, thus establishing point K.

3. With O_2 as center and O_2K as radius, draw an arc to normal O_2C, giving point L.

4. Repeat using successive centers O_3 through O_8, giving arcs LM, MN, MP, PQ, QR and RS. The curve ALMNOPQRS then represents the trace on the plane of the cross section of the folded surface passing through point A.

5. This same procedure may then be repeated to reconstuct the shape of the deeper horizons passing through points A_1 and A_2.

6. For the even deeper horizons passing through points A_3–A_6 a somewhat different approach is required. In order to maintain uniform thickness of the folded layers, their thicknesses must be marked off along the normals O_7S or O_8R. Swinging arcs through these marks into the core of the antiform produces angular, rather than rounded hinges. This is due to the fact that centers O_5 and O_6 lie above horizon A_3, and thus have no control on the shape of the curves below them. At even deeper levels O_3 and O_4 are also eliminated.

7. The trace of the hinge surface may be drawn as a smooth curve through the sharp hinges in the deeper parts of the antiform, and then extended at the bisector of arc PQ.

This method of reconstructing parallel folds may produce acceptable, even good results, and they get better as the quality and quantity of structural and stratigraphic control increases. There are, however, some important limitations, and an awareness of these should aid in constructing and interpreting such sections.

As described, this method will not work for plunging folds; in a vertical section, that is, in a plane not perpendicular to the axial direction, the traces of layers will appear with thickness which varies with attitude, even though the fold geometry may be strictly parallel. For similar reasons, the reconstruction of non-cylindrical folds should not be attempted by this technique. In Chapter 14, we will show how to construct the fold profile of plunging folds, and suggest ways of handling some types of non-cylindrical folds.

If non-parallel folds are present, even locally, there may be severe distortions if an attempt is made to force the structural data into a parallel mode.

Where measurement errors or local structural irregularities are present, these will be propagated throughout the reconstructed section. Reches and others (1981) have suggested a modification of the basic technique which may help avoid this problem. Several concentric arcs, using a number of adjacent centers are drawn on an overlay sheet to find the arc which gives the best fit to the structural and stratigraphic data.

At best, concentric circular arcs can only approximate the natural curves of parallel folds, although once again, the more attitude data available, the closer this approximation will be to reality. A large part of the problem arises because each arc segment utilizes only two dip angles. Mertie (1947) has explored quite a different approach involving no assumptions about the nature of the curvature, and which makes the best possible use of the existing data. The technique involves constructing a curve called an <u>evolute</u> which utilizes a number of adjacent dip measurements. Along this curve the center and radius of curvature of the traces of the parallel layers, called <u>involutes</u>, varies continuously. The method of constructing these curves is more involved than drawing tangent arcs, although by no means impossible, but it apparently has never been seriously used. It does represent a useful way of emphasizing just how limited concentric circular arcs are in approximating general curves (see Fig. 12.11).

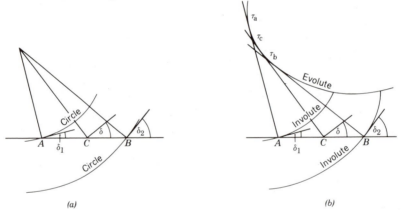

FIGURE 12.11 Alternative reconstructions of parallel folds (from Mertie, 1947): (a) circular arcs; (b) evolutes and involutes.

Finally, the appearance of the sharp cusps in the cores of the anticlines ought to warn us that there is something seriously wrong with the reconstruction at depth. To round off these cusps, as suggested by Badgley (1959, p. 34), is entirely cosmetic, and ignores completely the most serious defect of the Busk technique.

12.3 BALANCED CROSS SECTIONS

In the light of these limitations, it would be most useful to have some way of testing the results of our reconstruction for internal consistency and geologic reasonableness. In this section, we will describe just such a test. By way of introduction we make the following assumptions.

1. The deformation which produced the folds was two-dimensional in the plane of the cross section.

2. The deformation was isochoric, and as a consequence, area in the plane of the section is conserved.

3. Stratigraphic units originally had constant orthogonal thickness.

4. The fold geometry is parallel throughout.

If area is conserved, and bed thickness is constant, it follows that the length of any bed must also remain constant. Because of the requirement of the constancy of bed length, it follows that the length of trace of each folded surface must be the same from one bed to another. This leads to a simple test of consistency to the reconstructed shape of the parallel folds.

STEPS

1. Establish a pair of reference lines at either end of the section in regions of no interbed slip. These may be located at the hinges of major anticlines or synclines, or in regions well beyond the disturbed belt.

2. Measure the folded length of the traces of selected horizons between the two reference lines. This can be done most easily with a curvimeter. These lines should all be the same length.

3. If the lengths are not the same, the section must show a valid explanation of why they are not.

Sections which pass this test are termed balanced (Dahlstrom, 1969a; Hossack, 1979; Elliott, 1983). An important property of such sections, and indeed the principal reason for constructing them, is that the amount of shortening represented by the folds can be determined by comparing the original and deformed length of the folded traces. Note too that balance is a necessary, but not sufficient condition for the correctness of a section.

If this test for balance is applied to the parallel folds reconstructed by Busk's method of tangent arcs, it will be found that the bed length is not consistent (Fig. 12.12). A major part of this discrepancy lies in the cusp-shaped hinges in the cores of the anticlines. Clearly, these reconstructed forms can not represent the actual geometry of the deeper zones even approximately. This must mean that the parallel folding mechanism breaks down at depth, and that other mechanisms become important. The prediction of the geometry of these deeper structures from surface data alone is essentially impossible, and in many specific examples, it is still being debated. However, the main alternatives can be outlined.

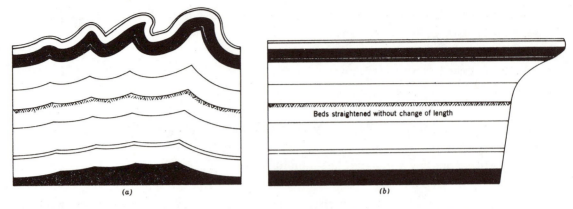

FIGURE 12.12 Inequal shortening with depth (from Carey, 1962): (a) reconstruction by the method of tangent arcs; (b) beds straightened without change in length.

The parallel folds may be cut at depth by thrust faults. If bed length is to remain constant, then the displacements on these faults should be consistent as well. However, as a matter of observation, displacements commonly change along the faults planes, and the faults may die out altogether (Fig. 12.13a). There are two ways of resolving this difficulty:

1. Interchanging fold shortening and fault displacement (Fig. 12.13b).

2. Upward imbrication (Fig. 12.13c).

In these illustrations, the fault traces are arbitrarily represented by straight lines. Such planar faults, together with constant bed length, require a substantial amount of interbed slippage in the vicinity of the faults which would alter the vertical ends of the blocks to the curves as shown. From these simple examples, it may be concluded that (Dahlstrom, 1969a, p. 747):

1. Faults with changing displacements are apt to be curved, usually concave upward, a feature which is confirmed by observation.

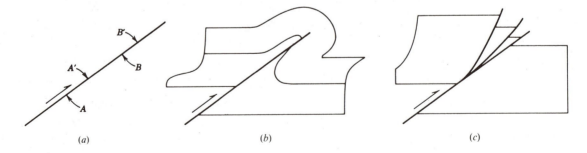

FIGURE 12.13 Thrust in parallel folds (from Dahlstrom, 1969a): (a) apparent anomaly in the amount of fault displacement (AA' is greater than BB'); (b) accomodation by folding; (c) accomodation by upward imbrication.

2. Interbed slip is a necessary part of the thrust faulting which accompanies the development of parallel folds.

3. Interbed slip can contribute to the change in displacement, and in extreme cases would become a type of imbrication itself.

Balanced cross sections are constructed by trial and error. After the purely geometric stage of reconstructing the parallel folds, the bed length is measured with a curvimeter, and areas with a planimeter. Adjustments are then made, and these are rechecked. The most sophisticated reconstructions which are available have been made in the foothill fold and thrust belt of the eastern Canadian Rockies, where abundant seismic, borehole and surface geologic data are available (for example see Balley and others, 1966).

By hand, the construction of accurate balanced sections is a tedious, time-consuming process. It is therefore not surprising that in the petroleum industry, where such sections are exceedingly important in the quest for oil-bearing structures, computer programs are now routinely used to produce them.

But what if the four assumed conditions do not hold?

The fold profile is the best approximation of the plane in which the deformation is most apt to have been two dimensional. In situations which have been looked at carefully, strain analyses indicates that extensions do occur in the direction of the fold axes, and, therefore, that the strain can not be plane. This means that there may be an area change in the plane of the section. Measured extensions indicate that longitudinal stretching in the range of 5-15% occur, and that 10% stretching is common. If the deformation is isochoric, this latter figure means that there will be about a 9% decrease in area in the plane of the section.

Volume changes are also known to occur. Strain measurements indicated that the formation of slaty cleavage commonly involves a volume reduction of 10-20%. Measurements of density increases suggests that volume reductions in the range of 0-10%, are not uncommon. Solutional transfer is the most troublesome to estimate quantitatively; but volume reductions in the range of 20-30% may occur. If a 10% reduction is taken as typical, in combination with a 10% stretching parallel to the fold axes, there will be a 15% reduction of area in the plane of the section, and this will certainly affect shortening estimates. These effects will reduce not only the area, but also the original length, and this will cause the shortening to be underestimated, and thus the calculated shortening will be on the conservative side.

Stratigraphic thicknesses are commonly not constant, especially over distances of tens of kilometres. Such changes can be taken into account. However, the method is relatively insensitive to normal changes in thickness. For example, from work in the Appalachians, Dennison and Woodward (1963, p. 671) noted that an error of 350 m

(1000 feet) in stratigraphic thickness on a regional scale introduces errors in the shortening calculation of only about 2.5-3%.

Finally, what if the folds are not parallel? The construction for nonparallel folds is more difficult, but it can be handled, as will be shown in a later section.

12.4 DEPTH OF FOLDING

Another approach to resolving the difficulties of the tangent arc method is to keep bed length constant, but allow small changes in thickness of each bed in order to keep area constant (Fig. 12.14). This requires that the folds change with depth, and these changes in form are termed disharmony.

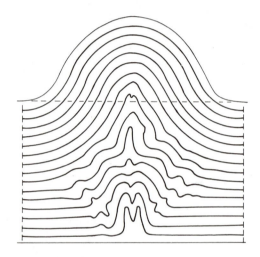

FIGURE 12.14 Reconstruction of folds at depth by maintaining original length and conserving area (from Goguel, 1952).

Another and much more fundamental structural change with depth is a direct consequence of this disharmony. How would the next lower bed of Fig. 12.14 be drawn? Cleary, the folded beds must have been completely detached from the underlying ones, and deformed essentially independently of them. This requires a shearing-off horizon, or decollement at the base of the figure.

The position of this basal detachment plane is commonly controlled by the location within the sedimentary sequence of weak layers, such as shale, or in extreme cases, salt and gypsum, or by the contact between the sedimentary rocks and an underlying rigid basement unit. In detail, the behavior of the lower part of the folded sequence depends on the mechanical properties of the rocks involved, Instead of tightly crumpled folds in the cores of the anticlines, cross-cutting thrust faults may form which root in the decollement zone. These faults may or may not break through to the surface, and if they form early, they themselves may be deformed by continued folding or by even later thrusting.

An illustration of some of the structures which may occur in the rocks above a decollement will be useful. Fig. 12.15 shows a historically important section through a part of the Jura Mountains. Surface mapping failed to reveal many of the structures and their complexities, and they were discovered only after the construction of a tunnel.

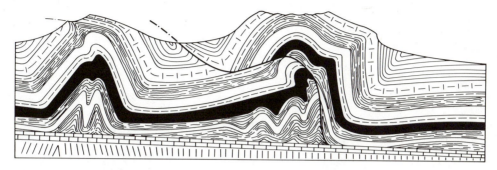

FIGURE 12.15 Section through a part of the Jura Mountains (after Buxtorf, 1916).

Finally, in terms of cause and effect, it is the position of the detachment plane, in combination with the amount of shortening, that largely determines the size and shape of the folds in the belt. The thicker the strata above the potential detachment horizon, the larger the individual folds, other things being equal. A comparison of two folds from the Jura Mountains illustrates this control (Fig. 12.16).

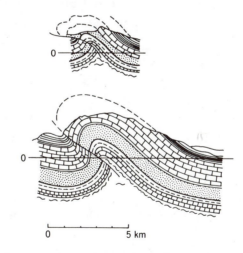

FIGURE 12.16 Two folds from the Jura Mountain with size proportion to thickness (from De Sitter, 1964)

The effect of folding is to make a packet of layered rocks shorter and thicker. If volume is conserved in the packet, the amount of material uplifted must exactly equal the decrease due to shortening. Since both the amount of thickening and shortening can be measured on a balanced section, the depth to the decollement can be determined.

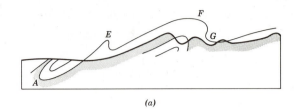

(a)

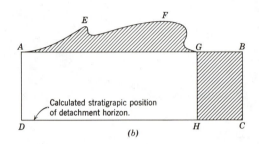

Calculated stratigrapic position
of detachment horizon.

(b)

FIGURE 12.17 Calculation of depth of folding (from Dahlstrom, 1969b).

CALCULATION (after De Sitter, 1964, p. 190; Dahlstrom, 1969b, p. 342)

1. The lateral shortening of a reference horizon between points of no
   slippage is determined by comparing the bed length as measured with a
   curvimeter and the lateral distance this same horizon occupies in the
   folded packet. In Fig. 12.17a, the bed length measured at the top of
   the Mississippian Rundel group in the Canadian Rockies is AEFG = AB.
   The distance GB represents the actual shortening.

2. The area of increased thickness is measured with a planimeter. In Fig.
   12.17b, this area lies between the straight line AB and the trace of
   the folded reference horizon AEFG.

3. The depth of folding follows directly from the relationship:

   Depth AD = (area uplifted)/(shortening)

   Note that the area GBCH is equal to the area measured in Step 2.

It should also be noted that the depth to the basal thrust is independently known,
the shortening may be found using this same formula. This is also an estimate of
the minimum displacement on the decollement thrust.

   This method of estimating the depth of folding has been used with good results
in the Jura Mountain, where the concept of the decollement originated. However,
these results are not without difficulties. How can a packet of sedimentary rocks
be deformed in a manner which is indepndent of the underlying material? Suggestions
have been made concerning the possible existance of related structures, such as im-

bricate slices in the underlying block. Besides raising several additional difficulties, there is now clear and compelling geophysical evidence that in the external zones of at least some mobile folds belts no such sub-decollent structsures are present, but the agents responsible for the deformation often remain in question.

It should be abundantly clear that parallel folding is a complicated process, and must involve other modes of deformation, including disharmony and shearing off. It does not follow, however, that every group of parallel folds has a single decollement thrust at depth. The necessary adjustments may take place locally and gradually rather than at a single horizon at depth. Many small-scaled examples of detachment structures can be found in the field.

More importantly, in the internal zones of mountain belts, the basement and cover rocks may both participate in the deformation. Synchronous folding and thrusting of the overlying, near-surface rock layers would be a certainty, and these would most likely involve, at least in part, the parallel mode.

Barnes and Houston (1969) have described a simple example which illustrates the principle involved. In a part of the Medicine Bow uplift in the North Rocky Mountains of Wyoming, a Precambrian basement complex is unconformably overlain by Paleozoic and Mesozoic sedimentary rocks. During the Laramide Orogeny, these layers were folded, presumably in response to distributive movement on microfractures in the basement unit (Fig. 12.18). Under these circumstances shearing off is not require. Compton (1967) described a similar example, and he was able to demonstrate actual slip on closely spaced fractures in a gneissic basement. In this case, up to 3700 m of overlying sedimentary rocks were deformed by folding, an interesting feature is the evidence of disharmony in the upper part of the sequence, particular in the cores of synclines. Dalstrom (1969b) has shown that in certain instances this upward increase in disharmony may actually lead to an upper detachment fault.

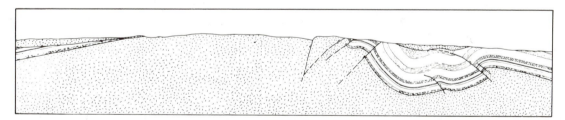

FIGURE 12.18   Laramide fold in the Northern Rocky Mountains (from Barnes and Houston, 1969).

12.5   KINK BANDS AND CHEVRON FOLDS

There is a special type of fold which has some interesting and important properties; the limbs are straight and the hinges sharp, and they commonly take two forms.

1. <u>Kink</u> <u>bands</u> are angular, step-like monoclines between two well developed parallel planes, called kink band boundaries, but which are actually hinge surfaces (Fig. 12.19a).

2. <u>Chevron</u> <u>folds</u> are symmetric folds with limbs of equal length (Fig. 12.19b).

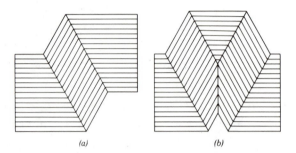

(a)   (b)

FIGURE 12.19 Kink bands and chevron folds.

Both types of folds maintain their shape in the direction of the trace of the hinge surface on the profile plane, a property of similar folds, but individual layers in the fold keep nearly constant orthogonal thickness, which is the characteristic of parallel folds. These folds typically develop in material with a prominant planar anisotropism such as crystals, slates, phyllites and schists, and thinly laminated sedimentary rocks, and they have also been produced experimentally in packs of thin cards.

The deformation within a kink band is essentially by flexural slip. The process, but not the kink geometry, can be modeled with a simple card-deck experiment: grip the deck firmly at both ends and without rotating the ends, shift the deck into a Z or S shape. Idealized kinks and chevron folds have sharp hinges, but due to the finite thickness of the laminations, there is invariably some curvature present in real examples.

As with other folds, the shape of these folds can be expressed in terms of the interlimb angle. Unlike more conventional folds where a whole spectrum of folds shapes can be observed, kinks and chevrons have a surprisingly restricted range of interlimb angles, usually in the range of 100°-120° for kinks and 50°-60° for chevrons. Gentle folds are not seen. Observations in the Juras indicate that some cases kinks are cut by thrusts and in others thrusts are involved in the kinking. This suggests that there is a Coulomb-like threshold for the formation of kinks, and once the process starts it is not halted until fully formed folds are produced.

The most commonly observed and described kink bands are small-scaled features. Typically, they occur singly or in parallel groups, but they may also occur in pairs in a fairly constant angular relationship. These pairs either:

1. Intersect in a complex zone; if the interlimb angles in both bands are 120°, chevron folds may develop in the area of intersection (Fig. 12.20a)

2. Mutually terminate in a decollent zone (Fig. 12.20b); Ramsay (1967, p. 454) illustrates an unusually clear example.

In both cases, a <u>conjugate</u> pair of bands produce a <u>box</u> fold.

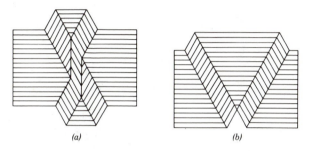

(a)   *(b)*

FIGURE 12.20   Intersecting kink band; (a) chevron folds; (b) local decollement (after Weiss, 1969).

In recent years large-scaled examples of these folds have also been recognized in the Appalachians (Faill, 1969), the eastern foothill belts of the Canadian Rockies (Dahlstrom, 1970, p. 364), and the Jura Mountains (Laubscher, 1977a,b). In the Rockies, the prerequisite for folds with smooth, rounded profiles, such as approximated by the Busk technique, seems to be the present of a dominant unit substantially more competent than the units above and below, such as a thick massive carbonate bed between shale units. On the other hand, kink bands form in well-bedded successions of fairly competent beds separate from one another by thin but effective slippage horizons. This correspondence between fold type and material properties supports the suggestion made by Laubscher (1977b, p. 341) that concentric folds and kinks bands may be end members of a spectrum of fold types.

Clearly, the extent that kinks, with their straight limbs and sharp hinges influences the fold style, the Busk technique will not adequately reconstruct the fold shape. In other ways, however, these folds behave as the more usual types of parallel folds, and this includes the present of a decollement zone. The depth to this zone can be calculated using the same length/area measurement as before, but it is even more simply determined by projecting the kink band boundaries downward to their intersection; for ideal geometry the decollement should be half way between the intersection of the two upper boundaries and the intersection of the two lower boundaries. In the Juras, the decollement associated with the larger box folds found in the way is essentially the same at that determined by the length/area method. Secondary planes of decollement are also present at higher levels, and these can be correlated with known shale horizons.

12.6 NONPARALLEL FOLDS

The breakdown of the ideal parallel fold geometry with depth, as implied by the disharmony-decollement relationships, is in any detailed form unpreditable, simply because it follows no geometrical rules that can be deduced from surface observations.

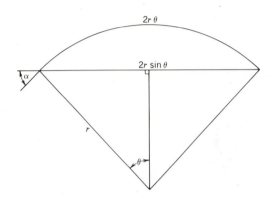

FIGURE 12.21 Shortening strain in a parallel fold.

There is a limit to the shortening possible in a parallel fold. Theoretically, bedding-plane slip must cease when the two limbs have been rotated into parallelism. In Fig. 12.21, the fold shape is assumed to be that of a circular arc. Then the arc length $\ell = 2r\theta$, where 2θ is the angle subtending the circular arc and r is the radius of curvature. The chord length $\ell' = 2r \sin \theta$. The maximum dip on the limb of the fold $\alpha = \theta$. Then the shortening strain associated with the folds is

(12.4) $$e_s = \frac{\sin \alpha}{\alpha} - 1 \qquad (\alpha \text{ in radians})$$

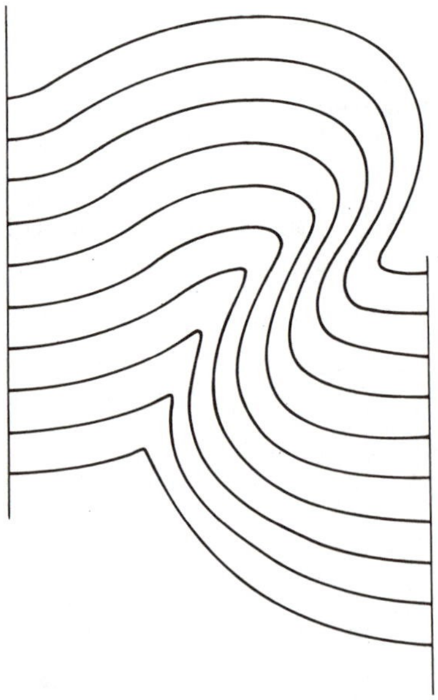

FIGURE 12.22 Sketch of a parallel fold modified by thinning of the overturned limb (after Busk, 1929, p. 57).

Thus when the limb are vertical, the shortening is equal to 36%. If the folds are deformed further, certain beds will begin to thin, and the fold will depart from strict parallelism. The closeness to the limit at which the thinning will occur depends on the mechanical properties of the rock material undergoing folding. After a certain degree of shortening, the parallel folds often become asymmetrical, and one limb becomes steeper than the other, and may finally become overturned. Although it may occur earlier, thinning is geometrically required at the point of overturning (Busk, 1929, p. 30). The less steep limb may still be parallel. In terms of the reconstruction by circular arcs, this nonparallelism is proved when correlation of a key horizon and the utilization of certain dip measurements is irreconcilable. The simplest approach is to make the necessary adjustments in the thinner limb by free-hand sketching (Fig. 12.22).

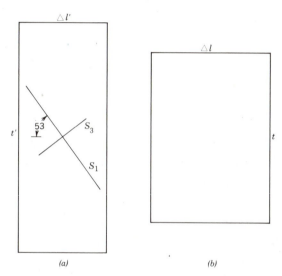

FIGURE 12.23 Correction to restore an increment of bed length

Once the folds depart from parallel geometry any estimation of the horizontal shortening becomes much more difficult. It is then necessary to apply a correction to the deformed length, thus determining the original bed length, and this requires information on the state of strain at as many points in the layer as possible. In Fig. 12.23, an increment $\Delta l'$ of the deformed bed is shown, together with the principal stretches derived from measurements. In terms of the original length of this same increment Δl and the extension parallel to the bedding,

$$\Delta l' = S\ \Delta l \qquad \text{or} \qquad \Delta l = \Delta l'/S$$

In other words, to recover the original length of this segment, we must multiply the deformed length by a correction factor $F = 1/S$. From Eq. 10.10,

$$(12.5) \qquad F^2 = \frac{1}{S^2} = \frac{\cos^2\phi'}{S_1^2} + \frac{\sin^2\phi'}{S_3^2}$$

Because of the problems of determining the full state of strain, it is useful to consider separately the distortion and the dilatation. This can be done by first combining the two terms on the right by using a common denominator, and making the following substitutions,

$$R_s = S_1/S_3$$

$$1+\Delta = S_1S_3$$

$$\cos^2\phi' = 1-\sin^2\phi'$$

the term $1/(1+\Delta)$ can be factored out, leaving

$$F^2 = \left(\frac{1}{1+\Delta}\right)\left(\frac{1}{R_s^2} + \frac{R_s^2-1}{R_s}\right)\sin^2\phi'$$

Applying this to Fig. 12.23, where $R_s = 2.7$ and $\phi' = 53°$, gives

$$F = 1.35[1/(1+\Delta)]^{1/2}$$

It may be difficult to determine the area strain Δ, which will not be zero if the strain has not been plane or if there has been a volume change, or both. This is the same problem faced in the construction of balanced sections, and without further information applying this result generally gives a minimum estimate of the original length, hence a minimum estimate of the shortening.

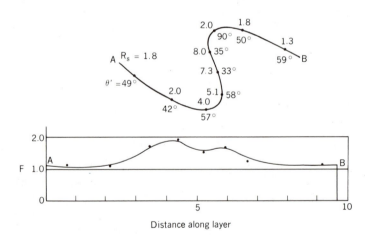

FIGURE 12.24   Shortening in a nonparallel fold: (a) folded layer and the strain parameters at several points; (b) graph of the correction factor as a function of distance along the layer (after Ramsay, 1969, p. 61).

PROBLEM

Given the folded layer of Fig. 12.24a, together with nine determinations of the isochoric strain ellipse, find the shortening of the layer.

STEPS (after Ramsay, 1969; see Fig. 12.24b)

1. Measure the arc length of the deformed layer, and determine the distance from point A to the points where the strain is known.

2. For each locality, calculate the correction factor and plot its value against distance. Draw a smooth curve through these points to give an estimate of the continuous variation of F with distance.

3. To find the original length of the layer from A to B, sum all the $\Delta\ell$ to give true original length,

$$\ell = \lim_{\Delta\ell \to 0} \sum_A^B F \, \Delta\ell = \int_A^B F \, d\ell$$

The value of this simple integral is the area under the curve which can be found with a planimeter. The result is 12.8 square units.

ANSWER

The length of the straight line AB is 6.1 units. The measured arc length around the folded layer is 9.7 units, and by removing the strain this is increased to 12.8 units. The shortening is then (6.1-12.8) = -6.7 units. Expressed as an extension: e = (-6.7/12.8) = -0.52. Because the area strain is not known, this is a mimimum shortening.

EXERCISES

1. Card deck model of a flexural slip fold on which a number of small circles have been stamped. Measure the orthogonal thickness, and verify that Eq. 12.4 holds.

2. The data given below were obtained along an east to west tranverse. Reconstruct the folds, calculate the depth of folding and invent disharmonic folds or thrusts or both to balance the profile. The area can be determined most simply by superimposing your construction on graph paper and counting squares.

Station	Distance	Elevation	Dip
A	0	650 m	30°E
B	900	800	41°E
C	2300	850	18°W
D	2550	750	37°E
E	3550	800	44°W
F	4900	1150	5°E
G	5350	1000	69°E
H	6900	650	8°W
I	8000	550	66°W

13
Similar Folds

13.1 INTRODUCTION

Several explanations of similar folds have been proposed (Ramsay, 1967, p. 43; Bayly, 1971; Matthews and others, 1971). The simplest of these can be illustrated with a card-deck model. A layer is represented by a band drawn on the edge of a card deck, and the deck is deformed by inhomogeneous simple shear; the band will be folded. This process is called shear folding. Because each card and the portion of the band marked on its edge remains undistorted during the displacements, and the thickness of the band measured parallel to the shear plane is constant. This is also the axial-plane direction, hence the folded layer has constant axial-plane thickness, and this is the distinguishing characteristic of and ideal similar fold.

During the process of this shear folding, the band on the edge of the cards plays no mechanical role. It simply serves as a marker of the patterns of inhomogeneous shear; this behavior is termed passive. Similarly in nature, a layer folded by shear folding can act only as a marker; it can not have mechanical properties which contrast significantly with the surrounding material, for if it did there would be a component of bending or buckling and an ideal similar fold would not be produced. Because most rock bodies are composed of contrasting lithologies, similar folds are probably rare, but they might be expected in rocks which are monomineralic, or nearly so. Salt, ice, and dunite are a few rocks where passive bands and streaks of impurities might be distorted into similar folds. They also might occur in rocks which are polymineralic, but statistically homogeneous, such as shale and granite.

Despite this rarity of ideal similar folds, a number of more general lessons can be learned from an exploration of the geometry shear folding. This chapter is devoted to several of them.

13.2 GEOMETRY OF SHEAR FOLDING

In describing shear folds it is convenient to establish a coordinate system directly related to the geometry of simple shear. The shear plane is usually taken as the ab plane, with the a-axis being parallel to the direction of shear. In the card-deck models, the face of the deck on which the drawings are made is the ac plane, and the edge of the cards on this plane is the a-direction. (This coordinate

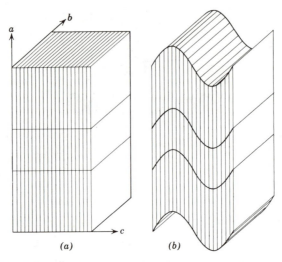

FIGURE 13.1 Card-deck model of a shear fold.

system has little or no meaning in other types of deformation, and should not be used to describe them).

If a layer is originally parallel to the bc plane, and if the form of the displacement curve is given by

(13.1) $a = A \sin c$

where A is the amplitude, then both bounding curves of the layer will have this same form (Fig. 13.1). In simple shear no change occurs in the b-direction, and therefore these folds will be cylindrical. Note that the angle of shear is given by the slope of the displacement curve.

Other original attitudes and displacement curves are also possible. For example, the layer may be inclined to the c-axis at some angle ϕ (Fig. 13.2a). If the displacement curve is the same, then the traces of the folded layers on the ac-plane will have the form (after Ramsay, 1967, p. 426),

(13.2) $a = A \sin c + c \tan \phi$

This case brings out two additional features:

1. The axial plane thickness is not generally equal to the orthogonal thickness of the original layer.

2. The hinge of the fold may be shifted laterally from the maximum and minimum points on the displacement curve.

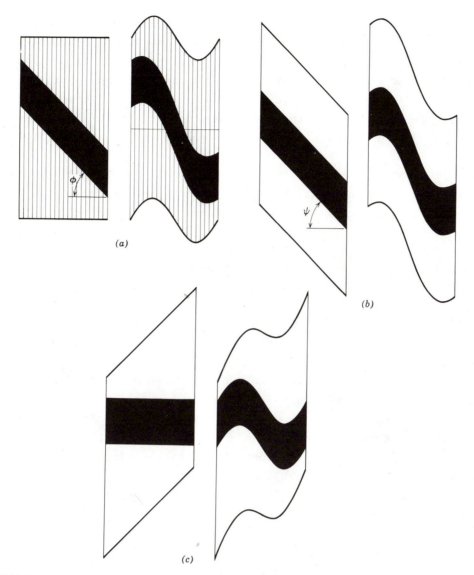

FIGURE 13.2 Symmetric and asymmetric shear folds: (a) symmetric displacement
of an oblique layer; (b) a component of homogeneous simple shear
destroys symmetry; (c) a special combination of homogenous and
inhomogeneous shear produces a symmetric fold.

This same result may be obtained somewhat differently. A band originally
parallel to the c-direction (as in Fig. 13.1) may be rotated into an inclined
orientation by homogeneous simple shear and then distorted by inhomogeneous simple
shear of the form given by Eq. 13.1. The total displacement curve, and the resulting fold will have the form

(13.3)   a = A sin c + c tan ψ

where ψ is the angle of shear of the homogeneous component. The order of the stages
of deformation may be reversed or the two components may act simultaneously during a
more complex pattern of simple shear with the same result.

Finally, an original inclination (as in Fig. 13.2a) may be eliminated by homogeneous simple shear, and the displacement curve of Eq. 13.1 then superimposed to produce the fold of Fig. 13.2c. The total displacement curve is the same as in Eq. 13.3, but because the slope of the original band and the slope of the line defining the homogeneous simple shear are opposite, the resulting fold has the simpler, symmetrical form of Eq. 13.1. It should also be noted that the strain distribution in this fold will be quite different that the identical appearing fold of Fig. 13.1.

Three objections have been raised concerning this shear folding mechanism. First, folds often termed <u>similar</u>, but which do not possess the ideal similar geometry have been shown to result from an entirely different mechanism--the homogeneous flattening of originally parallel folds (Ramsay, 1967, p. 441). This difficulty is avoided by reserving the term for the special class of fold shapes, rather than applying it to all folds with pronounced thinning of the limbs, a characteristic of folds with weakly convergent and divergent isogonal patterns.

The second objection, and a more serious one, is directed at the mechanical feasibility of the systematic reversal in the sense of shear as required by the examples given here. However, under certain circumstances, single-sense inhomogeneous simple shear is also capable of producing well-developed similar folds (Fig. 13.3). For a fold of the type given by Eq. 13.1, the slope at the limb inflection point is the maximum, for example, if $A = 1$, $\alpha_i = 45°$, if $A = 2$, $\alpha_i = 60°$, and so forth. For single-sense simple shear to produce such a fold, the initial angle of inclination of the band $\phi \geqslant \alpha_i$, and $\psi \geqslant \phi$, but of opposite slope. This mechanism seems to operate in glacier ice (Ragan, 1969b).

In three dimensions, the relationship between shearing and fold geometry is even more involved. In shear folds, the axial plane always coincides with the ab plane, and the hinge line of the fold and the fold axis are always parallel the intersection of the shear plane and the surface being folded. The simplest case

FIGURE 13.3 Single-sense shear to produce a well-developed similar fold.

occurs when the passive layer is parallel to the b-axis (Fig. 13.4a); this is, in effect, the case examined above.

On the other hand, if the original layer is inclined to the b-axis at an angle β, the fold axis will also be inclined at this same angle β, and the orientation of the fold axis can not be used to determine the a-direction (Fig. 13.4b); clearly, there is no reason to expect that the b-direction should necessarily be parallel to the passive layer.

Also in this case, since the profile plane of the fold is no longer parallel to the ac plane, the fold shape will always be a subdued version of the displacement curve. The amplitude A_f of the fold measured in the profile plane is related to the amplitude A_d of the displacement curve by

(13.4) $$A_f = A_d \cos \beta$$

A limiting case occurs when β = 90°, and A_f = 0; that is, when the layer is parallel to the a-axis no fold develops at all (Fig. 13.4c).

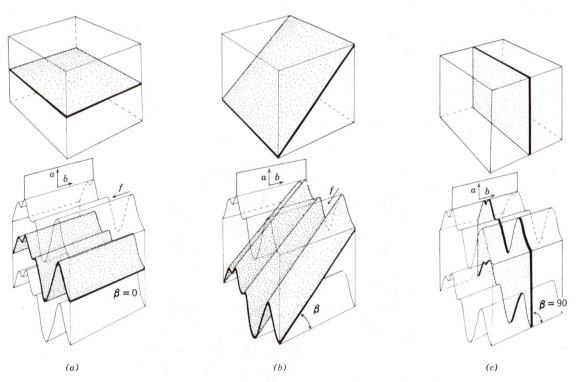

(a)   *(b)*   *(c)*

FIGURE 13.4 Layers variably inclined to the b-axis (from Ramsay, 1967, p. 425-456): (a) original layer parallel to the b-axis; (b) layer inclined to the b-axis at angle β; (c) layer perpendicular to the b-axis.

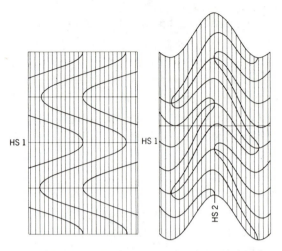

FIGURE 13.5 Card-deck model of superposed folds in two dimensions.

13.3 SUPERPOSED FOLDS IN TWO DIMENSION

Card-deck models can also be used to illustrate the effects of shear folding on previously existing folds. The form of the first folds is, as before, simply drawn on the edge of the pack, the deck deformed by inhomogeneously (Fig. 13.5). During this deformation, the originally straight traces of the first hinge surface HS1 behave in the same fashion as layer boundaries of the previous examples. At the same time, the traces of the first folds are complexly recurved. Two features are noteworthy.

1. The hinge points of the first folds do not coincide with the points of maximum curvature of the second folds.

FIGURE 13.6 Example the problem of the analysis of superposed folds.

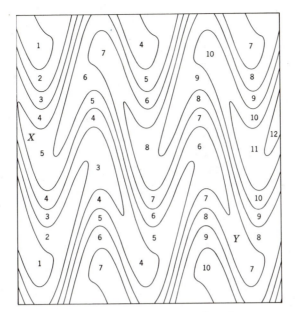

FIGURE 13.7 Tracing and numbering.

2. The trace of the hinge surface of the second displacement curve HS2
   does not pass through the points of maximum curvature on the limbs of
   the now twice folded bands, but alternates from one side to the other
   as it passes from limb to limb.

Given such superposed folds, it is possible to unravel the two stages of
deformation using the simple rules of shear folding (see Fig. 13.6).

ANALYSIS (after Carey, 1962)

1. Identify the first generation folds: on a tracing of the superposed
   folds, number the layers in sequence (the relative ages may not, and
   need not be known). If continuity is lost in highly attenuated zones
   it is usually possible to work around them; if not, a second or even
   third sequence may be started. Once all units are labeled, special
   patterns will identify the cores of the first folds. For example, at
   point X in Fig. 13.7 the numbers run from 3 4 5 4 3, and at point Y
   from 9 8 7 8 13. Without relative ages they can not be distinguished,
   clearly one of these cores represents an anticline and the other a
   syncline.

2. Involuted hinge surfaces of the first folds: wherever the core layers
   form an apex, the trace of the folded hinge surface of the first fold
   HS1 must pass into the next layer. Since the first hinge points can
   not be accurately located, the point of maximum curvature is used as a
   close approximation. Through these approximately located hinge points
   draw in the traces of the first hinge surface using solid, dashed or

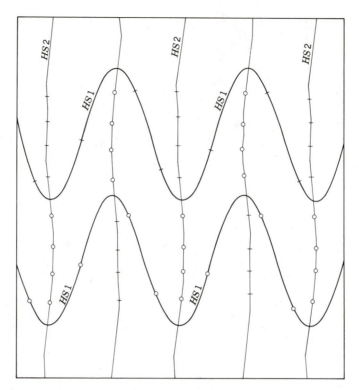

FIGURE 13.8 Traces of hinge surfaces HS1 and HS2.

dotted lines to indicate the degree of confidence in their locations. These traces can then be marked with zeros or cross-bars depending on whether the index numbers are minimum or maximum in the cores of the folds (Fig. 13.8).

3. Hinge surfaces of the second folds: a second set of traces can be drawn through the hinge points of the folded limbs. These appear as a series of roughly straight and parallel lines. Using crosses and zeros to mark sequences which rise or fall it will be seen that groups of symbols alternate along these traces. If difficulty is encountered in following the involuted HS1 traces though strongly attenuated zones or across widely spaced areas, they may be completed using the following clues.

   a. Involuted HS1 traces of like sign must join.

   b. The sense of curvature of the HS1 traces must be the same where they cross a particular HS2 trace, that is, all must be either concave up or down.

   c. The folded HS1 traces cross the straight HS2 traces only at points where the sign changes occur. This fact may also be use to determine the number of HS1 traces to be inserted across gaps.

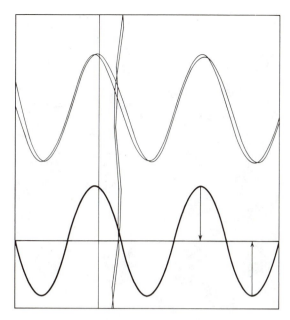

FIGURE 13.9 Determination of the shape of the second folds.

4. The directrix and form of the second folds: by collecting all the HS1 and HS2 traces on a single sheet, and by visually averaging both, the displacement curves and the directrix responsible for the second folding can be extracted from the complex pattern (Fig. 13.9).

5. Form of the first folds: the hinge surface of the first folds are assumed to have been planar. Therefore the effect of the second folding can be eliminated by shifting the patterns parallel to the displacement curve derived from the original superimposed folds (Fig. 13.9). This shifting is most easily accomplished by:

   a. Drawing a series of closely spaced lines on the original fold pattern parallel to the directrix of the second set of folds.

   b. On a tracing sheet, draw a second set of parallel lines with the same spacing.

   c. Overlay this sheet on the fold pattern and mark off the points of intersection of the folded points with the first line, and then shift the tracing according to the reverse displacement curve and repeat for a second line. Repeat this for all the lines.

The pattern of the first folds can be drawn by connecting points across these guidelines. The first attempt is apt to be somewhat crude. Irregularities due to drafting and positioning errors may be filtered out by repeating the tracing process illustrated in Fig. 13.9. The smoothed versions will appear as in Fig. 13.10.

FIGURE 13.10 Unfolding the second folds by straightening out the HS2 traces.

13.4 WILD FOLDS

In both theory and concept, this process of superimposing folds upon folds could be repeated any number of times. To model such multiple folds with the aid of card decks would require that the convoluted patterns of earlier experiments be transfered onto a deck with a different orientation relative to the a–direction. Unfortunately, any attempt to perform such an experiment met with severe practical difficulties which mount exponentially beyond the second set. However, such patterns are easily produced with the aid of a computer–driven plotter. The technique consists simply of adding sine curves of varying amplitudes and wave lengths alternatively along the x–and y–axes. Fig. 13.11 shows typical runs for three and four sets of superposed folds.

Irregular folds are characterized by irregularities of the axial planes, discontinuities and rapid variations in the thickness of bands (Fleuty, 1964, p. 477). The most disordered types also show a wide variation in attitudes of hinges and hinge surface. Such folds are particularly common in migmatic gneisses, where they give the appearance of stirred porridge (De Sitter and Zwart, 1960, p. 253), and they are sometimes called <u>wild</u> folds (Kranck, 1953, p. 59; Berthelsen and others, 1962). Except for their perfectly periodic character and continuity, certain

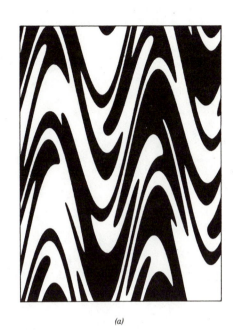

(a)   (b)

FIGURE 13.11 Multiple superposed folds: (a) three sets; (b) four sets.

aspects of these computer generated patterns are similar to the wild folds found in nature, suggesting that their appearance may be more a matter of complexity than irregularity.

13.5 SUPERPOSED FOLDS IN THREE DIMENSIONS

Card-deck models can also be used to illustrate superposed folds in three dimensions. The first folds are represented by a cylindrical surface cut across the deck; deformation of the deck by inhomogeneous simple then refolds this surface. Although the method requires special preparations, including the cutting and deforming of the cards, a number of interesting and informative experiments can be performed, and these are well worth pursuing further (see O'Driscoll, 1964). A simple example will indicate the approach and its potential. If original upright folds are deformed by a second set of upright folds trending at right angle to the first, a series of domes and basins result (see Fig. 13.12). Other angles between the first and second folds can be simulated by first homogeneously shearing the deck in a horizontal direction; the domes and basins are then asymmetrical and en echelon.

Because of their likeness to the patterns caused by the intersection of two sets of waves, these are called <u>interference</u> patterns. If instead of a single folds

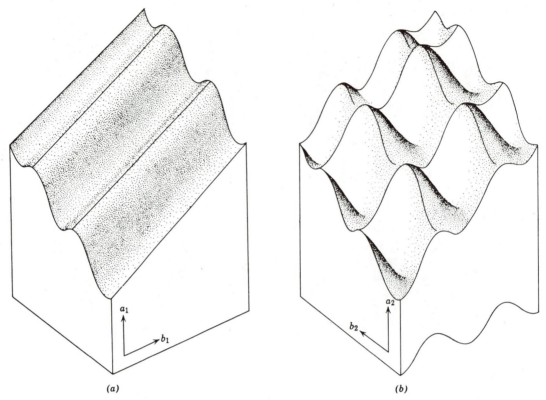

FIGURE 13.12   Superposed folds in three-dimensions (after O'Driscoll, 1962, p. 166).

surface, the deformation operated on a multilayered block, and if an exposed plane cuts horizontally through the superposed folds produces an outcrop pattern of the interference structures. The characteristics of this type of pattern depends on the relative orientation and size of the first and second folds. If, as in the model, the first folds are horizontal and upright, and further, if the sizes of both sets of folds are about the same, the resulting patterns can be described directly in terms of the attitudes of the first folds. Within this framework, several types of patterns can be distinguished.

1. If the first folds are horizontal and upright, and the trends of the two folds are perpendicular, the pattern of domes and basins alternate with a high degree of symmetry and regularity (Fig. 13.13a).

2. Inclined first folds are reflected by dome-basin patterns which no longer have simple rounded forms. This reflects the difference in dip on the limbs of the first folds (Fig. 13.13b).

3. Inclined first folds with overturned limbs result in characteristic cresent and mushroom patterns (Fig. 13.13c).

4. If the first folds are recumbent or reclined, and the trends of the two sets of folds are parallel, the patterns do not differ in kind from the

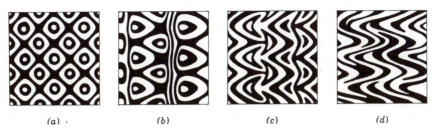

FIGURE 13.13 Interference patterns of superposed folds (from Ramsay, 1967, p. 531).

two-dimensional patterns which are so easily modeled with card decks (Fig. 13.13d).

Other relative orientations give patterns which are transitional with these four, and many can be illustrated by means of card-deck experiments as described by O'Driscoll.

A full discussion of interference pattern, together with many excellent illustrations, is given by Ramsay (1967, p. 518f). His classification scheme has recently been expanded and extended by Thiessen and Means (1980).

13.6 FOLDS IN SHEAR ZONES

A special type of superimposed fold is produced when preexisting folds are cut by a shear zone. Generally, all preexisting lines are rotated toward the a-direction during simple shear. The convergence of such lines is particularly striking at high shear strains. One especially interesting effect involves folds which outside the shear zones show only a slight variations in the orientations of their hinge lines (Cobbold and Quinquis, 1980; Ramsay, 1980). In contrast, inside the shear zone they show an extreme variation, with the result that a sheath fold is produced (Fig. 13.14).

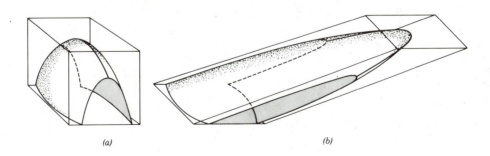

FIGURE 13.14 Sheath fold (after Ramsay, 1980, p. 93).

EXERCISES

1. Print a row of small circle across the end of a card deck and then
   produce the following types of shear folds:
   a. A symmetrical sinusoidal fold with reverse sense of shear (Fig. 13.1)
   b. Add a component of homogeneous simple shear to produce an asymmetrical fold
      (Fig. 13.2).
   c. A symmetrical fold by single sense shear (Fig. 13.3).

2. In the the light of these experiments write a short statement concerning the
   indeterminant relationship between final fold form and strain distribution.

3. Draw the profile of a similar fold on the edge of a deck of cards with the trace
   of its hinge surface oblique to the direction of shear. Homogeneously deformed
   the deck to produce a superimposed fold. Note the thickness variation in the now
   twice folded layer, the location of the new hinge point and the location of the
   hinge points of the first fold.

4. Using Fig. X13.1 remove the effects of the F2 folding to give the form of the F1
   folds. As will soon become apparent, the F1 folds in this problem are highly
   regular and it is only necessary to proceed to the point where their form can be
   confirmed with some confidence rather than attempting a complete and detailed
   reconstruction.

FIGURE X13.1

14
Folds and Topography

14.1 MAP SYMBOLS FOR FOLDS

If exposures are good, it should be possible to locate and measure the attitudes of both the hinge and hinge surface of a fold. On the other hand, if these features can not be directly observed, then the attitude information and the location of the hinge surface must be found from the map pattern. In any case, the fold is then depicted on a map with a line marking the trace of the hinge surface, together with symbols indicating the attitude of the hinge and hinge surface and the direction of closure (Fig. 14.1). This chapter is concerned with indirect methods by which this may be accomplished.

14.2 OUTCROP PATTERNS

Just as structural planes intersect the earth's surface to give characteristic outcrop patterns, equally distinctive though more complex patterns result from the

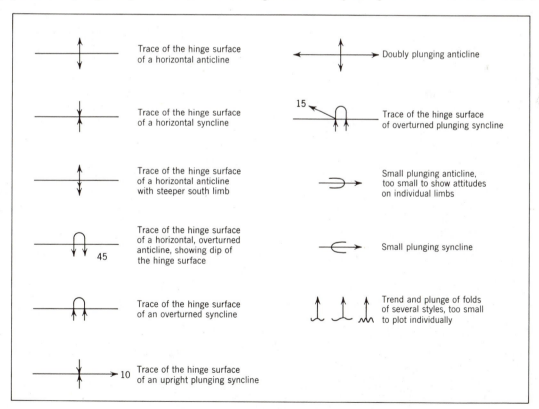

FIGURE 14.1 Map symbols for folds.

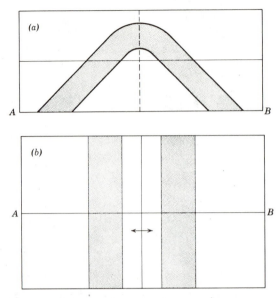

FIGURE 14.2 Horizontal upright antiform: (a) fold profile; (b) map pattern on a horizontal surface.

intersection of folds with the topographic surface. Horizontal folds have the simplest type of outcrop pattern. On a horizontal exposure plane, the pattern of such a fold is essentially the sum of the patterns of the inclined limbs, that is, a series of parallel outcrop bands (Fig. 14.2). If the relative age or correlation of the units, or the attitude at several points is not known, it may not be possible to interpret the pattern as being part of a fold at all. However, once these are known, the existence of a fold becomes clear. If the same fold is exposed in an area of some topographic relief, the inclination of the opposing limbs are immediately evident, and the interpretation of a fold can be made with some confidence. If the topography is such that a hinge intersects the surface, the existence of the fold becomes obvious (Fig. 14.3).

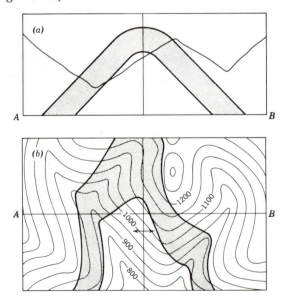

FIGURE 14.3 Horizontal upright antiform: (a) fold profile; (b) map pattern in an area of topographic relief.

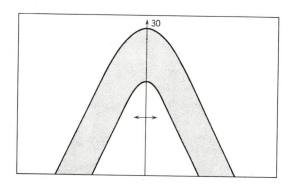

FIGURE 14.4 Map pattern of a plunging upright antiform.

For plunging folds, a converging pattern is characteristic and unmistakable (Fig. 14.4). However, on the basis of the pattern alone it is not possible to distinguish antiforms and synforms. Again, if the dip at several points or the direction of plunge is known the type of fold may be immediately determined. Some of this information is supplied if the plunging fold is exposed in an area of topographic relief (Fig. 14.5).

In all these cases, given attitude data, vertical cross sections could be constructed. However, only if the fold is horizontal does such a section give authentic information about the fold geometry. In all other cases the shape of the folded surfaces, the thickness variations in the folded layers, and the interlimb angles are distorted in vertical sections. If the aim is to portray fold geometry, such vertical sections are useless and they must be avoided.

14.3 DOWN-PLUNGE VIEW OF FOLDS

By viewing map patterns of tilted, but not folded strata in the down-dip direction an important visual simplification is achieved (Chapter 3). This same approach may be used to even greater advantage in viewing the map patterns of plunging folds. By turning the map, and adopting a view so that the line of sight is in the direction of the plunging fold axis, the profile of the fold is actually seen. As before, the principle is that the distortions at the earth's surface are elimi-

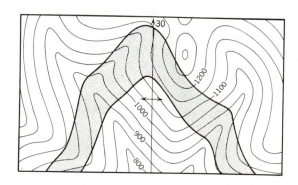

FIGURE 14.5 Plunging upright fold exposed in an area of topographic relief.

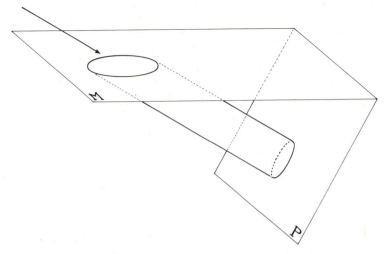

FIGURE 14.6 Down-plunge view of a circular cylinder: M = map, P = profile.

nated by this down-plunge view. This is easily seen in the case of an inclined cylindrical pipe of circular cross-section (Fig. 14.6). If the pipe is cut horizontally, its trace will be an ellipse. If this ellipse is then viewed in the
direction of the inclined axis of the pipe, it will again appear to be circular.
Fig. 14.7 shows a geologic map of several upright plunging folds. In a down-plunge
view, antiforms and synforms are simply and directly seen as such. The map pattern
also indicates that disharmonic folds are also present, and the down-plunge view
automatically includes them in their proper place in the structure. In contrast, a
vertical section mechanically constructed along a line SS' would fail to represent
the small plications in the cores of the folds.

If the folds are inclined, the correct shape of the plunging folds is seen only
if the line of sight is parallel to the fold axis, not when it is parallel to the
trace of the hinge surface. Fig. 14.8 illustrates this rule. As shown on the geo-

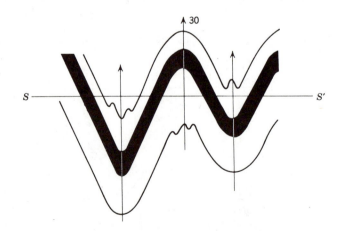

FIGURE 14.7 Map pattern of upright plunging folds (from Mackin, 1950).

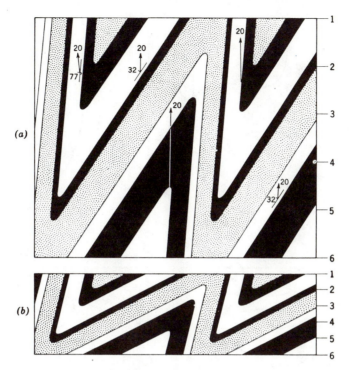

FIGURE 14.8 Gently-plunging-steeply inclined folds (from Mackin, 1950): (a) map pattern; (b) profile plane.

logic map (Fig. 14.8a), the folds plunge due north at 20°. Though not shown, the hinge surface trends N15°E, and dips steeply to the west. With a line of sight parallel to the fold axis, the true shape of the folded layers is revealed; the result of this view is also shows on the constructed profile (Fig. 14.8b). Note carefully that if the folds are viewed parallel to the strike of the hinge surfaces, that is, toward N15°E, the folds falsely appear upright, with apparent thinning on the steep eastern limbs of the antiforms.

The down-plunge view reveals another important feature of fold patterns. The folds of Fig. 14.7 have vertical hinge surfaces, and thus the trace of this surface also connects points of greatest curvature of the outcrop pattern, a fact that can be readily seen in the down-plunge view. This coincidence holds only when the folds are exposed in areas of negligible relief. If the relief is significant there will, in general, be a discrepancy between the real and apparent hinge points (see Figs. 14.3b, 14.5). For inclined plunging folds exposed on horizontal plane surfaces, the line connecting the points of greatest curvature of the outcrop pattern and the trace of the hinge surface always depart. The degree of departure depends on the geometric nature of the folded surfaces, and the effect is compounded by the topographic relief. For similar folds, the two lines are parallel but not coincident (Fig. 14.9a), and for parallel folds the line are neither coincident nor parallel (Fig. 14.9b). If this relationship seems difficult to accept, a down-plunge view of the folds of Fig. 14.9 will immediately confirm its validity.

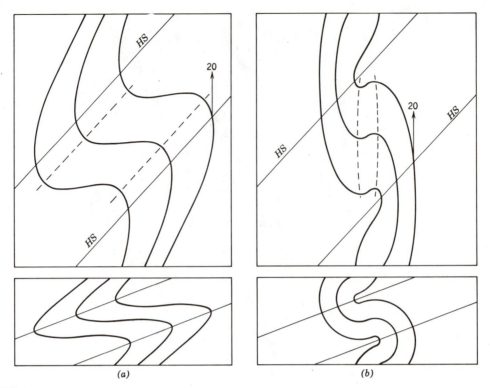

FIGURE 14.9 Map and profile views demonstrating the lack of correspondence between the trace of the hinge surface HS and the line connecting the points of greatest curvature of the outcrop pattern; both sets of folds plunge 20° due north: (a) similar folds; (b) parallel folds (after Schryver, 1966).

14.4 FOLD PROFILE

The true form of cylindrical folds seen in the down-plunge view may also be constructed from a geologic map. There are two different, though equivalent, construction techniques. The first is the more straightforward one, and applies to areas of negligible relief. It involves constructing the foreshortened map pattern with the aid of a grid.

CONSTRUCTION (after Wegmann, 1929)

1. On the geologic map, draw a square grid with one coordinate direction parallel to the trend of the fold axis (Fig. 14.10a).

2. Viewed down-plunge, the grid spacing across the line of sight, i.e, perpendicular to the axial trend, remains unchanged (1, 2, 3,...). In the direction of the axial trend the spacing is reduced to s sin p, where s is the original grid spacing, and p is the plunge. This new grid spacing may be computed, or found graphically (Fig. 14.10b).

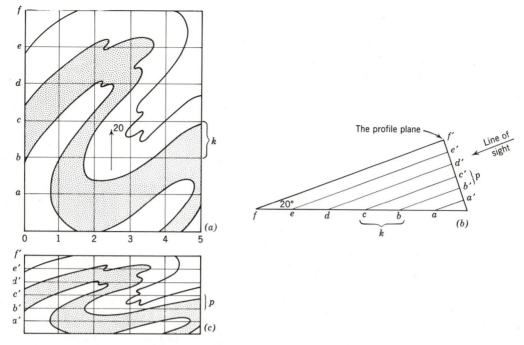

FIGURE 14.10 Construction of profile of plunging folds: (a) geologic map with grid; (b) down-plunge view, and method of calculating grid spacing on profile, (c) foreshortened grid on the profile plane.

3. A second grid is then constructed representing the down-plunge view: the spacing 1, 2, 3,... is the same as originally constructed on the map, while a', b', c',... is the foreshortened one.

4. The fold pattern is then transferred from the map grid to the profile grid point by point and the folds then sketched in (Fig. 14.10c).

The resulting fold profile involves no speculation; no line appears on it that is not also on the map. Contrast this with the construction of a vertical cross section of parallel folds in which surface attitudes are projected in directions at right angles to the fold axis. Both profiles and vertical sections require the projection of data. For the profile, however, the data is projected parallel to the fold axis, which is a direction of minimum change in cylindrical folds, whereas the vertical section requires projection in the direction of maximum change.

The second technique, using an orthographic construction, is more involved, but it has the advantage of being adaptable to more complex situations.

CONSTRUCTION (after Wilson, 1967)

1. Establish folding lines FL1 parallel and FL2 perpendicular to the trend of the fold axes, preferably outside the map area (Fig. 14.11a).

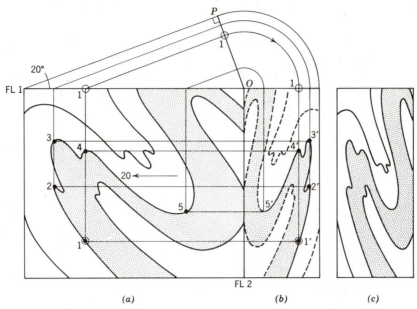

FIGURE 14.11 Construction of the fold profile by orthographic methods: (a) map with folding lines parallel and perpendicular to the trend of the fold axes; (b) an "up-structure" profile view; (c) the true profile by reversal.

2. Folding about FL1, draw a vertical section parallel to the fold axes. In this section, the line OP represents the profile plane in edge view. A series of selected point on the map are projected to FL1, thence to OP using the angle of plunge.

3. Folding about FL2, these same points are projected from both the map and from OP to fix their location on the profile plane (see details of the projection of point 1 on the map to 1' on the profile on Fig. 14.11b).

4. After a sufficient number of points have been transferred in this manner, the form of the folds on the profile plane can be completed by connecting appropriate points.

5. The result is an "up-plunge" view, but this can be easily converted to a down plunge view simply by reversing it (Fig. 14.11c).

The advantage of this procedure over that of the grid method is that a fold profile can also be constructed when the map pattern is influenced by topographic irregularities.

CONSTRUCTION

1. As before, construct FL1 parallel and FL2 perpendicular to the trend of the plunging folds.

2. Instead of projecting directly to the profile plane via FL1, the topography must first be taken into account. This is done by adding a series of scaled elevation lines to the edge view using the contour interval of the topographic map (Fig. 14.12a). The points on the outcrop traces of the folded surfaces are then projected across FL1 to their corresponding elevation lines.

3. These points are then projected to the profile plane OP, using the plunge angle. This final projection can be made easier if the use of the circular arcs is avoided. By constructing line OB to bisect the angle POR, the points can be projected along the line of plunge directly to this bisector, and then to the profile, thus saving one step (Fig. 14.12b). The fold profile is sketched in as before from points projected from the edge view and from the map.

With the true shape of the fold displayed, the trace of the hinge surface can be added to the profile by connecting hinge points. Points on the outcrop trace of the hinge surface can then be found by projecting hinge points back to the map.

Stauffer (1973) has devised an alternative method for finding hinge points on the map, called the elliptical-arc technique, which involves constructing ellipses

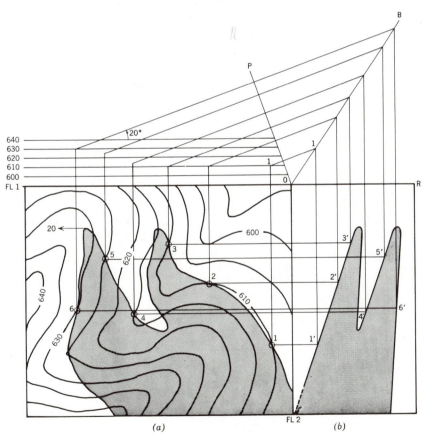

FIGURE 14.12 Fold profile constructed from outcrop pattern in an area of topographic relief: (a) geologic map; (b) fold profile.

which are tangent to the outcrop trace of a folded surface in the vicinity of the hinge zone. From these ellipses, in combination with the attitudes of the hinge and hinge surface, the hinge points can then be found.

Strictly, the construction of a fold profile requires that the fold axes of the entire area be constant in trend and plunge, or at least have negligible variability. If the axes are not parallel, the folds are not cylindrical, and no single direction of view or projection exists. However, if the plunge angle changes progressively over an area, it is possible to draw an approximate profile, and thus to depict the general fold style. One method is to draw a series of overlapping strip profiles for small areas where the plunge angle is essentially constant, and join them to make a a composite section. A second method is to adapt the orthographic construction used above.

CONSTRUCTION

1. As before establish folding lines FL1 and FL2 (Fig. 14.13a).

2. Project the data points to FL1 and for each, plot the associated plunge angle. Using the tangent-arc method of §12.2, curved plunge lines are constructed to project the points to the profile plane.

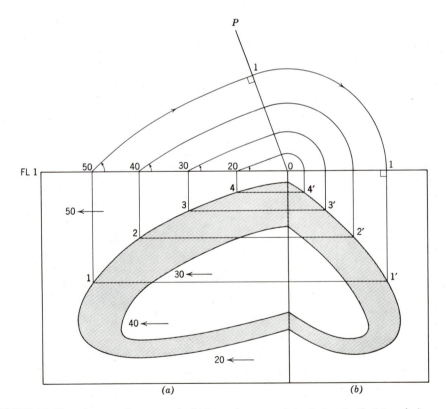

FIGURE 14.13 Approximate profile of noncylindrical fold: (a) map; (b) profile.

3. From the map and the edge view OP, points are then projected to the profile plane, and the form of the folds are completed by sketching between the control points (Fig. 14.13b).

Computer programs are now becoming available which can construct down-plunge
sections from the data on geologic maps (see Kilby and Charlesworth, 1981).

14.5 ATTITUDE OF THE HINGE SURFACE

With two points of known location and elevation, an apparent dip of a planar
hinge surface can be found. The fold axis is also an apparent dip and with two such
dips known its true dip and strike can then be found with the methods of §1.6. If
the location of only a few hinge points are known, and the hinge surface is strictly
planar its outcrop trace can be reconstructed by the method of §3.5.

EXERCISES

1. The folds depicted in Fig. X14.1 plunge 30° due east. With the aid of a
down-plunge view add the appropriate structural symbols to the map.

FIGURE X14.1

2. The fold in Fig X14.2 plunge 25° due north. Construct a profile of the fold and
   add the appropriate symbols to the map.

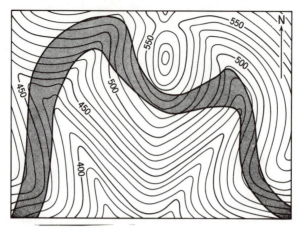

FIGURE X14.2

3. From the geologic map of the Coatesville–West Chester District, Pennsylvania
   (Fig. X14.3), assuming that the attitude of the fold axes is 15°/245°, construct
   a profile. Concentrate on the main elements of the pattern rather than on
   details. Noting the order of superposition describe the main structures in the
   map area. In the light of your results do you think it necessary to interpret
   any part of the map pattern as being due to superposed folding? For two opposing
   views on this question, see McKinstry (1961) vs Mackin (1962).

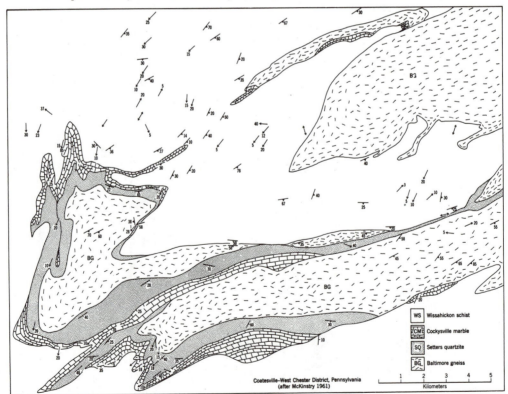

FIGURE X14.3

15
Structural Analysis

15.1 INTRODUCTION

Problems involving the angular relationships between lines and planes may be solved with the methods of descriptive geometry, although the advantages of using stereographic projection should now be obvious. However, if certain problems are to be solved graphically then the use of the stereonet is indispensable. The three-dimensional structural geometry of a rock mass, especially if complex, is one of these. The same basic techniques may also be applied with profit to much simpler situations, and this is a convenient way to introduce the method.

15.2 S-POLE AND BETA DIAGRAMS

In cylindrical folds, the hinge zones may be too smooth to allow accurate field measurement of the attitude of the hinge line, or the folds may be too large or incompletely exposed. If attitudes along the folded surface can be measured, the orientation of the fold axis may be determined by a simple stereographic plot of the data.

PROBLEM

With the following data, determine the attitude of the fold axis.

1. N68°E, 30°NW	4. N35°E, 35°SE
2. N60°E, 45°NW	5. N41°E, 50°SE
3. N88°E, 16°N	6. N20°E, 20°E

METHODS

There are two different, though equivalent plotting techniques:

1. <u>Beta diagram</u>: plot each measured plane as a great circular arc. These all intersect at a point called the β-axis (Fig. 15.1a).

2. <u>S-pole diagram</u> (also called the π diagram): plot the poles of the measured planes. These define a great circle, and the pole of this plane is the β axis (Fig. 15.1b).

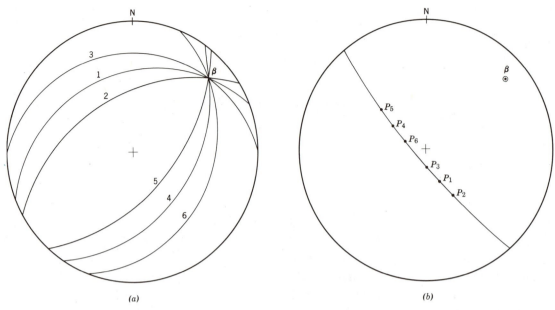

FIGURE 15.1 Stereographic plots of the attitudes around a cylindrical fold:
(a) beta diagram; (b) S-pole diagram.

ANSWER

Both methods give the attitude of the fold axis as 10°/049°.

The basis of these methods is the collection of all the measured attitudes at a single point. This has both advantages and disadvantages. The obvious advantage is that only in this way can we determine the orientation of the fold axis. The disadvantage is that these attitudes are removed from their context on the map. The best of both worlds is always to use the stereographic plots in conjunction with the geologic map.

15.3   FOLD AXIS AND AXIAL PLANE

The reason for carefully distinguishing between the hinge line and the fold axis may now be appreciated. The β-axis = fold axis characterizes the relationship of any two attitudes, and therefore all attitudes. This axis has no specific location in the fold, only orientation. In cylindrical folds, the hinge lines and fold axis are parallel, but they refer to quite different aspects of the fold. In simple cylindrical folds, there is a similar relationship between the planar hinge surface of the fold, and the axial plane, and there is an interrelationship between both pairs of features, as a simple example will illustrate.

PROBLEM

Given the geologic map of an overturned, plunging antiform (Fig. 15.2a), determine the attitude of the fold axis and the axial plane.

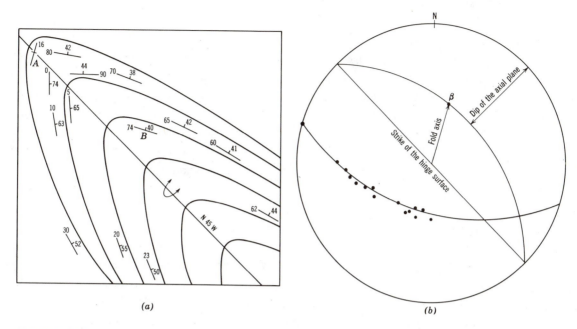

(a) (b)

FIGURE 15.2 The attitude of the axis and axial plane from a map: (a) geologic map of an overturned plunging fold; (b) stereogram showing construction of the axis and axial plane.

CONSTRUCTION (Fig. 15.2b)

1. An S-pole diagram of the measured attitudes yields the β-axis.

2. With this direction known, a profile may then be constructed to locate the trace of the hinge surface if it can not be found by more direct means. The trace of this surface is added to the map.

3. Add to the stereogram the strike of the planar hinge surface which is also parallel to the axial plane. As the fold axis is parallel to the axial plane, the axis is, in effect, an apparent dip of that plane. Therefore the great circle through the β-intersection and the strike of the hinge surface gives the dip of the axial plane.

4. For an overturned fold, the orientation of the fold axis may be estimated from the geologic map by inspection.

   a. The axial trend parallels the strike of the vertical attitude (Point A).

   b. The plunge angle equals the dip of the plane whose strike is perpendicular to the vertical attitude (Point B).

15.4 EQUAL-AREA PROJECTION

In practice, stereographic plots of structural lines or planes are never as perfect as illustrated in Fig. 15.1. Irregularities, departures from ideal geometry, and measurement errors all contribute to a scatter of points. If the scatter is small, it is generally possible to visually locate the best-fit point or great circle within acceptable limits, but with only a few points the confidence will be low. A larger sample is then required. This in turn raises several problems. The first is that plotting a large number of points takes a good deal of time. Some devices have been invented to help speed the process and reduce the tedium (for example see Draper, 1984).

Also, with a larger number of scattered points on the net, a practical problem of treating and evaluating the data arises. There are several alternatives. First, the data may be treated statistically, and the best fit calculated. It is always advantageous to examine the data visually. Computer programs are available for handling large quantities of data, with output in either printed or plotted form. This approach is essentially the older, completely graphical method made efficient by the computer. The most common method of presenting such data, whether processed by computer or by hand, is to contour the density of the plotted points.

The visual evaluation of plotted data, contoured or not, requires a special type of net. If a series of randomly oriented lines are plotted as points on the Wulff net, the resulting distribution would not be statistically random. There would tend to be a concentration in the center of the net, that is, the random lines would falsely show a weak preferred orientation in the vertical position. The reason for this is that an area on the net of, say, $10° \times 10°$ in the center of the net is smaller that the same angular area at the margin. To overcome this a different type projection is needed. The method used is called the Lambert equal-area, or simply the equal-area projection. The geometric basis of this projection is given in Fig. 15.3a which shows a vertical diametral plane through the reference sphere of radius R. Line ZO' is the vertical diameter, and OP is an inclined line in this diametral plane. The point P' is the projection of P on the projection plane. The distance d from the center of the projection circle O' to P' is, from the diagram,

$$(15.1) \qquad d = O'P' = O'P = 2R \sin(\theta/2)$$

where p is the inclination of the line, and $\theta = 90° - p$. By this same method, the radius of the projection circle is

$$r = 2R \sin(90°/2) = 2R/\sqrt{2}$$

It is convenient to make this radius equal to the radius of the reference circle,

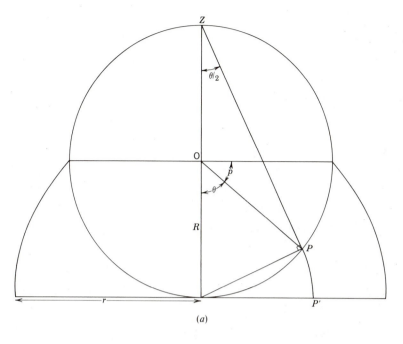

(a)

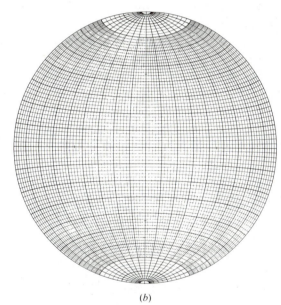

(b)

FIGURE 15.3 Lambert equal-area projection: (a) method of projection; (b) equal-area or Schmidt net.

that is, make d = 2R, when p = 0°. This is accomplished by dividing Eq. 15.1 by $2/\sqrt{2}$, giving,

(15.2) $\qquad d = R\sqrt{2} \, \sin(\theta/2)$

With this result, a series of curves may be drawn corresponding to the small and great circle on the Wulff net. The result is the equal-area or Schmidt net (Fig. 15.3b). The techniques of plotting and manipulating data on this net is identical to that used on the Wulff net. The only practical difference between the two nets is that small and great circles on the sphere do not project as circular arcs on the Schmidt net.

15.5 CONTOURED DIAGRAMS

Once the point diagram has been prepared on this net, the densities are counted out. A wide variety graphical counting methods have been devised (Stauffer, 1966; Denness, 1970, 1972; see also Turner and Weiss, 1963, p. 58f). The principle common to all techniques is the use of a counting area, usually 1% of the total surface area of the net, to determine the density of points over the entire surface of the net.

The method used here is one of the simplest yet devised, and it applies reasonably well to all situations. A special counting net is required which is completely subdivided into small triangles (Fig. 15.4). Six of these triangles form a hexagonal area which is equal to one percent of the total area of the net. In addition to ease of use, this counting net has the advantage of a fixed relationship between the total number of points and the counted density. Each point is counted three times (except for a small discrepancy caused by the semicircular areas at each end of the six spokes).

COUNTING PROCEDURE (Fig. 15.5)

1. Superimpose the point diagram and a second tracing sheet on the counting net. At the center of each hexagon, the total number of points

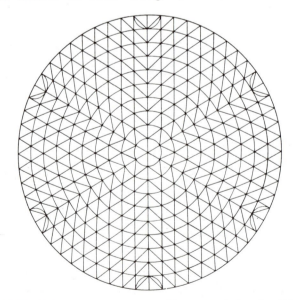

FIGURE 15.4 Counting net (from Kalsbeek, 1963).

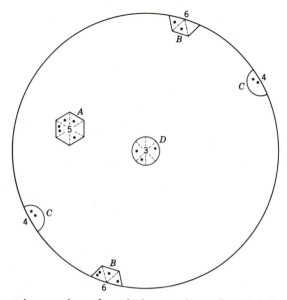

FIGURE 15.5 Counting point densities using the Kalsbeek counting net.

within that hexagon is written (see Fig. 15.5, Point A). For the main body of the diagram there will be numbers at the center of each overlapping hexagon. For parts of the diagram with no points, the hexagons may be left blank, rather than noting a zero for each.

2. At the circumference of the net, the points in each half hexagon on one side of the net are combined with the complementary half on the opposite side, and this number is written on both sides of the net along the primitive (see Point B).

3. Points at the ends of the spokes are counted using the complementary half circles (Point C). At the very center, the small 1% circle is used (Point D).

Following the counting out process, the tracing sheet bearing the numerical densities expressed as the number of points per one percent area is removed from the counting net, and contours of equal density are then drawn. To facilitate comparisons of diagrams with different numbers of total points, contours are drawn in percentages of total points per 1% area of the net. Therefore, the numbers posted during the counting process must be converted to percentages. In the special case of exactly 100 points, each number will, of course, also be the required percentage figure. If 50 points are plotted, each point represents 2% of the total, and the posted numbers are doubled, and so forth.

CONTOURING

1. Within the main body of the diagram, contours of equal density are drawn as shown at Point A of Fig. 15.6a. It is usually easiest to locate the area of greatest concentration and work outward.

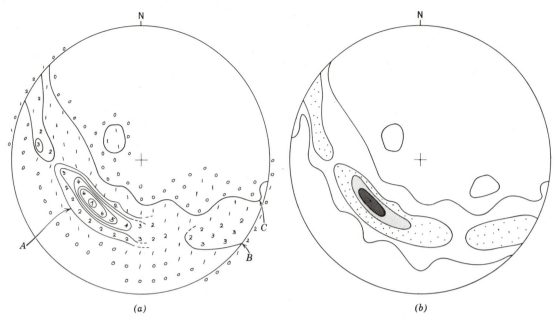

(a) *(b)*

FIGURE 15.6 Contouring a diagram: (a) counted densities of 50 points and preliminary contours; (b) completed diagram; contours 2-4-8-12% per
1% area, maximum 14%.

2. For contour lines which approach the primitive, the counts along the
   edge are used. When a contour line intersects the primitive it must
   reappear exactly 180° opposite (see Point B).

3. When a contour line should be strictly drawn intersecting the primitive, but it is clear that it immediately loops back again, it is permissible to avoid actual contact with the primitive (Point C).

4. When the preliminary contouring is complete, several modifications may
   be made to improved the appearance of the diagram (Fig. 15.6b):

   a. The maximum found during the counting may not be the true maximum
      of the diagram. The greatest concentration can be found by return-
      ing the point diagram to the counting net. Using the central 1%
      circle, shift the diagram around until the largest number of
      points lies within the circle.

   b. All the contour lines may not be necessary to show the pattern,
      for example, if the spacing is very close, some intermediate lines
      may be eliminated. The values of the remaining contours in the
      final diagram are indicated in the legend in the form 2-4-8-12%
      per 1% area, maximum 14%.

   c. The area of maximum concentration is often blackened. Although
      usually unnecessary, patterns may be used for the areas of lesser
      concentration. Particularly effective is the used of stipple pat-

terns graded so that the areas of greater concentration have a denser appearance. Line patterns detract from the visual effect, and should be avoided.

In order to convey as much objective data as possible to the reader, it is useful to supply both the point diagram and its contoured equivalent.

15.6 INTERPRETATION OF DIAGRAMS

Pattern is the key to interpreting a point diagram and its contoured counterpart. The real equivalents of the ideally perfectly linear and perfectly planar patterns are:

1. A point maximum is a axially symmetrical clustering of points about a single direction.

2. A girdle is a grouping of points in a band along a great circle.

For folds, we may choose to construct a β–diagram which produces a point maximum, or an S-pole diagram which produces a girdle pattern. There are several compelling reasons for adopting the S-pole diagram.

1. In the β diagram, the number of intersections $N = n(n-1)/2$, where n is the number of individual great circles. This number rapidly rises with the number of circle, and such a large number of intersections is apt to give the impression of a large sample size, and therefore a false sense of confidence in the result. It also involves much more work to produce a β diagram. For several hundred individual great circles, which is not a particularly large number of measurements, the number of intersections becomes impossibly large.

2. As a result of inevitable scatter, spurious concentrations of intersections may result. This will be especially true in open or tight folds, that is, where the angle between the limbs is small. These spurious intersections will not be randomly distributed about a mean position, and they may exceed in number of significant β-points (Ramsay, 1964).

3. Perhaps the most important advantage is that the S-pole diagram, if based on a representative sample of the attitudes of the structure, gives information concerning the shape of the folded surface, the interlimb angle and the attitude of the axial plane.

An instructive approach to understanding S-pole diagrams is to follow the patterns as if progressive develops during the folding. Consider the cylindrical folding of a single layer. Before folding the poles of the horizontal layer would plot as a concentration of points at the center of the net (Fig. 15.7a), that is, the poles would statistically define a vertical line. If the diagram were constructed parallel to the profile plane, there would be a point maximum at each end of a diameter of the net. As the layer is folded about a horizontal axis, the original vertical poles are spread into a fan. In terms of pattern, whether projected horizontally or vertically, the original point maximum spreads into a <u>partial</u> girdle (Fig. 15.7b). With further folding, the girdle continues to spread (Fig. 15.7c). Finally, with rotation of the limbs into parallelism, a <u>full</u> girdle develops (Fig. 15.7d).

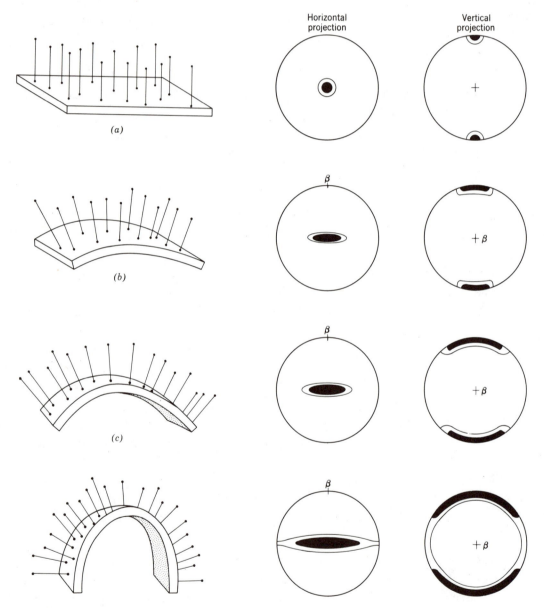

FIGURE 15.7 Development of the S-pole diagram during folding; note that the same stereographic diagrams would result from both antiforms and synforms: (a) statistically horizontal layer; (b) layer bent through 45°; (c) layer bent through 90°; (d) layer bent though 180°.

If the fold shape is dominated by planar limbs (see Fig. 11.3a), the S-pole pattern will consist of a point maximum associated with each limb, and the interlimb angle will be the supplement of the angle between these two maxima. On the other hand, if the fold shape is dominated by a uniformly curved hinge zone (see Fig. 11.3b), the density of points within the girdle will be uniform, and the interlimb angle will be the supplement of the angle between the two extreme poles in the girdle. Most folds have shapes and patterns between these two extremes.

It will also be noted that symmetrical folds have symmetrical patterns, both in terms of location and concentration of points (Fig. 15.8a,b). Conversely, the pattern of asymmetric folds are also asymmetric; for such folds a large number of variations in the patterns are possible. Fig. 15.8c illustrates a simple example: the overally shape of the contours are symmetrical, but the point maxima within the girdle have noticably different values. The stronger one marks the pole of the dominant limb.

For purposes of introduction, the folds illustrated above are horizontal or upright or both. The axis and axial plane can, of course, have any attitude, and this will be reflected on the diagram. Several plunging and inclined folds are shown in Fig. 15.9.

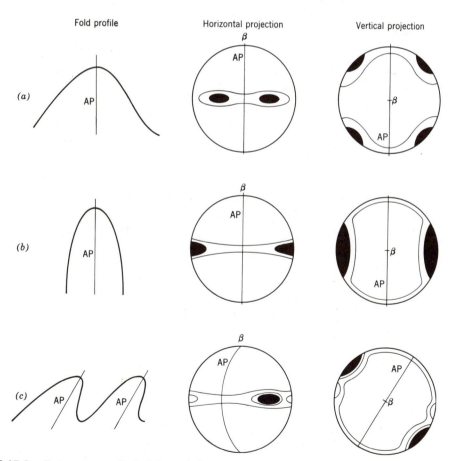

FIGURE 15.8 Patterns of folds: (a) symmetrical open fold; (b) symmetrical isoclinal fold; (c) asymmetrical fold with inclined axial plane.

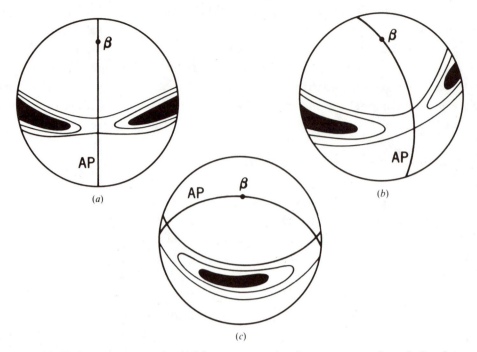

FIGURE 15.9 Folds with different attitudes axes and axial planes.

15.7 SUPERPOSED FOLDS

The S-pole diagram may also be viewed as a test for the homogeneity of the fold axes in the area being examined. As such, the diagram can be used to decide if, and in what direction, a fold profile can be drawn. On the other hand, the pattern may not be interpretable; the scatter may be such that no clearcut S-pole girdle is present. Such areas are inhomogeneous with respect to axial directions. This will be the general case in rock masses that have undergone two or more episodes of folding.

The approach in areas of polyphase folding is to seek smaller, homogeneous subdivisions for which the data does yield interpretable diagrams. An artificial example will suggest the approach that is used.

PROBLEM

From a geologic map of an area in which the rocks have undergone two episodes of folding (Fig. 15.10), determine the geometrical relationship between the two sets of folds.

ANALYSIS

1. Subdivide the map area into smaller subareas, each of which contains structures that are statistically homogeneous, that is, subareas characterized by cylindrical folds. These subdivisions may be located by

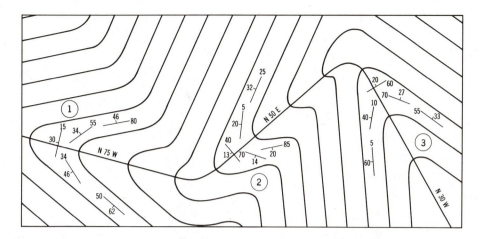

FIGURE 15.10 Idealized map of superposed folds; subareas 1, 2 and 3 are recognizable by the segments of the apparent traces of the hinge
surface which are rectilinear.

trial and error, or by the recognition of the rectilinear nature of the
apparent traces of the hinge surfaces, or by other structural evidence.

2. A Plot of the data from each subarea then yields the orientation of the
   folds in each homogeneous part of the structure (Fig. 15.11). Changes
   from one subarea to the next can then be determined by comparing
   diagrams.

3. Synoptic diagrams are useful in illustrating these variations, and in
   obtaining information about the second folds.

   a. Beta intersections of the axial planes from the three sub-areas
      define the axis of the second folds (Fig. 15.12a).

   b. The axes of the three subareas lie on a single great circle, which
      indicates a special type of dispersal of pre-existing fold axes

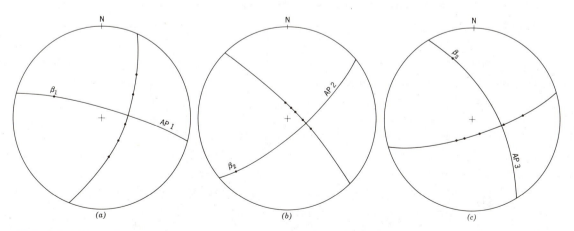

FIGURE 15.11 Stereograms of the orientation data from subareas 1, 2 and 3.

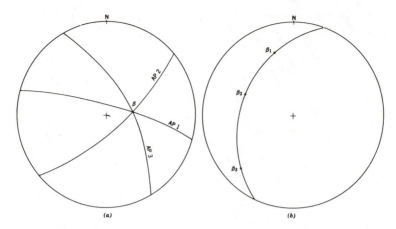

FIGURE 15.12   Synoptic diagrams: (a) axial planes; (b) fold axes.

and linear structured during the second deformation (Fig. 15.12b). This pattern is characteristic of similar folding.

Turner and Weiss (1963, p. 178–179) give an extended and more realistic example of this same type of analysis which is well worth examining in detail.

In general, results of this type, together with information on the type of folding, permits individual hinge lines to be traced through the superposed folds (Stauffer, 1968).

15.8   SAMPLING PROBLEM

The validity of interpretations based on point diagrams and their contoured equivalents depend in large part on the data being representative, that is, accurately reflecting the range and relative abundances of attitudes. Because of problems of limited exposure, and inability to explore the third dimension, it is difficult to justify any particular collection of data as being a strictly representative sample. Usually all that one can do is to conscientiously try to avoid introducing any obvious bias.

To illustrate this sampling problem, we will consider a situation which frequently arises when attempting to determine and describe the distribution of joints of various orientations in a rock mass for geotechnical purposes. Joints are cracks and fractures in rock along which there has been negligible movement. If joints are essentially planar and parallel so that they form <u>sets</u>, they are said to be systematic. Commonly several sets of systematic joints are present. Joints which belong to no set are referred to as random joints, and if only random joints are present they are said to exhibit a random pattern (though it is uncertain whether any group of joints are strictly random in the statistical sense).

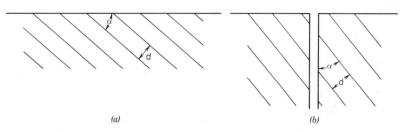

FIGURE 15.13 Joint sets in cross section: (a) vertical section through a set of joints perpendicular to the plane of the section which intersect the surface at angle α; (b) vertical section through a drill hole showing a joint set intersecting the hole at angle α (after Terzaghi, 1965, p. 295).

To evaluate the relative frequencies of various joint orientations observed at a particular locality, the dip and strike of each joint is measured, and each is then represented on a Schmidt net by plotting its pole. If such a joint survey is carried out in an area where the slopes are so irregular that exposures with a great variety of orientations are available for examination, the resulting joint-orientation diagram is likely to be a reasonably representative sample of the joints in the area. However, if the investigator must rely on observations made on nearly two-dimensional exposures, the orientational data is unlikely to provide even approximately correct information concerning the abundance of the joints of all sets present at the locality. Fig. 15.13 illustrates the effect of the angle α between a exposure plane and a joint. If the average spacing between joints is d, then the number of joints N_α with a particular orientation encountered along a sampling line of length L will be,

$$N_\alpha = (L/d)\sin \alpha$$

Because the number of intersections between an outcrop surface and joint planes with a given spacing is proportional to the sine of the angle of intersection, an idealized plot of the poles of joints of all possible orientations may be constructed (Fig. 15.14). In this diagram, the successive contours, starting at the circumference, are the loci of poles of joints which intersect the horizontal surface such that sin α = 0.1, 0.3, 0.5, 0.7, 0.9 respectively. Similarly, the relative densities of poles in the zones bounded by these isogonic (equal-angle) lines are also shown. As can be seen in Fig. 15.14a, joints with poles near the isogonic line for sin α = 1.0 (the primitive) will be about 10 times as abundant as poles near the isogonic line sin α = 0.1. The results of a survey from a vertical drill hole exhibit a similar but more serious deficiency (Fig. 15.14b); in this case there is a disproportionately large number of near-vertical poles and very few near the primitive. If both of these surveys were made in the same body of rock, it would be easy to conclude that joint systems developed at depth was quite different that the one observed at the surface, when in fact the differences are due entirely to the sampling problem.

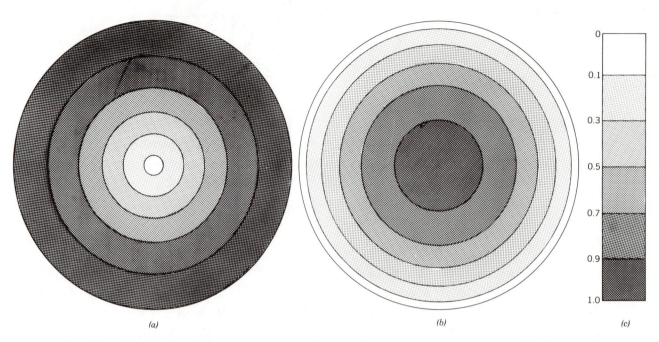

FIGURE 15.14   Idealized contoured diagram of the poles of random joints: (a)
as measured on a horizontal surface; (b) as measured in a
vertical drill hole (after Terzaghi, 1965, p. 296).

The accuracy of such a diagram can be increased by replacing the number of
joints N_α intersected at an angle α, by a value N_{90} representing the number of
joints with the same orientation which would have been observed on an outcrop with
the same dimensions intersecting the joints at an angle of 90°, where,

$$N_{90} = N_\alpha /\sin \alpha$$

However, no adequate correction can be made for low values of α because in real rock
the number of intersections is significantly affected by local variations in spacing
and continuity if α is small. Further, no correction whatsoever can be applied if α
= 0. Hence even a corrected diagram fails to indicate the abundance of horizontal
and gently dipping joints.

In this illustration, the outcrop surface was assumed to be horizontal, but it
should be clear that there will be a <u>blind</u> <u>spot</u> in the vicinity of the pole to the
exposure plane whatever its attitude. For similar reasons, the results of the
survey of joints in a drill core will not adequately sample the existing joints; the
problem is more severe because there is a blind zone for joints which are parallel
to the drill hole. In both cases, an adequate sample of such planes requires that
joints on other exposure planes or in other drill hole directions, and the data so
obtained, appropriately weighted, are then combined into a collective diagram
(Terzaghi, 1965, p. 298f).

For exactly the same reasons, the measurement of the orientations of cleavage planes made on a thin section with a universal stage are subject to bias; this is called the Schnitteffekt (see Turner and Weiss 1963, p. 226f)

EXERCISES

1. With the attitude data given on the map of Fig. X15.1, construct both a beta and an S-pole diagram. What is the trend and plunge of the fold axis?

2. Using the map of Fig. X15.2, determing the nature and orientation of the structure.

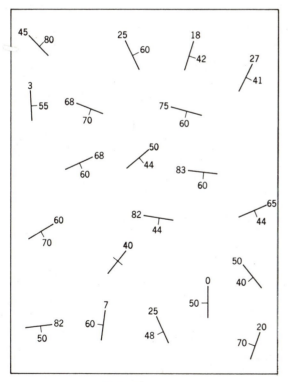

FIGURE X15.1

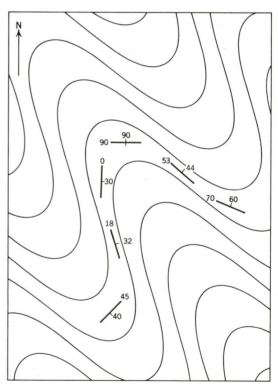

FIGURE X15.2

3. With the attitude data given on the map of Fig. X15.3, construct a point diagram and its contoured equivalent. Then determine the following:
   a. Trend and plunge of the fold axis.
   b. Attitude of the axial plane.
   c. Sketch the style of the folding.
   d. Approximate interlimb angle.
   e. Number of intersections on a beta diagram.

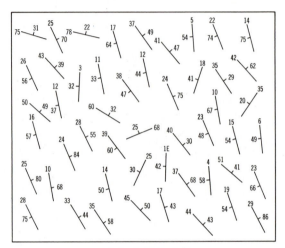

FIGURE X15.3

16
Linear and Planar Structures in Tectonites

16.1 INTRODUCTION

The term fabric includes the complete spatial and geometrical configuration of all the components that make up a rock. It is an all encompassing term that describes the shapes and characters of individual parts of a rock mass and the manner in which these parts are distributed and oriented in space (Hobbs, Means and Williams, 1976, p. 73). These components and their boundaries are elements of the fabric, and they may be either planar or linear. A description of the manner in which these elements and the boundaries between them are arranged in space constitutes a statement of the fabric of the body. Something of the nature and orientation of these fabrics may be determined using the methods of structural analysis.

16.2 ISOTROPY AND HOMOGENEITY

A rock with randomly oriented fabric elements will have the same physical and geometrical properties in all directions, and is therefore isotropic. Such rocks are rare in nature; usually the best that can be said is that a rock mass is statistically isotropic on some specified scale.

Almost all rocks show some degree of preferred orientation, and therefore are anisotropic. Such rocks are of considerable interest for the processes of formation and deformation are often recorded by these anisotropic fabrics. Such rocks are termed tectonites.

If any two identically oriented, equal-volume samples taken from a rock mass are identical in every respect, the mass from which they came is said to be homogeneous. At best, some rock masses are only quasi-homogeneous, that is, the distribution of the various different mineral components is only approximately uniform. Samples from such a mass that are large compared with the grain size will then be statistically indistinguishable, and the mass is said to be statistically homogeneous. A region of a rock body which is homogeneous with respect to the orientation or pattern of orientation for a given fabric element is a termed a fabric domain. In attempting to describe a particular fabric, it is important to insure that the sample is from such a domain.

16.3 ANALYSIS OF PLANAR AND LINEAR FABRICS

At the outcrop or in the hand specimen, planar or linear structures which make up the fabric are visible as traces on exposure faces. If the structure is simple and well developed, there may be no problem in determining its nature and attitude. However, when the traces are faint, or several different traces are present on the same exposure faces, it may be difficult to tell whether a planar or linear structure is present merely by inspecting several two-dimensional exposure faces. If the attitudes of the traces and the exposure plane on which they occur are measured, the data can be fitted into a three-dimensional picture with the aid of the stereonet.

TABLE 16.1   CLASSIFICATION OF FABRICS (after Den Tex, 1954)

---

1. Planar structures
   a. Planar parallelism of planar fabric elements (Fig. 16.1a)
   b. Planar parallelism of linear fabric elements (Fig. 16.1b)
2. Linear structures
   a. Linear parallelism of linear fabric elements (Fig. 16.1c)
   b. Linear parallelism of planar fabric elements (Fig. 16.1d)
3. Composite structures
   a. Combined structures: two or more planar and/or linear structures in combination.
   b. Complex structures: two fabrics marked by either linear or planar fabric elements only.
      (1) Linear + planar fabrics marked by linear elements
      (2) Linear + planar fabrics marked by planar elements

PLANAR STRUCTURES

If a planar structure is present, then each trace is an apparent dip of that plane, and the method used to find the true dip from two apparent dips may be used. However, more that two points are required to demonstrate that the structure is, in fact, a plane, and the more points that are used the more certain is its existence. With only two points, a plane that satisfies the measurements can be found, and this then used to check the results on a exposure face (natural or artificial) perpendicular to the plane.

PROBLEM

The following measurements could have been taken from five different outcrop faces, or from an oriented specimen on which five non-parallel faces had been cut. Determine the attitude of the planar structure.

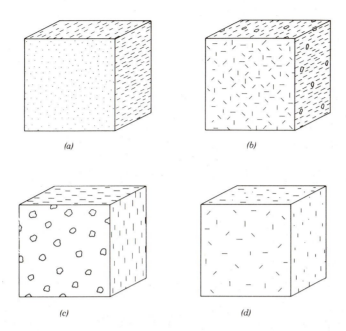

FIGURE 16.1 Planar and linear structures (from Oertel, 1962): (a) planar structure marked by planar fabric elements; (b) planar structure marked by linear fabric elements; (c) linear structure marked by linear fabric elements; (d) Linear structure marked by planar fabric elements.

Face	Attitude	Pitch of the trace
1	N40°E, 21°SE	76°S
2	N75°W, 32°N	20°NW
3	N80°E, 32°S	76°E
4	N30°W, 65°W	44°NW
5	N12°W, 45°W	86°N

CONSTRUCTION (Fig. 16.2)

1. Using the pitch angle, plot the point representing the line of the measured trace on each exposure face. The great circle of the exposure plane is, of course, used in locating this point, but it need not be added to the diagram.

2. Rotate the five points representing all the trace points until they fall on the same great circular arc. This great circle represents the structural plane defined by the measurements.

ANSWER

The planar structure has an attitude of N50°W, 55°SW.

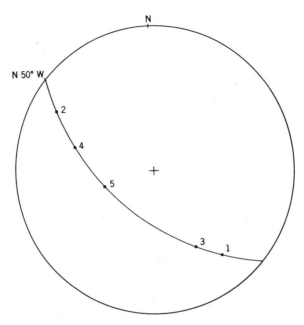

FIGURE 16.2   Attitude of a planar fabric based on outcrop traces.

LINEAR STRUCTURES

If the structure is linear, traces will we present on all faces except the one perpendicular to the line. As shown in Fig. 16.3, a rod-shaped fabric element cut by an oblique exposure plane P will expose an ellipse. The long axis of this ellipse is an apparent lineation of the true linear structure, that is, the trace marked by the long axis in an orthographic projection of the linear structure onto an exposure face. The true direction of the line lies in the plane N which is normal to the exposure plane and also contains the trace (if you have trouble visualizing this, make such an oblique cut on a dowel or pencil). Normal planes can be constructed from measurements on exposure faces; the intersection of any two will ideally fix the attitude of the line. In practice more points are needed to confirm the nature of the structure and to increase confidence in the result.

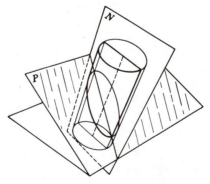

FIGURE 16.3   Relationship between an inclined exposure plane P and its intersection with a rod-shaped element; the axis of the linear
element lies in the plane normal to the exposure plane N that
contains the trace (from Lowe, 1946).

PROBLEM

From the following measurements determine the attitude of the linear structure.

Face	Attitude	Pitch of trace
1	N12°W, 85°E	45°N
2	N45°E, 90°	44° SW
3	N30°W, 65°W	48°N
4	N53°E, 85° SE	34° SW
5	N46°W, 18° SW	36° NW

CONSTRUCTION (Fig. 16.4a; after Lowe, 1946)

1. Plot as pairs of points the apparent lineation and the pole of each exposure plane.

2. Through each trace point and the pole of the corresponding exposure plane, draw in a great circular arc.

3. These great circular arcs intersect at point L which represents the linear structure.

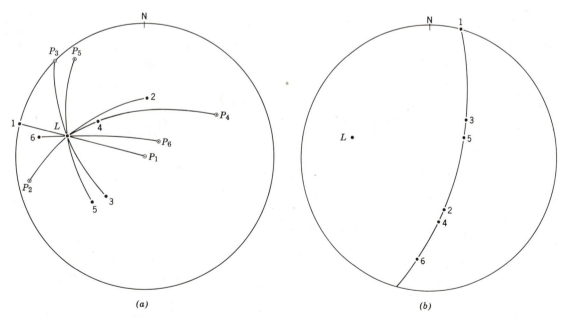

FIGURE 16.4 Attitude of linear structure based on an analysis of exposure traces: (a) Lowe's method; (b) Cruden's method.

ANSWER

The attitude of the linear structure is 26°, N76°W.

This method of locating the line of intersection of the N planes is similar to that used in the β diagram, and has the same disadvantage. As we have seen in §5.7, when the planes intersect at small angles, the location of the intersection point is strongly influenced by small variations in the measured angles. In contrast to the idealized data of this sample problem, it may be difficult to accurately determine the orientation of the line.

An alternative method avoids this problem, and has the additional advantage of being amenable to numerical treatment.

CONSTRUCTION (Fig. 16.4b; after Cruden, 1971)

1. As with Lowe's method (Steps 1 and 2), locate the trace points and the poles of the exposure planes.

2. Locate the great circular arc which passes through each trace–pole pair and plot its pole.

3. These new pole points themselves define a great circle which represents the plane perpendicular to the linear structure.

ANSWER

The pole of the best fit great circle found by this method is identical to the intersection found by Lowe's method.

The technique for identifying planar and linear structures and determining their attitude may be combined in cases where two or more traces are present on exposure faces.

16.4 COMPLEX STRUCTURES

In some rocks, fabric elements of just one shape may be arranged to give more than one structure, even though there is only a single trace on an exposure face. A simple experiment may help to see how this is possible. A collection of pencils scattered randomly on a table top is analogous to the case of a planar structure marked by linear elements (Class 1b, Table 16.1). Similarly, a parallel alignment of pencils on the table top is analogous to linear structure marked by linear elements (Class 2a). Now if these aligned linear elements are slightly dispersed so that all orientations within a small angle of azimuth of, say, 30° are represented, the result is a configuration intermediate between the two end member classes. It possesses both a dominant linear fabric (the still strong alignment) and a subordinate planar fabric (the tendency to spread in the plane of the table top). A similar pattern involving planar elements can be generated which are intermediate between Classes 1a and 2b.

The geometric analysis of such fabrics depends on the fact that in a given series of random exposure planes, certain faces will be more favorable oriented for observing the traces of the internal structure than others. Specifically, those exposure planes parallel to a linear structure, and those perpendicular to a planar structure will exhibit the best developed traces. Conversely, those planes perpendicular to the linear structure and those parallel to the planar structure will show no traces at all.

PRINCIPAL LINEAR, SUBORDINATE PLANAR STRUCTURE

Such complex structures occur when, for example, linear elements are statistically arranged with linear parallelism, but with a deviation into a plane. The deviation that produces this subordinate structure means that the principal structure cannot be as well developed.

Fig. 16.5a shows the stereographic plot of the poles P of exposure planes which contained measurable traces ℓ. Exposure faces oriented so that no trace is visible on them must also be examined, but they are not plotted. The measured planes should be well distributed in space for this type of analysis to yield a reasonable estimate of the structure.

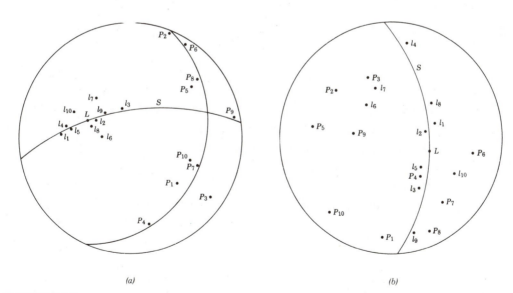

(a)   (b)

FIGURE 16.5 Complex structures (from Den Tex, 1954): (a) principal linear with subordinate planar structure; (b) principal planar with subordinate linear structure.

ANALYSIS

1. The points representing the traces cluster about a center L that
   represents the principal linear structure.

2. Note the tendency for the poles of the trace-containing exposure planes
   to be distributed along the great circle 90° from the linear structure
   L.

3. The trace points are as well spread out along another great circle s,
   which contains six of the 10 ℓ-points, as well as the center L.   This
   is the subordinate planar structure.

Construction of the linear structure using intersection great circles drawn
through the respective poles and traces would have approximated the position of L,
but it could not have detected the spreading that marks the subordinate planar
structure.

PRINCIPAL PLANAR, SUBORDINATE LINEAR STRUCTURE

A similarly complex structure may result from planar elements oriented in such
a manner as to mark a planar structure, but with deviations that are controlled by a
tendency to remain parallel to a line within the plane. Fig. 16.6b shows the plot
of such a case.

ANALYSIS

1. The majority (7 of 10) trace points exhibit a tendency to lie close to
   a great circle s.

2. There a similar, though less pronounced tendency for the poles P also
   to lie along this same great circle. Six of 10 poles are less than 45°
   from s and are roughly symmetrical with respect to it. The s plane
   is therefore the principal planar structure.

3. Within s, there is a tendency for the trace points to cluster aroung a
   center L. The 4 remaining poles are located approximately 90° from L.
   The point L is, therefore, a subordinate linear structure lying within
   the principal plane.

16.5 L–S TECTONITES

The complex fabrics of the last section are members of a continuous spectrum of
fabrics which range from perfectly linear, or L–tectonites, to perfectly planar, or

S-tectonites. Intermediate members of this series have intermediate relative strengths of the linear and planar components. A useful way of characterizing fabrics of this spectrum is to note that they generally have three mutually perpendicular planes of symmetry, the plane of S is one of these, and the other two are parallel and perpedicular to L.

For field identification, it has been found useful to divide these fabrics into five discrete intervals (Flinn, 1965; Schwerdtner, and others, 1977):

$$L, \ L > S, \ L \simeq S, \ L < S, \ S \ \text{tectonites}$$

The importance of this descriptive approach is that it may be possible to correlate the fabric geometry with the shape of the strain ellipsoid.

Other types of fabrics, such as the combined structures, will generally have lower symmetries, either monoclinic or triclinic. These can not be included in the L-S tectonite scheme, and their fabric components have to be described separately.

EXERCISES

For the following sets of measurements, analyze the data for the structure involved.

Exposure Plane	Pitch of trace 1	Pitch of trace 2
1. N80°W, 30°N	20°W	
N50°E, 80°N	30°W	
Horizontal	N46°W	
N 5°E,10°S	40°S	
N72°E,20°S	80°W	
2. N15°W, 70°E	80°N	
N52°E,50°SE	40°NE	
N 0°, 45°E	74°N	
N86°E,60°S	30°E	
N43°E,50°W	25°NE	
N52°E,35°N	14°NE	
3. N30°E, 30°W	15°N	85°N
N45°W, 20°SW	30°SE	70°SE
N20°W, 60°E	52°N	5°S
N25°E, 40°E	36°S	28°N
N80°W, 70°E	70°E	40°W
N50°E, 55°SW	50°SW	15°NE
4. N70°W, 30°S	28°E	80°W
N60°W, 10°E	35°E	90°
N90°W, 20°N	55°E	73°E
N20°W, 40°E	26°N	none
N15°W, 45°W	45°S	30°N
N55°E, 57°SW	85°S	15°N
N50°E, 90°	20°SW	60°NE
N40°W, 30°NE	60°N	35°N
5. N50°W, 50°NE	55°NW	
N90°W, 30°N	90°	
N20°W, 30°N	27°N	
N10°W, 70°E	40°N	
N48°E, 26°NE	46°NE	
N20°W, 70°E	68°N	
N32°E, 52°W	38°N	
N 4°E, 70°W	10°N	
N85°W, 45°N	90°	
N85°E, 65°S	none	
N20°E, 75°W	60°N	
N45°W, 20°NE	46°NW	
N90°W,52°S	none	

17
Drill Hole Date

17.1 INTRODUCTION

The exploration of the underground by diamond drilling and the recovery of core samples is an important technique for the geologist, especially in engineering and mining projects. A good overview of many practical aspects of drilling and coring is provided by Goodman (1976, p. 127-157). The information gained from a drilling program depends on a number of factors: the number of holes drilled, the orientation of the holes, the core recovery, and the structures seen in the cores. Because drilling is expensive, it is important to gain the maximum structural information from a minimum number of holes.

We will consider here two type of structures and the information that can be obtained about them from a study of recovered cores. The simplest cases involve planar structures. The method of determining the attitude of such planes will be described in some detail. With this as background, some of problems involved in interpretation the structure of folded rocks are introduced.

17.2 ATTITUDE OF A PLANE FROM AN ORIENTED CORE

Special devices or techniques are available which allow the in situ orientation of the core to be recovered (Zimmer, 1963; Goodman, 1976, p. 142f). We assume here that the method of fixing this orientation takes the form of a mark made along the top edge of the inclined cylindrical core parallel to the axis of the drill hole (see Fig. 17.1a); other marking schemes can be converted to this one, or the construction easily adapted to other conventions.

APPROACH

The first step is to describe the attitude of the planar structures in the core as if the core axis were vertical with the index mark in the direction of the trend of the drill hole. In this local reference frame, the angle between the bedding or other structural plane and the core axis, called the core-bedding angle θ, is converted to the local dip angle $\delta' = 90° - \theta$. The trace of an inclined structural plane in a cylindrical core will appear as an ellipse, and the major axis of this ellipse marks the local dip direction. This is easily identified by locating the low point on the trace (see Fig. 17.1b). The bearing β' of this dip direction rela-

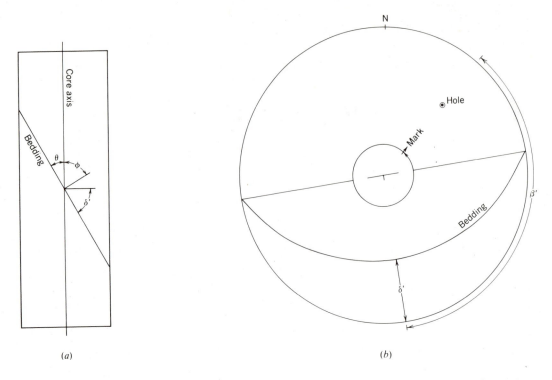

(a)   (b)

FIGURE 17.1   Analysis of an oriented core for the true attitude of a plane:
(a) cross section of a core showing angles describing bedding
inclination; (b) the core in local reference frame showing
orientation of dip direction.

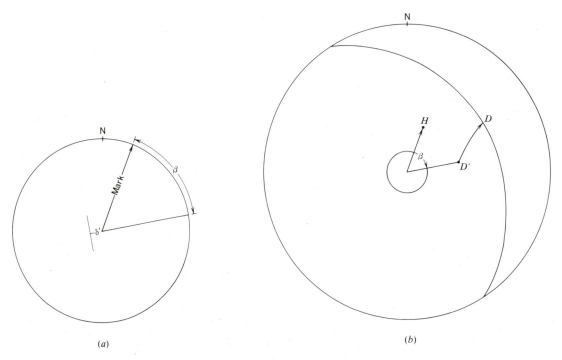

(a)   (b)

FIGURE 17.2   Attitude of a plane from a recovered core: (a) specifying the
local dip direction; (b) stereographic construction of attitude.

tive to the index mark is then measured. With this information, the
determination of the true strike and dip of the structural plane is
essentially the problem of two tilts in reverse.

PROBLEM

An oriented core recovered from a drill hole with an attitude of 55°/020°,
yielded the following measurements: $\theta = 40°$, $\beta' = 60°$ measured clockwise
from the index mark (see Fig. 17.2a). Determine the true attitude of the
structural plane.

CONSTRUCTION (Fig. 17.2b)

1. Plot point H representing the inclined drill hole and point D' representing the line of dip of the plane in the local reference frame using
   the angles $\delta' = 90°-\theta = 50°$ and $\beta'= 60°$.

2. By rotating the vertical core so that its axis coincides with H, the
   point D' moves 40° along its small circle to the point D marking the
   true dip of the plane.

ANSWER

The true strike and dip of the plane in the core is N33°W, 26°E.

17.3 ONE DRILL HOLE INTERSECTING A PLANE

During routine drilling, marking devices are not used and it is not possible to
keep the core from rotating in the drill pipe during recovery. As a result, only
three pieces of structural information will be available from a single recovered
core: (1) the distance along the hole to a recognizable structural plane, (2) the
angle between the core axis and the bedding, and (3) the attitude of the drill hole.

In the special case of a vertical hole, the local dip is the true dip, but because of the rotation, the strike direction is unknown. We can imagine all the possible orientations by rotating the core at its original location through 360°. During this rotation, the structural planes in the core will be tangent to a right circular cone. The intersection of this cone with the earth's surface will be circle.
There are two special cases. If the structural planes are vertical, the cone
degenerates to a line, and if they are horizontal, it degenerates to a plane. Only
in this latter case is the attitude uniquely define by a single drill hole.

If the single drill hole is inclined, the local dip angle no longer represents
the true dip. Again, the true attitude will be tangent to the surface of the cone
generated by rotating the core about its axis. Depending on the relationship be-

tween the drill hole inclination and the local angle of dip of the structural planes, this cone will intersect the earth's surface as an ellipse, parabola or hyperbola. As before, the attitude is completely determined in the special case when the drill hole is normal to the plane.

In both the cases of vertical or inclined holes, the construction of the appropriate conic section at the earth's surface would permit all possible attitudes to be delimited, but this requires drawing these sections accurately, and this is not a practical approach, especially since there is a much easier way. A simple problem will introduce this alternative method which uses the stereonet. In plotting the orientation of the bedding it will be convenient to use the angle between the axis of the core and the pole to the bedding, called the <u>core-pole</u> angle (see Fig. 17.1b).

PROBLEM

A drill hole inclined 40° due north intersects a plane with attitude of N50°W, 50°SW. What will the core-pole angle be in the recovered core?

CONSTRUCTION (Fig. 17.3a)

1. Plot points H representing the inclined drill hole and P representing the pole to the known plane.

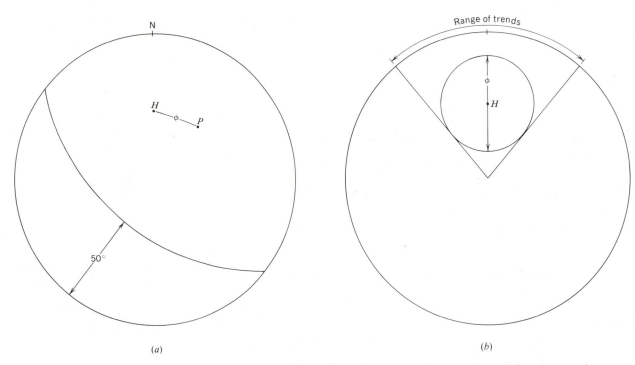

(a)   (b)

FIGURE 17.3 Relationship between bedding and core-pole angle: (a) core-pole angle from known attitude; (b) problem of the attitude from a single hole.

2. The angle between H and P, measured along the common great circle is the core-pole angle ϕ (see Fig. 17.1b).

ANSWER

The core-pole angle is $\phi = 30°$.

The inverse of this problem will introduce the basic construction used in solving all drill hole problems without core orientation graphically.

PROBLEM

The core recovered from a drill hole with attitude 40°/000° has a core-pole angle $\phi = 30°$. What can be said of the true attitude of this plane?

CONSTRUCTION (Fig. 17.3b)

1. Plot the point H representing the inclined drill hole.

2. The possible poles to the structural plane lie at a constant angular distance from H, that is, on a small circle. Points on this circle may be located by rotating H to a series of great circular arcs and counting off 30° along each, as shown. The locus of these points describes a small circle. (A more efficient method of drawing small circles is detailed in §17.4).

ANSWER

As can be seen, the poles have a 40° range of trend and plunge angles range from 10° to 70°, and, of course, the corresponding planes have a range of dips and strikes. Without further information, there is no basis of locating the particular point on the small circle which is represents the true pole, hence the true dip and strike remains unknown.

Additional information can be obtained from several sources. One possibility is that the true attitude of the plane may be partially known, that is, either the strike, or the dip may be known.

PROBLEM

A drill hole with attitude 30°/160° intersects bedding planes and the core-pole angle is 25°. The true strike of beds is known to be N50°E. What are the dip possibilities?

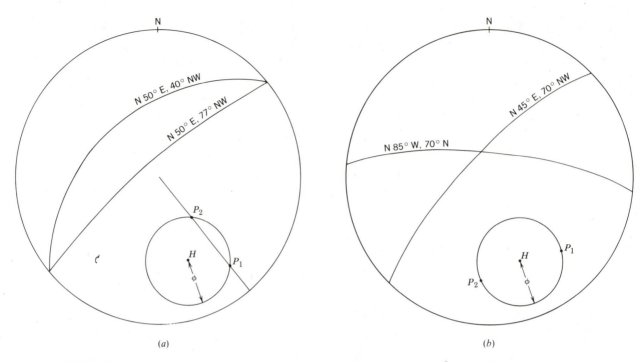

(a)   (b)

FIGURE 17.4   Single hole and partial attitude known: (a) strike known: two
possible dips; (b) dip known: two possible strikes.

CONSTRUCTION (Fig. 17.4a)

1. Plot the point H representing the inclined drill hole, and draw in the
    small circle with angular radius ϕ = 25°.

2. Draw in a radius of the net perpendicular to the known strike direction
    intersecting the small circle in two points. These points represent
    the possible poles to the bedding.

ANSWER

The plunge angles of these two poles correspond to dips of 40° and 77°.

A similar construction, based on known dip angle, also yields two possible
poles to bedding, and the two corresponding strike directions can then be determined
(see Fig. 17.4b).

Another piece of information can be obtained if two structural planes are present, and the attitude of one of them is known. This allows the orientation of the
core to be fixed, and the attitude of the unknown plane can then be determined.   As
in the case of the oriented core, use is made of the local reference frame to
describe the observed relationship between the two planes.

PROBLEM

A drill hole plunges 70°/000° intersects two planes. One of these is known to have a true attitude of N40°E, 40°SE. Measurements made on the core yield the local attitude of the second plane: the core-pole angle ϕ = 40°, and the local bearing, relative to the dip direction of the known plane, β' = 65°, measured clockwise. What is the attitude of this second plane?

CONSTRUCTION (Fig. 17.5)

1. Plot the drill hole H, the pole P_1 of the plane whose attitude is known, together with the great circle of the local reference plane whose pole is H.

2. The dip direction of the known plane in the local reference frame is located by drawing a great circle through points P_1 and H, to intersect the local reference plane at point D_1; the local strike S_1 is 90° along this same great circle (Fig. 17.5a).

3. The dip direction of the second plane is then located β' = 27° + 38° = 65° from D_1 along the great circle in a clockwise sense to locate point D_2. The local strike direction S_2 of this plane is then located 90° away (Fig. 17.5b).

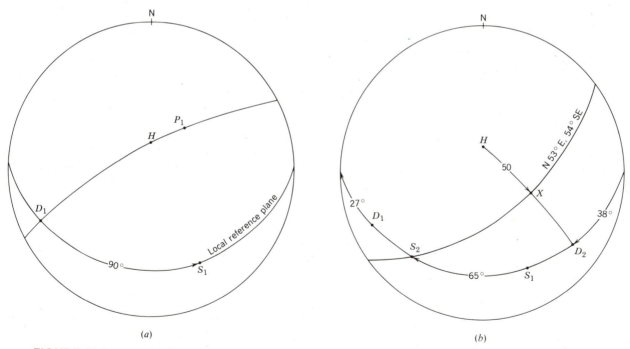

(a)   (b)

FIGURE 17.5 Single core containing two planes, the attitude of one being known (after Goodman, 1976, p. 146): (a) stereographic plot; (b) attitude of the second plane.

4. Along a great circular arc through points H and D_2, measure $\theta = 50°$
   from H toward D_2 to locate a point X on the second plane. S_2 is a
   second point on this plane, and a great circle is then drawn through
   both.

ANSWER

The attitude of the second plane is N53°E, 54°S.

17.4 CIRCLES IN STEREOGRAPHIC PROJECTION

To show that any circle on the sphere projects as a circle, we will consider
the general case of a small circle about an inclined axis. Any small circle is the
line of intersection of the surface of a sphere and a circular cone with vertex at
the center of the sphere (Fig. 17.6a). Label the axis of the cone OV and its vertex
angle LOM. The small circle on the sphere has a diameter of LM and the point V is
at its center.

PROOF (Fig. 17.6a)

1. The projectors of this circle on the sphere to the projection plane
   form the cone LZM.

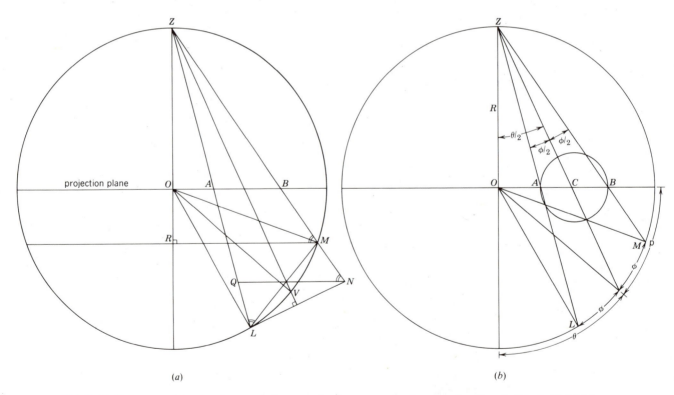

(a)   (b)

FIGURE 17.6 Small circles: (a) proof they project as circle (after Phillips,
1963, p. 24); (b) derivation of expressions for the center and
radius of a small circle (after Macelwane and Sohon, 1932, p.
136).

2. The line LM is the trace of a circular section of this cone and it is oblique oblique to the line ZV. Therefore, the section through LN, which is perpendicular to ZV, must be an ellipse.

3. Line QN is constructed to be inclined to ZV at the same angle as LM, and therefore it marks the trace of a conjugate circular section of the cone.

4. Line MR is constructed parallel to the projection plane. Then angles ZMR and ZLM are equal because they both subtend the arcs equal to ZM.

5. Because QN and LM are conjugate, the angles ZLM and ZNQ are also equal.

6. Therefore QN is parallel to MR, and also to the projection plane. Parallel sections through a cone are similar, and therefore the section at AB in the projection plane is a circle.

With this result any small circle may be drawn easily on the stereonet using a simple construction.

CONSTRUCTION (Fig. 17.6b)

1. Construct the section of the cone LOM corresponding to the required small circle.

2. Project points L and M to the plane of projection using the point Z, thus locating points A and B.

3. The center C of the small circle on the projection plane is found by bisecting the segment AB, which can be then completed with a compass. Note that C does not lie on the axis of the cone.

Algebraic expressions for the location of the center of the small circle and its radius may also be obtained. From Fig. 17.6b,

$$OA = R \tan(\theta/2 - \phi/2) \qquad \text{and} \qquad OB = R \tan(\theta/2 - \phi/2)$$

where R is the radius of the net, θ is the supplement of the plunge of the cone axis, and 2ϕ is the vertex angle of the cone. The distance from O to the center of the small circle is then

$$c = R[\tan(\theta/2 + \phi/2) + \tan(\theta/2 - \phi/2)]/2$$

and its radius is

$$r = R[\tan(\theta/2 + \phi/2) - \tan(\theta/2 - \phi/2)]/2$$

Substituting the identities

$$\tan\frac{1}{2}(\theta+\phi) = \frac{\sin\theta \pm \sin\phi}{\cos\theta + \cos\phi}$$

and

$$\tan\frac{1}{2}(\theta-\phi) = \frac{\sin\theta \pm \sin\phi}{\cos\theta + \cos\phi}$$

into these two expression and rearranging, we obtain

$$c = \frac{R\sin\theta}{\cos\theta + \sin\phi} \quad \text{and} \quad r = \frac{R\cos\phi}{\cos\theta + \sin\phi}$$

Because $\theta = (90°-p)$, these two can be written in a form which is more convenient for geological purposes.

(17.1a)
$$c = \frac{R\cos p}{\sin p + \cos\phi}$$

(17.1b)
$$r = \frac{R\sin\phi}{\sin p + \cos\phi}$$

If the inclination of the drill hole is small, or the core-pole angle large, or both, the small circle may overlap the primitive, and it is then necessary to construct its opposite the upper hemisphere. This involves projecting the intersection of the cone about the opposite of the line of the drill hole to the projection plane using the projection point Z (see Fig. 17.7). This second circle will also overlap the primitive and its points of intersection with the primitive are diametrically opposite those of the first circle. Although it is in two parts, the small circle is now complete in the lower hemisphere.

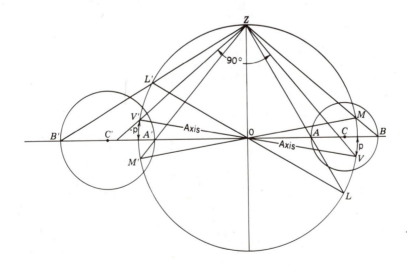

FIGURE 17.7 Construction of the opposite of a small circle.

An even simpler way is to determine the center and radius of the second small circle by substituting -p in Eqs. 17.1 (the negative sign indicating an upward inclination of the cone axis).

17.5 STEREOGRAPHIC SOLUTION OF THE TWO DRILL HOLE PROBLEM

Solutions to problems involving two differently oriented drill holes are obtained by drawing the small circles representing the core-pole cones associated with each hole. The points of intersection of these circles are possible poles to bedding.

PROBLEM

Recovered cores from two inclined drill holes yield the following data on the plunge and trend of each hole, and the associated core-pole angle:

$$P_1 = 70° \qquad P_2 = 40°$$
$$t_1 = 110° \qquad t_2 = 240°$$
$$\phi_1 = 60° \qquad \phi_2 = 40°$$

CONSTRUCTION (Fig. 17.8a)

1. For each hole plot the point representing the line of inclination of the hole, and construct the associated small circle whose radius is the core-pole angle.

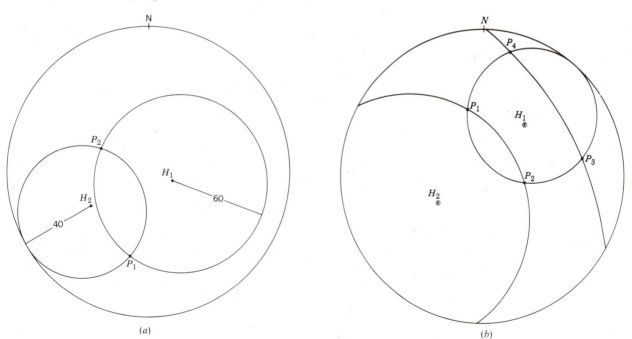

FIGURE 17.8 Two drill holes: (a) two possible attitudes; (b) four possible attitudes.

2. Read off the plunge and trend of the points common to the two small circles.

ANSWER

The attitudes of these poles and the corresponding dips dip directions of the planes are:

$$1.\ 29°/192°\quad \text{or}\quad 61°,\ N12°E$$
$$2.\ 50°/296°\quad \text{or}\quad 40°,\ S64°E$$

and these are the two possible attitudes of the intersected planes.

If one or both of the small circles is partly in the upper hemisphere there is the possibility of two additional points of intersections, and it is therefore necessary to construct additional arcs using the method of Fig. 17.6.

PROBLEM

Two inclined drill holes and their corresponding cores yield the following data:

$$p_1 = 40°\qquad p_2 = 50°$$
$$t_1 = 40°\qquad t_2 = 240°$$
$$\phi_1 = 40°\qquad \phi_2 = 70°$$

CONSTRUCTION (Fig. 17.8b)

1. As before, plot the two drill holes and draw in the associated small circles.

2. Because the small circle associated with the second hole is partially in the upper hemisphere, its opposite is also constructed.

ANSWER

The four possible poles to bedding and their corresponding dips and dip directions are:

$$1.\ 40°/347°\quad \text{or}\quad 50°,\ S13°E$$
$$2.\ 56°/097°\quad \text{or}\quad 34°,\ N83°W$$
$$3.\ 7.5°/014°\quad \text{or}\quad 82.5°,\ S14°W$$
$$4.\ 18°/079°\quad \text{or}\quad 72°,\ S79°W$$

If a recognizable marker horizon is present in both cores, it will generally be possible to determine the attitude of the structural plane uniquely. The presence of the marker at known distances along the two holes permits two points on the marker horizon to be found and from these an apparent dip.

PROBLEM

Drill hole 1 is inclined 60° due north; it encountered a marker bed at 28.5 m, and the core-pole angle is 52°. Hole 2, located 100 m due south and at the same elevation, inclined 50° due west. The same marker was encountered at at 52.5 m, and the core-pole angle was 68°. What is the attitude of the marker bed?

STEREOGRAPHIC CONSTRUCTION (Fig. 17.9a)

1. Plot the two inclined drill holes and their corresponding core-pole small circles.

2. The points of intersection of these circles are the two possible poles to the marker bed. The corresponding strike and dip pairs are: N30°E, 30°W and N68°E, 80°S

ORTHOGRAPHIC CONSTRUCTION (Fig. 17.9b)

1. Locate the two drill holes and plot their trends in map view.

2. At each location, establish a folding line parallel to the surface projection of the drill hole, and plot the plunge angle and inclined distance to the marker.

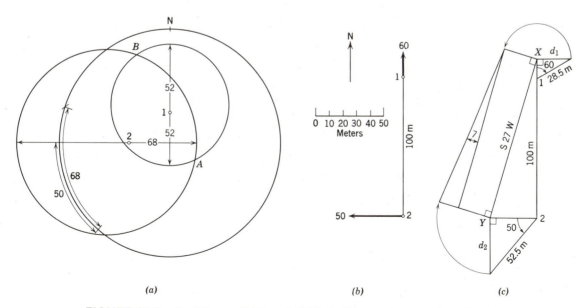

(a) (b) (c)

FIGURE 17.9 Problem of two drill holes and a marker horizon.

3. Locate the surface projections of these two points on the marker (points X and Y). The depths to the marker at each of these points d_1 and d_2 also appear in these sections.

4. Using XY as a folding line, plot the depths at each end, and thus determine the inclination of the a line on the marker in this direction.

ANSWER

The apparent dip in the direction of the line XY is 7°, S27°W, and this is compatible only with the plane defined by N30°E, 30°W, which is therefore the correct attitude.

17.6 ANALYTICAL SOLUTION OF THE TWO DRILL HOLE PROBLEM

For each pole common to two intersecting small circles, there are three unknowns--the three direction cosines (see Eqs. B.6). Correspondingly, three relationships involving these three unknowns quantities are required for a mathematical solution. The first is the identity

$$(17.2) \qquad L^2 + M^2 + N^2 = 1$$

The other two are obtained from the expression for the angle between two lines in terms of direction cosines,

$$\ell_1 L + m_1 M + n_1 N = \cos \phi_1$$

$$\ell_2 L + m_2 M + n_2 N = \cos \phi_2$$

where the direction cosines with subscripts identify the orientation of each drill hole and the capital letters denote the direction cosines associated with the as yet unknown pole to the bedding. Rearranging these two expressions, we have

$$\ell_1 L + m_1 M = \cos \phi_1 - n_1 N$$

$$\ell_2 L + m_2 M = \cos \phi_2 - n_2 N$$

Using Cramer's Rule, we express L and M in terms of determinants of these two equations.

$$(17.3a) \qquad L = \frac{\begin{vmatrix} (\cos \phi_1 - n_1 N) & m_1 \\ (\cos \phi_2 - n_2 N) & m_2 \end{vmatrix}}{\begin{vmatrix} \ell_1 & m_1 \\ \ell_2 & m_2 \end{vmatrix}} = \frac{\begin{vmatrix} \cos \phi_1 & m_1 \\ \cos \phi_2 & m_2 \end{vmatrix} - N \begin{vmatrix} n_1 & m_1 \\ n_2 & m_2 \end{vmatrix}}{\begin{vmatrix} \ell_1 & m_1 \\ \ell_2 & m_2 \end{vmatrix}}$$

$$(17.3b) \quad M = \frac{\begin{vmatrix} \ell_1 & (\cos \phi_1 - n_1 N) \\ \ell_2 & (\cos \phi_2 - n_2 N) \end{vmatrix}}{\begin{vmatrix} \ell_1 & m_1 \\ \ell_2 & m_2 \end{vmatrix}} = \frac{\begin{vmatrix} \ell_1 & \cos \phi_1 \\ \ell_2 & \cos \phi_2 \end{vmatrix} - N \begin{vmatrix} \ell_1 & n_1 \\ \ell_2 & n_2 \end{vmatrix}}{\begin{vmatrix} \ell_1 & m_1 \\ \ell_2 & m_2 \end{vmatrix}}$$

In these equations there are five determinants to be evaluated; these are

$$D_1 = m_2 \cos \phi_1 - m_1 \cos \phi_2$$

$$D_2 = m_2 n_1 - m_1 n_2$$

$$D_3 = \ell_1 \cos \phi_2 - \ell_2 \cos \phi_1$$

$$D_4 = \ell_1 n_2 - \ell_2 n_1$$

$$D_5 = \ell_1 m_2 - \ell_2 m_1$$

With these, we then write Eqs. 17.3 as

$$(17.4a) \quad L = (D_1 - ND_2)/D_5$$

$$(17.4b) \quad M = (D_3 - ND_4)/D_5$$

Substituting these back into Eq. 17.2 and collecting terms gives the quadratic in N,

$$(D_2^2 + D_4^2 + D_5)N^2 - 2(D_1 D_2 + D_3 D_4)N + (D_1^2 + D_3^2 - D_5^2) = 0$$

Using the roots of this equation in Eqs. 17.4, we then have two sets of direction cosines, one for each of the two intersections representing possible poles to bedding. The plunge and trend of these poles can be determined by using Eqs. B.7.

In the case of four possible attitudes, the plunge $-p$ and trend $t+180°$ are used for the small circle which overlaps the primitive to find the direction cosines of the other two poles. Charlesworth and Kirby (1981) give a similar development and several additional applications.

17.7 THREE DRILL HOLES

In any situation, cores from three drills holes completely fix the attitude of a structural plane. If a marker is encountered, the attitude can be determined by the simple method of the three point problem, but its attitude could already have been found with only two differently oriented drill holes. Without a marker, the attitude can be determined by finding the unique pole common to the three core-pole cones.

PROBLEM

Three drill holes have the following orientations and core-bedding angles in the recovered core. What is the possible attitudes of the bedding?

1. $60°/000°$; $\phi = 51°$
2. $50°/270°$; $\phi = 67°$
3. $55°/045°$; $\phi = 38°$

APPROACH

It is straight forward extension of previous methods to plot the points representing the three inclined drill holes and to construct the intersecting small circles representing the core-pole cones (Fig. 17.10)

ANSWER

The point common two the three circles gives the unique pole $60°/120°$.

The effort of plotting three small circles, especially if one or more of them overlap the primitive, is time-consuming process, and use of the analytical method becomes an attractive alternative.

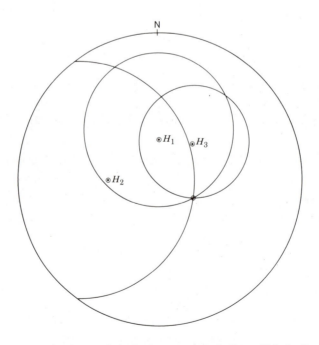

FIGURE 17.10 Three holes: (a) holes H_1 and H_2; (b) holes H_2 and H_3.

17.8 INTERPRETATION OF FOLDS

The deciphering of the structure of folded terranes from drill cores is considerably more difficult. If the wave length of the folds is large and enough holes are drilled it should be possible to represent the structural geometry by structure contours or formline contours (see Chapter 18).

If the beds are sharply folded on a small scale, then each hole will intersect planes with a variety attitudes. Given certain conditions it may then be possible to determine something of the attitude of these structures. In order to show the basis of the method, consider the following situation. The attitudes of the axes of a series of small-scale cylindrical isoclines are 40°/110°, and these folds are intersected by a drill hole with attitude 36°/208°. At each point along the recovered core, there will be a measurable core-pole angle, and for each of these a small circle about the plunging drill hole may be constructed (Fig. 17.11a). Of all these possible circle, one is unique; it is the circle tangent to the S-pole great circle, and for this the core-pole angle $\phi = \phi_{min}$.

The inverse of this problem, that is, the coring of such folds without knowledge of the fold attitudes will give the same series of small circles, of which one can be identified as the smallest, but without further information a unique solution is not possible.

One way of obtaining a solution is to examine cores from several differently oriented holes, or, since in practice it is common for drill holes to curve with depth, from an examination of cores from a number of points along the hole whose variable attitudes are known. If the curve of the deflected drill hole lies in a vertical plane, and the minimum small circles are drawn at a number of point, two great circles may be drawn which are tangent to them. In Fig. 17.11b, such small circle associated with points H_1, H_2 and H_3 along the hole are shown; note that for an additional point H_4 the hole is locally parallel to S-pole great circle, or, equivalently, perpendicular to the fold axis, and the minimum small circle has degenerated to a point.

In this example, the two possible fold axes have quite different attitudes: 40°/110° and 40°/325°, and it may be possible to reject the spurious one if the general structural trend of the area is known, even approximately. If, in addition to curving in the vertical plane, a horizontal component of curvature is also present, then the bilateral symmetry exhibited by Fig. 17.11b will be destroyed and a unique solution can be obtained.

Other structural information may also be taken into account. If slaty cleavage is seen in some of the rocks, then two different planes will be present in the cores. Such cleavage often has a reasonably constant attitude, and its attitude may be known from surface exposures. With the attitude of the cleavage planes known,

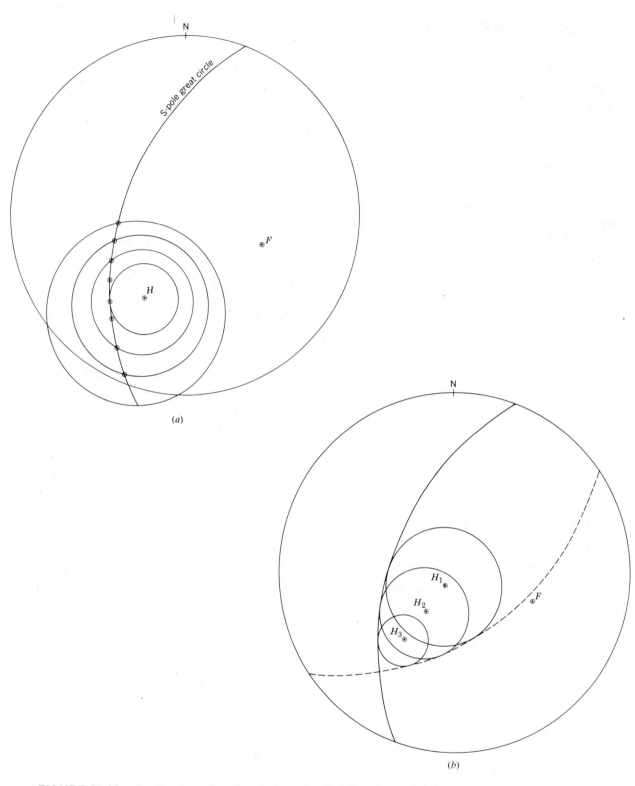

FIGURE 17.11 Analysis of cylindrically folded beds: (a) stereogram with four of small circles representing possible core-pole cones; (b) stereogram illustrating the determination of two possible orientations of the fold axis from a number of ϕ_{min} small circles (after Laing, 1977, p. 673-674).

the attitude of the variable bedding could then be determined at any point using the construction of Fig. 17.5. Further, the intersection of cleavage and bedding could be found, and this related to the fold axis. Details of these and other closely related techniques, and some important limitations are discussed by Laing (1977).

EXERCISES

1. Two vertical holes are drilled. H_1 encountered a marker horizon at a depth of 65 m. H_2, located 120m to the N30°W, encountered the same marker at 33 m. What is the dip, and what are the strike possibilities?

2. In a vertical drill hole the core-bedding angle is 20°. In a second hole, inclined at 50°, N45°E, the core-bedding angle is 15°. What are the possible attitudes?

3. A vertical hole encountered a marker at 14.8 m, and the core-bedding angle is 60°. A second hole inclined at 30°, N20°E is located 60 m due east. The marker was found at 33.6 m along the drill hole and the core-bedding angle is 45°. What is the attitude?

4. Three drill holes intersect a prominent planar structure. From the following information, what is the attitude of this plane?

H_1: 82°, S37°W, ϕ = 17°
H_2: 61°, S21°E, ϕ = 34°
H_3: 50°, S07°E, ϕ = 30°

5. With a recognizable marker horizon why do two differently inclined holes give more information than two parallel holes?

18
Structural Contours

18.1 INTRODUCTION

A <u>structure</u> <u>contour</u> is a line of equal elevation drawn on a structural surface; such contours are used to depict the form of the surface. In the simplest case, a structure contour drawn on a planar surface is a straight line and is equivalent to a line of strike. On curviplanar surfaces, they appear as curved lines, which are everywhere tangent to the strike of the surface. As an imaginary line connecting points of equal elevation, structure contours are partially analogous to topographic contours. The visualization of the features portrayed by both type of contours follow the same rules. In addition, however, structure contours have several unique properties: the surface represented by contours may overhand, or it may be broken by faults. In certain circumstances, it may also be useful to use an inclined, rather than horizontal datum plane (Conolly, 1936).

18.2 CONTOURING

The available factual information concerning the configuration of a structural surface usually takes the form of a series of points of known elevation. The construction of a structure contour map from this raw data involves two general steps.

OBJECTIVE INTERPOLATION

The choice of the contour interval depends on the amount of relief present on the surface, the map scale, and the spacing of the data points and the accuracy of their locations and elevations. Intermediate elevations points are then estalished which correspond to this chosen interval, and tentative contour lines are drawn through them. The location of these intermediate points may be found by eye, especially if only a very few contours must be interpolated, or with special devices (Marsh, 1960; Schweinfurth, 1969), or by a simple construction.

PROCEDURE

1. Connect two adjacent elevation points with straight line (for example, points A and B of Fig. 18.1a), which are taken from the map of Fig. 18.2a).

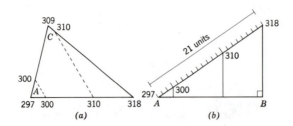

FIGURE 18.1 Linear interpolation of intermediate elevation points.

2. By assuming that the slope between these two points is a straight line on the surface, the location of points of intermediate elevation can be easily found (Fig. 18.1b).

   a. Draw a line perpendicular to AB, through point B. This establishes two sides of a right triangle.

   b. The difference in elevation between A and B is 21 m (= 318-297). Using any convenient scale draw the hypotenuse of this triangle 21 units long through point A.

   c. Along this hypotenuse count off the vertical distances to even multiples of the contour interval. Here, the interval is chosen to be 10 m, so that points at 3 units and 13 units along the line represent the of 300 m and 310 m contours. The locations of these points on the ground are found by dropping perpendiculars back to the line AB.

The locations of tentative contours between other pairs of points can be also be found by this linear interpolation procedure.

The intermediate points found by this method and tentative contour segments through them are only a first approximation of the form of the structural surface. For a curviplanar surface, curvilinear contours must, of course, be used. An examination of the map may reveal advantageous places to start drawing, for example, in areas where close spacing of data points gives greater control, or where the steep slopes require a number of contours to be interpolated. Working in bands, curved contour lines are drawn which agree with the known elevations. In the absence of other information, these contours should be as smooth as possible, and with a spacing which is as nearly equal as possible. In general, this will require that the contours do not pass exactly though the interpolated elevations, but this is understandable since the assumption of uniform slope is generally false.

The contour pattern should progressively evolve during the work. It will be found that altering the position of one contour line requires that several adjacent

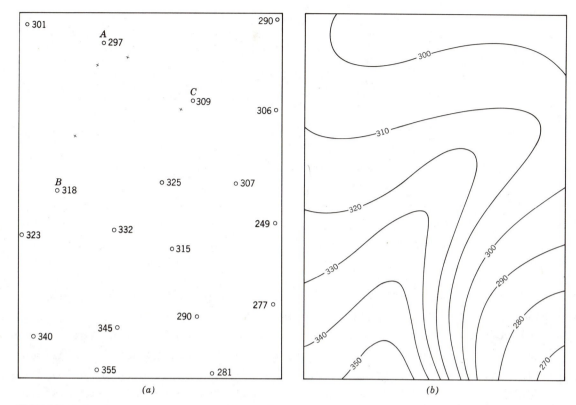

FIGURE 18.2 Method of contouring: (a) map showing the elevations; (b) the contoured equivalent.

ones must also be shifted.   The final pattern at this stage of objective contouring is the one which technically accounts for the known elevations, but introduces no features not demanded by them.   The map will be easier to read if certain contours are shown with a heavy line, such as every fifth one, and if some or perhaps all are labeled with their values (Fig. 18.2b).

Computer programs are widely available which perform this entire contouring procedure using linear or other criteria.   Results may be displayed on a terminal and the final map then produced on either a line printer or a graphics plotter.

CONTOURING LICENSE

The structure contour map resulting from this process can still only be an approximation of the true structural picture.   It can always be improved upon by the skill and imagination of the geologist, and, of course, by adding more data points. The latitude in drawing the contours that allows imaginative expression is termed contouring license.   Because the data reveal only a small part of the whole picture, there is always a question of how to represent the surface between these known points, especially when the spacing is large.   The ability to make an intelligent interpretation in such a case is part of becoming a capable geologist.

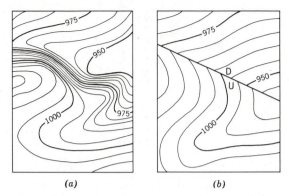

(a)   (b)

FIGURE 18.3 An example of contouring license: (a) objectively contoured map; (b) an interpretation involving a fault (after LeRoy and Low, 1954, p. 23).

When drawing contours, certain features may become evident that suggest several alternatives to the data. Fig. 18.3a shows the result of objective contouring; the steep irregular zone in the central part of the map seems out of place. By a more careful contouring and paying attention to the maintenance of nearly equal contour spacing and gradual changes in the strike direction, the presence of a fault suggests itself (Fig. 18.3b). In the second drawing, the contouring should start on both sides of the suspected area and proceed independently toward it.

The wider the spacing of the control points, the greater the latitude for interpretation. Fig. 18.4a shows a routine projection of contours into a covered area for which no data is available. Fig. 18.4b shows quite a different structure in

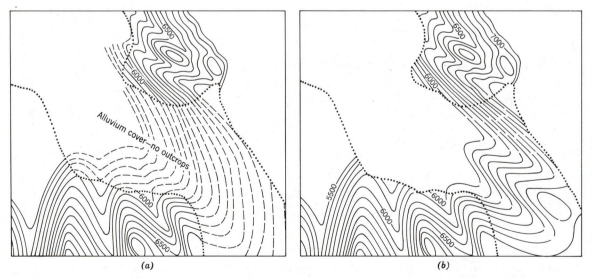

(a)   (b)

FIGURE 18.4 Projection of data into an unknown area (from Low, 1957): (a) routine extrapolation; (b) projection based on rhythmic character of the known folds.

this same area; the same rhythmic folds that occur in the known area are continued into the covered portion. There is no direction evidence for the extra fold, but imagine its importance if it were an oil trap.

STRUCTURE CONTOURS AND FAULTS

Unlike the topographic surface, faults separate structural surfaces into two or more parts; the discontinuity in such surfaces also causes a discontinuity in the contours. Three different situations may arise:

1. Vertical faults appear to offset the contour lines (Fig. 18.5a).

2. Normal faults produce a gap between the lines (Fig. 18.5b).

3. Reverse or thrust faults cause an overlap of contours; the hidden contours are often shown by dashed lines (Fig. 18.5c).

The fault surface itself, especially if it has a curved shape, or is extensively developed, may also be represented by structure contours; they might be differentiated by the use of dashed or dotted lines, as in Fig. 18.5d. Such contoured fault surfaces together with the structural surfaces they disrupt help define the geometry of the fault system, and which in turn may lead to information concerning the type and amount of displacement.

STRUCTURE CONTOURS AND COMPLEX FOLDS

Structure contours are a powerful method of expressing the form of gently warped structural surfaces. More complex forms may also be represented by such contours. There are two special cases. If the fold has a vertical limb, the contours

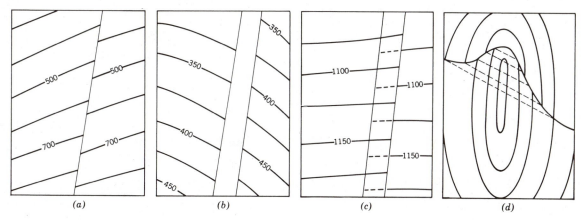

FIGURE 18.5 Structure contours disrupted by faults: (a) vertical fault; (b) normal fault; (c) reverse fault; (d) faulted fold with contours on the fault.

will merge into a line that represents the map trace of that limb. If the fold is overturned, the hidden underside may be indicated with dashed contours.

Structure contours have also be used to study complexely refolded rock (Berthelsen, 1960). As subsurface data in such areas are rare, the necessary information must be collected from surface mapping. Accurate measurements of the locations of the structural surface and its attitude must be made. Clearly, being confined to the earth's surface, only a relatively small part of the chosen horizon may be confidently contoured. Great topographic relief extends this range.

18.3 FORM LINE CONTOURS

In some areas a recognizable key horizon may be lacking. If a structure contour map is to be drawn in such cases it must be constructed from measured attitudes alone. Because no particular horizon is being represented, absolute values can not be given to the contour lines. However, the configuration of the structure can be shown by the pattern and spacing of the contours. Such contour lines drawn without reference to a datum are called form line contours.

For any given contour interval i, the appropriate spacing s between contour lines representing an inclined plane of known dip is given by

$$s = i \cot \delta$$

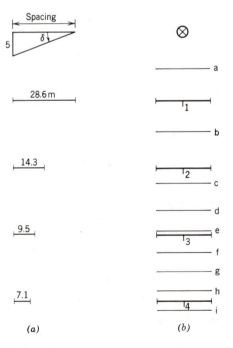

FIGURE 18.6 Spacing of form line contours.

In Fig. 18.6a, the spacing for dip angles from 1-4°, for a 5 m contour interval are shown. It is convenient to plot these spacing to the scale of the map along the edged of a card to speed the actual contouring. Given two or more dips along a line, the problem is to draw contours spaced to quantitatively express the variable inclination of the surface (Fig. 18.6b).

PROCEDURE (Fig. 18.6)

1. Because the form line contours have no absolute value, they may start anywhere. In this simple example, contour a is arbitrarilly located.

2. The next measurement in the down-dip direction is 1°, suggestive that a 1° spacing be tried. It is found that this distance extends equally on both sides of the dip location, and therefore the second contour b is drawn. Note that this assumes that the average dip over this distance is 1°. Since it range from approximately 0.1° to 1.5°, and there is no evidence of an abrupt change, this must be nearly correct.

3. The next dip is 2°. However, if a 2° spacing is tried, it is clear that the average dip over this distance is less than 2°, though it is certainly greater than 1.5°. This requires a spacing between 1° and 2°, and closer to the latter. By trial and error a balance between the spacing and the average inclination is obtained; this can be done fairly accurately by judging the fractional spacing by eye.

4. This balancing process continues in this manner until all the required contours are drawn.

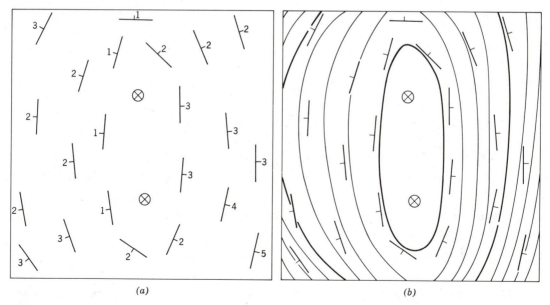

(a)   (b)

FIGURE 18.7 Form line contours: (a) map with attitudes; (b) final map with contours.

In contouring the attitude data on a map (Fig. 18.7a), proceed in much the same manner. The intial contour is drawn with an arbitrary location, but as closely parallel to the measured strike lines as possible. It is often found convenient to start by defining the highest part of the structure in this way. Subsequent contours are then drawn with a spacing which is appropriate to the local dip angles. The final map will then appear as in Fig. 18.7b.

18.4 ISOPACH AND ISOCHORE MAPS

A closely related type of map is based not on the elevations on a certain horizon, but on the thickness of the stratigraphic unit. Such a map is called an isopach map, and the contours are lines of equal thickness. Isopach maps are very important in studying the regional variation in the thickness of stratigraphic units, as an aid to understanding the paleogeography at the time of deposition. A slightly different approach is based on the vertical distance between the top and bottom of the strata in question. This measurement is the "depth" from the top to the bottom. Such a map is called an isochore map. The importance of this type of map lies in the fact that it defines more accurately the present condition of the strata, rather than the thickness which is largerly a function of primary conditions. Estimates of the volume of a given unit can be made more easily from an isochore map. Also, given an isochore map of a unit and a structure contour map drawn on its upper surface, a second structure contour map can be constructed for the bottom of the unit by subtraction. If the dip angles are small, the isopach and isochore maps, are for practical purposes, equivalent. Both types of maps are based on the areal distribution of data points, and the procedure for producing them are identical with those described for structure contours.

EXERCISES

1. The map of Fig. X18.1 shows a series of points where the top of a particular formation and its isochore interval (in brackets) are known. Draw three contour maps:
   a. A structure contour on the top of the formation.
   b. An isochore map of the formation.
   c. A structure contour map on the base of the formation.

2. Fig. X18.2 is a map of measured surface attitudes; the area has negligible relief and no mappable horizons. Show the structure with form line contours.

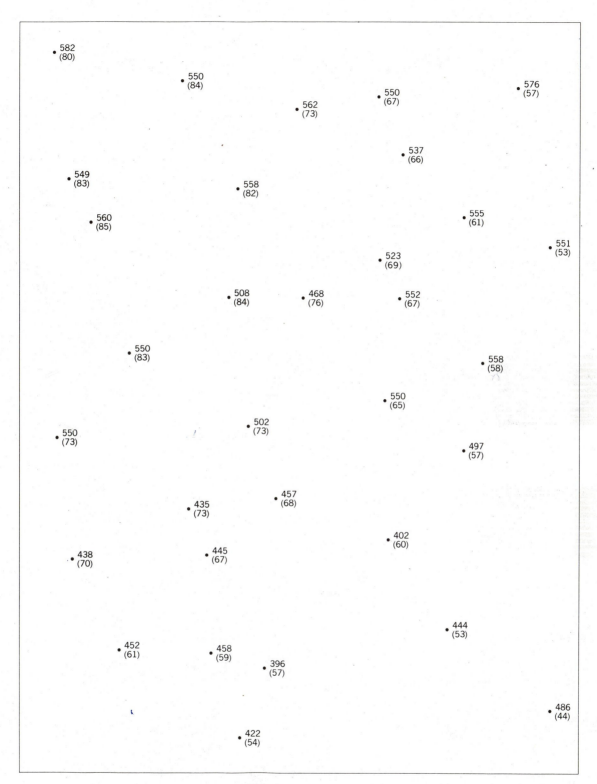

FIGURE X18.1

FIGURE X18.2

19

Maps and Cross Sections

19.1 THE GEOLOGIC MAP

A variety of structural techniques have been described in previous chapters. In the main, the approach has been one of dissecting the geologic map and examining its parts. The map is, however, more than the sum of these geometric parts, and it remains to consider some of the more collective features.

Properly done, the map is an exceedingly important tool in geology. The graphic picture it gives of the location, configuration and orientation of the rock units of an area could be presented in no other way. Essential as the map is, however, it is not without limitations, and if it is to be of maximum use these limitations must be fully understood. The most important point to realize is that geologic maps generally record both observations <u>and</u> interpretation. In part, the element of interpretation is due to a lack of time and complete exposure; it is almost never possible to examine all parts of an area. If a complete map is to be produced, this lack of observed continuity then requires interpolation between observation points, and such interpolation is, to a greater or lesser degree, interpretive.

To distinguish between observation and interpretation several devices may be adopted. Most commonly, special symbols are used to identity several degrees of certainty in the location of lithologic contacts (Fig. 19.1); additional map symbols can be found in Compton (1962, p. 334). The choice of these symbols depends both on the ability to locate the boundaries in the field and on the scale of the map. Although other conventions could be adopted, one common rule of thumb is that a solid line is used if the contact is known and located to within twice the width of the line (Kupfer, 1966). Accordingly, a very thin, carefully drawn pencil line 0.1 mm

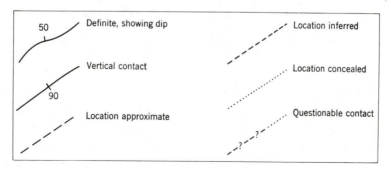

FIGURE 19.1 Map symbols for lithologic contacts.

wide covers 1 m on a 1:10,000 map, and would be appropriate for a contact known within 2 m on the ground, and a thin ink line 0.3 mm is appropriate for a contact known within 6 m. Clearly, the detail that can be shown on a map is scale dependent.

If contacts are represented by less certain lines, it is important that their inferred locations make geometric sense; if a contact line crosses a valley it should obey the Rule of V's according to the inferred attitude. It is quite misleading to show uncertain contacts as if they were all vertical, though many examples of this practice can be found.

Factual and interpretative data may also be distinguished by considering the two aspects more or less separately. An outcrop map is one method of presenting field observations in a more objective way (Fig. 19.2a). Another way of conveying the essential information of an outcrop map, but without actually drawing in the boundaries of the exposed rock masses, is to show abundant attitude symbols, which then serve two functions: to record the measured attitude, and to mark the locality where the attitude can be measured.

However, even an outcrop map or its equivalent can never be entirely objective, for several reasons. What constitutes an exposure of rock is itself subject to some interpretation. For purposes of mapping, a thin rocky soil at the top of a low hill might be considered to be an outcrop by a worker in a poorly exposed area. In contrast, a geologist working in mountainous terrain would probably not give such exposures a second glance. Differences such as these will certainly affect, and may even control the accuracy and completeness of the mapping.

Even with these limitations, it is, of course, important to strive for as high a level of objectivity as possible, and to discuss the problems and difficulties involved in this quest in the text which accompanies the map.

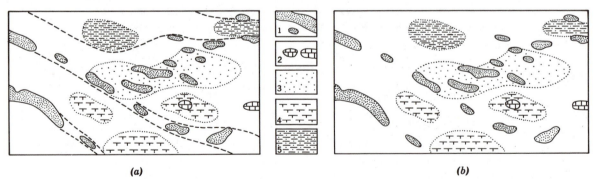

FIGURE 19.2 Hypothetical maps; lithologies: 1 = sandstone, 2 = limestone, 3 = sandy soil, 4 = limy soil, 5 = clayey soil: (a) outcrop and soil map showing facies interpretation (after Kupfer, 1966); (b) interpretation based on recognition of melange; the blocks of sandstone and limestone are shown by outcrops and soils, the clay soil and covered areas are underlain by melange matrix (after Hsu, 1968).

There is another and much more fundamental reason why geologic maps are inevitably interpretive. Even the simplest rock mass is extremely complex, and a complete physical and chemical description of a single outcrop might take years, and questions concerning the origin of the rock would almost certainly remain. Clearly, such detailed studies are rarely feasible. The question then arises—what observations are to be made and recorded? The process of deciding what is important is guided in at least two ways. First, observations are made which have proven in the past to give results. Routine descriptions of attitude, lithology, visible structures and so forth are an important preliminary stage; some check lists have been published to facilitate this type of field description. However, the creative part of field study involves asking critical questions and then attempting to find the answers. These questions are formulated on the basis of knowledge, intuition and imagination. In this search for understanding, the older, often well-established approaches may actually be a barrier which must be broken through if progress is to be made. For example, the interpretive aspect of the map of Fig. 19.2a is based on an application of the so-called laws of superposition, original horizontality, original continuity and faunal assemblage (see Gilluly and others, 1968, p. 92, 103). There are, however, rock bodies of largely sedimentary materials which do not obey these laws: a melange is an example (Hsu, 1968). French for mixture, the term melange is applied to a mappable body of deformed rocks consisting of a pervasively sheared, fine-grained, commonly pelitic matrix with inclusions of both native and exotic tectonic fragments, blocks or slabs which may be as much as several kilometres long (Dennis, 1967, p. 107). Fig. 19.2b is an interpretive map based on the recognition of a melange.

Furthermore, the identification of even well-exposed rock is not always so straight forward that all geologist agree. And, as progress is made, concepts change. The most dramatic way of illustrating these changes is to compare two maps of the same area made at different times. One of the most startling examples, given by Harrison (1958, p. 228), involves a part of the Canadian Shield. An earlier map was made at a time when "granites" were thought be entirely magmatic in origin. Later, a map was produced after it was realized that metamorphism and metasomatism could produces many of these same rocks. The result is that there is little in common between the two maps. This is an extreme example, but most geologic maps still reflect, to a greater or lesser degree, the prejudices of their authors and their times.

As with many things, progress in mapping is an evolutionary process. Each step along the way is, at best, an approximation. These steps, because they are incomplete, necessarily involve some interpretation on the part of the investigator. Just as in the making of a geologic map, so too does the use of the map as an aid to understanding the structure and history of an area involve several stages of development. The first step does not constitute making structural interpretations, but is rather a repetition of the experience and thinking of the original observer. This step is indispensible in gaining a complete understanding of both the map and the area it represents, and the facility to do this can only be achieved with practice. Two attitude toward maps greatly increases their usefulness:

1. Regard any geologic map as a progress report. Improvement can always be made by further work based on the original mapping, either by the study of new exposures, or a more detailed study using new concepts and techniques.

2. Develop a critical outlook toward the lines and symbols on the map. By refusing to accept them completely, especially those that are clearly interpretive, and by adopting a questioning attitude toward the nature of the various map units and structures, new questions may arise that can be answered directly from the map, or from a visit to the area.

19.2 OTHER TYPES OF MAPS

Lithologic map units, and even different structural elements are often shown in color on geologic maps, in combination with the normal symbols printed in black. However, a carefully prepared black and white structural map is often superior to a colored one. On such maps the lithologic units should not be represented by purely geometric patterns, such a parallel rulings and other such patterns. Such patterns fail to express the variously curved lines of strike of the deformed rocks. It is both easier and conveys considerably more information to draw the lines of strike freehand. Further, certain features can be depicted in this way which would be most difficult otherwise. For example, the transition between directionless and foliated rocks can be expressed by a parallel change in the map pattern. The two contrasting maps of Fig. 19.3 illustrates the value of this approach.

In addition to surface geologic maps, there are a number of other types which may be constructed. Maps may, of course, be drawn wherever rock are exposed, as in a mine. A structure contour map is a type of geologic map. Similarly, an isopach map is a geologic map, with the zero contour being the undergound equivalent of a lithologic contact. An isopach map is also a picture of the structure of the lower boundary of a formation at the time the upper boundary was horizontal. A paleogeologic map portrays the distribution of rock units immediately below the

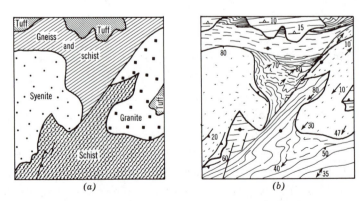

FIGURE 19.3 Two black and white structures maps of the same area (from Balk, 1937).

surface of an unconformity. A worm's eye map is a picture of the unconformity as seen from below; Levorsen (1960) gives a number of examples. Palinspastic map restore the rock units to their relative positions before structural displacement. Although difficult to draw, such map are important because they introduce stages of historical development into the description of the geology of an area.

19.3 GEOLOGIC HISTORY

After describing the geometry of a rock mass, the next step is to work out the time sequence by which that geometry developed. This concern for history includes both the local chronology, and the dating of these event in terms of the geologic time scale. The dating is largely a matter of paleontology and radiometric measurements. The local sequence of events, however, can be worked out without reference to either the geologic time scale or to absolute time.

There is a geologic feature, not previously discussed, which is of great assistence in dating structures and bracketing periods of deformation. An unconformity is a surface of erosion or nondeposition that separates younger from older rocks. There are several important types (Dennis, 1967, p. 159). An angular unconformity is characterized by an angular discordance between the two sets of strata (Fig. 19.4a). In contrast, a parallel unconformity (sometimes called a disconformity) is marked by an evident break between the two parallel strata (Fig. 19.4b). A nondepositional unconformity is a surface of nondeposition; physical evidence of this surface may not be evident, and paleontologic evidence may be needed to demonstrate the time gap (Fig. 19.4c). A heterolithic unconformity (also called a nonconformity) describes the situation where the older rock are non-stratified (Fig. 19.4d).

In determining the local chronology it must be kept in mind that several events may have been synchronous; for example, deposition may occur during folding and faulting. A further complication is that a given structure may be the result of several episodes of movement. Nevertheless, though it may be quite involved, the sequence can be worked out using rather simple geometrical relationships. The following criteria are self-evident:

1. Folds are younger than the folded rocks.

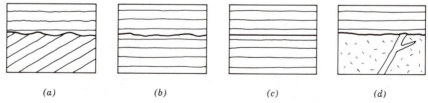

(a)   *(b)*   *(c)*   *(d)*

FIGURE 19.4 Important types of unconformites: (a) angular unconformity; (b) parallel unconformity; (c) nondepositional unconformity; (d) heterolithic unconformity.

2. Faults are younger than the rocks they cut.

3. Metamorphism is younger than the rock it effects.

4. The erosion represented by an unconformity is younger than the overlying rocks and older than the overlying ones. This is strictly true
   only for a small area; erosion and deposition at widely separated local-
   ities may be synchronous.

5. Intrusive igneous rock are younger than the host rocks. This is especially clear where they are in cross-cutting relationships. A similar
   rule holds for other types of intrusions such as salt domes and
   sandstone dikes, with the qualification that the act of intrusion is
   younger though the materials may be older or younger.

An elementary, hypothetical map and the mimimum number of events in chrono-
logical sequence derived from it illustrates this approach (Fig. 19.5).

19.4 THE STRUCTURE SECTION

One of the problems in reading a geologic map is to perceive the structures portrayed on its two-dimensional surface in their proper three-dimensional setting.
Several techniques for doing this, especially the powerful down-structure method of
viewing maps, have been described in earlier chapters. Vertical structure sections,
though they have their limitations, are also useful in helping to work out and
depict the structural relationships at depth, particularly when the structures are

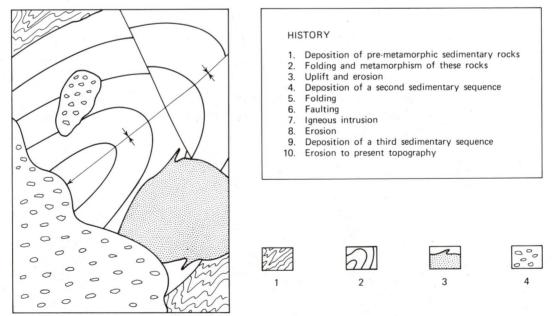

HISTORY

1. Deposition of pre-metamorphic sedimentary rocks
2. Folding and metamorphism of these rocks
3. Uplift and erosion
4. Deposition of a second sedimentary sequence
5. Folding
6. Faulting
7. Igneous intrusion
8. Erosion
9. Deposition of a third sedimentary sequence
10. Erosion to present topography

1   2   3   4

FIGURE 19.5 History from a geologic map.

diverse and no single down-structure direction exists. The line of the section is chosen so as to reveal particular geological relationships. The section is oriented so that the right-hand end is its more easterly end or is oriented due north (Compton, 1962, p. 45). Once the location and orientation of the section are fixed, the technique for contructing a vertical section is straight forward, and, in general, consists of two parts:

1. The topographic profile along the chosen line of the section.

2. Structural data, such as contacts, attitudes, and so forth, which are added to the line representing the topographic surface and then extrapolated into the underground.

TOPOGRAPHIC PROFILE

The edge of a piece of paper is laid the full length of the chosen section line (Fig. 19.6). Points of intersection of the topographic contours and the section are marked along this edge. Other features, such as the crests of hills or the locations of streams should also be marked, even though a contour line is not

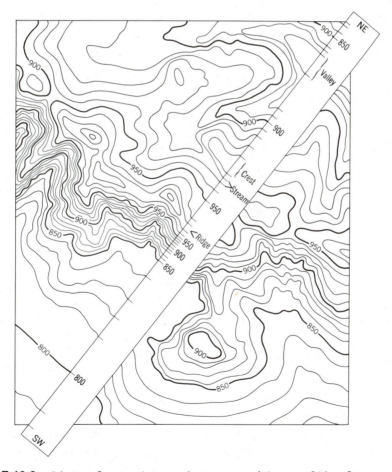

FIGURE 19.6 Line of section and topographic profile from a map.

present. The elevations of the contours must also be indicated; every contour may be marked, especially if they are widely spaced, or if closely spaced, only those which mark change in slope directions may be used.

A series of elevations lines are then drawn on a second sheet of paper with a spacing equal to the contour interval and plotted at the same scale as the map. The topographic points along the section line are then transferred from the edge of the marked paper, which now represents the line of section, by projecting the contour marks to the corresponding elevation lines, and the series of points so located is joined with a line representing the topography. If the spacing of the contours is wide, the map may have to be consulted to assist in sketching in topographic details (Fig. 19.7).

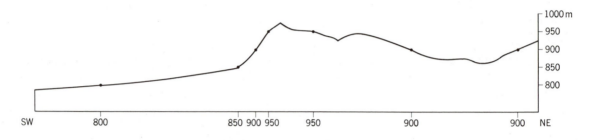

FIGURE 19.7 Topographic profile along the line of section.

In constructing the topographic profile, it is an easy matter to exaggerate the vertical dimension by enlarging the vertical scale while keeping the horizontal scale the same. The effect of such an alteration of the vertical scale produces profound geometrical changes, some of which are examined in the next section. Generally, only unexaggerated sections should be constructed for serious structural work, because only with them are the true geometric relationships preserved. If more space is required to plot a wealth of structural data, the whole section should be enlarged uniformly. The effects of vertical exaggeration are detailed in a later section.

STRUCTURE SECTION

The second step is to add the structural data to the constructed topographic profile. The various contacts and attitude points can be marked at the same time as the topographic elevations; if there are abundant details which must be transferred it may be less confusing to make two separate plots.

Map data on either side of the section line may be projected to the section; the allowable distance of projection depends on the constancy of the trend of structural features normal to the section line. Where the structures plunge, the map patterns may be projected to the plane of the vertical section, essentially the same constructed used for fold profiles. This may also be used to show the structure which existed above the section be have been removed by erosion. When this is complete, all the surface information will have been utilized.

If subsurface information is available, it should then be added. This includes direct geologic data on lithologies and contacts obtained from drill holes, and indirect data from geophysical surveys, such as prominant reflection surfaces.

In extrapolating the surface data downward into the underground, several different approaches may be used. In folded sedimentary rocks, and making some reasonable assumptions the form of the folds may be reconstructed with the Busk tangent-arc method (Fig. 12.11). Even if the plane of the section is not perpendicular to the axes of folds, up- and down-plunge projections may still be used as an aid in depicting the structure of folds.

Lithologic boundaries with no regular features can only be projected downward using the surface attitude; this is only a first approximation and will, at best, be valid only for relatively shallow depths. One example of such an unpredictable boundary is the contact of a discordant intrusive igneous body. Alternatively, such contacts might be shown schematically.

After taking these rather purely geometrical steps, the making of further predictions will depends on a thorough understanding of the various processes of folding, thrusting and so forth, and this can only come with experience. As on the geologic map, the various lines of the structure section should indicate the degree of certainty in location. Questionable areas should be so indicated or even left blank. The predictions will not everywhere have the same confidence at the same depth. It follows that the lower limit of the structural representation should have an irregular boundary. For example, it might be possible to portray a thousand metres of uniformly dipping sedimentary rocks with reasonable accuracy; on the other hand, the nature of the rocks and structures under a thin sheet of alluvium might be completely unknown.

If structure sections are prepared in black and white, lithologic symbols can be used to indicate the different rock units in much the same manner as with black and white structure maps. A variety of symbols are used, and a few of these are shown in Fig. 19.8; Compton (1962, p. 334f) gives many more. The meaning of any such symbols used must, of course, be identified on the section, either by labeling the units directly or by including a list in the legend.

In its final form, the section whould be labeled with geographical coordinates because knowning the orientation is as important as its location. The line of the section should also appear on the accompanying map. Prominant topographic features should be labeled to assist in orienting the reader. The scale, especially if different than the map, is also important.

19.5 OTHER TYPES OF SECTIONS

A number of variations are possible. Composite sections can be drawn by projecting data to the section plane from some distance, or by combining in one section
several different lines that meet at angles. This is generally done to show greater
diversity than would be possible along a single, straight line of section.

One way of suggesting three dimensions is to use multiple sections. Two grouping are common, but any number of combinations are possible. A coulisse diagram is
a group of parallel sections drawn and arranged serially to take advantage of some
special point of view, such as along the strike of a fault, or in the direction of a
fold axis. Fence diagrams may be thought of as two intersecting coulisse diagrams
giving the appearance of an egg crate.

Time, rather than geographic location, may be the basis for a series of structure sections. As with palinspastic maps, increments of deformation are subtracted
from the observed structural geometery, and thus progressively earlier stages in the
historical development are illustrated. Fig. 19.8 is an abridged version of a
famous group of such sections; an examination of the originals and the maps on which
they are based is well worth the effort.

19.6 VERTICAL EXAGGERATION

It is a very common practice to draw cross sections with the vertical scale
enlarged relative to the horizontal scale; that is, to stretch the section vertically while leaving the horizontal dimension unaltered (see Suter, 1947). This practice is especially common in stratigraphic or geomorphic sections where more space
is needed to plot vertical details or to accentuate certain features which would
otherwise be obscure. The result is known as a vertically exaggerated section, and
the degree of the stretch is defined by an exaggeration factor V, where,

$$(19.1) \qquad V = \frac{\text{vertical scale}}{\text{horizontal scale}}$$

For example, if the horizontal scale is 1/50,000 and the vertical scale is 1/10,000,
the vertical exaggeration factor is V = 5.

Largely because of continued exposure to such sections most geologists tend to
think in terms of them, and unexaggerated sections often have an "unnatural" appearance. It is therefore vital to understand the detailed geometric implications of
vertical exaggeration. This will both aid in deciding whether to draw such sections
or not, and in interpreting the exaggerated sections of others.

As a result of this vertical stretching, both the angle of dip and thickness
are systematically distorted in the exaggerated section. A useful way of consider-

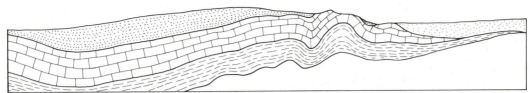
Early folding near margin of sedimentary basin with simultaneous deposition of coarse clastics in the marginal trough.

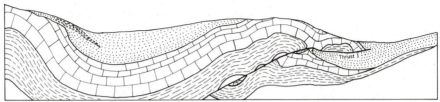
Thrusting and continued folding. Rocks are carried toward the trough. Deposition of coarse clastics continues.

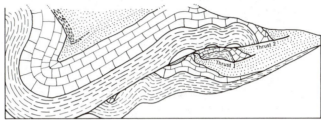
Further folding with formation of new thrust.

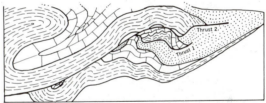
Thrusts are folded, as main syncline becomes recumbent.

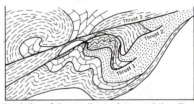
Involution of the syncline and renewed thrusting.

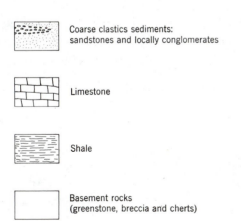

Coarse clastics sediments:
sandstones and locally conglomerates

Limestone

Shale

Basement rocks
(greenstone, breccia and cherts)

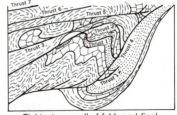
Tightening up all of folds and final
imbricate thrusting.

FIGURE 19.8 Diagramatic sections showing progressive development of complex folds and thrusts (after Ferguson and Muller, 1949).

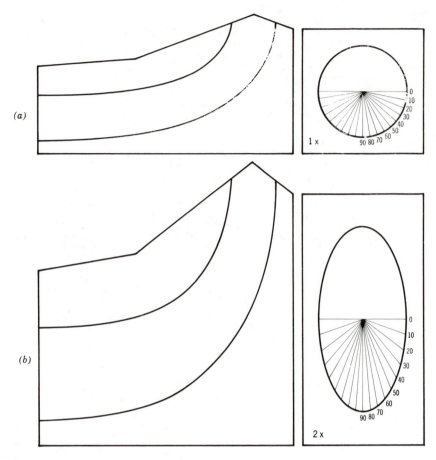

FIGURE 19.9 Natural and vertically exaggerated structure sections (in part after Wentworth, 1930).

ing these changes is to consider the vertical exaggeration as an artifically intro-duced strain. For example, vertical exaggeration can be described by a strain ellipse (see Fig. 19.9), and this emphasizes the profound geometric changes which accompanies such exaggeration.

In exaggerating a section, horizontal dimensions are unchanged and vertical dimensions are multiplied by the exaggeration factor. In Fig. 19.10, the distance w is unchanged while the distance d becomes Vd. From this geometry, we obtain two relationships.

$$w = d/\tan \delta \qquad \text{and} \qquad w = Vd/\tan\delta'$$

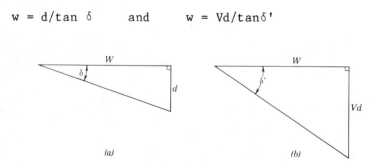

FIGURE 19.10 Effect of vertical exaggeration on dip angle: (a) geometry before exaggeration; (b) after exaggeration.

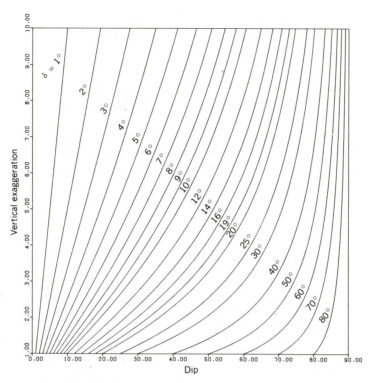

FIGURE 19.11   Graph of angle of dip (or slope) with vertical exaggeration.

Equating these two expressions and rearranging we then have

(19.2) $\tan \delta' = V \tan \delta$

where δ and δ' are, respectively, the original and exaggerated dips.   In Fig. 19.11 this equation is graphically represented for selected values of δ over a range of exaggeration factors V = 2–10.

An examination of this graph makes clear the effect of vertical exaggeration on the dip (or slope) angle.   In general, all angles are steepened, but small angles are effected relatively more, with the result that small differences in inclination are accentuated.   It is this property which is of advantage in depicting subtle variations in slope in the presentation of geomorphic information.   On the other hand, differences between steep dips are diminished.   For example, at V = 10 planes dipping at angles of 30° and 60° apppear with inclinations of 80° and 87°.   Thus the important distinction between normal and thrust faults is lost, and this is a serious disadvantage.

The effect of exaggeration on thickness is also of interest.   The basic situation is shown in Fig. 19.12.   From this geometry we obtain expressions involving the original thickness t and the exaggerated thickness t'.

$$w = t/\sin \delta \qquad \text{and} \qquad w = t'/\tan \delta'$$

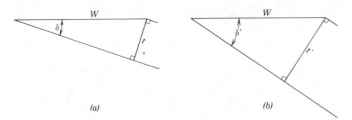

FIGURE 19.12 Effect of vertical exaggeration on thickness: (a) geometry before exaggeration; (b) after exaggeration.

Again equating and rearranging, we have

(19.3) $t' = (\sin \delta'/\sin \delta)t$

Fig. 19.13 gives a graphic representation of this equation for values of V = 2-10. As can be readily seen, t' varies with the dip. The limiting cases occur when the layer is horizontal, and the thickness is multiplied by the exaggeration factor,

$$t'_0 = V \, t$$

and when the bed is vertical and its thickness remains unchanged,

$$t'_{90} = t$$

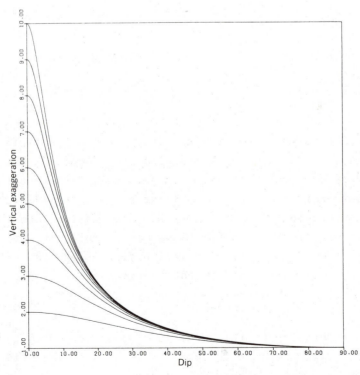

FIGURE 19.13 Graph of exaggerated thickness as a function of dip.

Most of the variations in thickness are confined to a relatively narrow range of shallow dip angles. For example, at V = 10,

$$t_0' = 10 \quad \text{and} \quad t_{25}' = 2.3$$

Thus a layer of uniform thickness but variable inclination within this range will appear to have approximately a four-fold variation in thickness, and it is easy to see that small real variations in thickness would be masked.

From these considerations it should be clear that vertically exaggerated cross section severely distort the form and orientation of geologic structures, thus tending to destroy the very information the structure section seeks to show—the true geometric relationships at depth. Therefore, they should not be used for serious structural work. For those few sitation where a vertical exaggerated section may aid a presentation, the smallest possible exaggeration factor should be used. In addition, there is an important responsibility to keep the reader informed of the degree of exaggeration used. This can be accomplished several ways:

1. Include a bar scale for both the horizontal and vertical dimensions.
2. Quote the actual exaggeration factor.
3. Include a protractor of exaggerated dip angles.
4. Include a natural-scale section in addition to the exaggerated one for easy comparison.

This latter approach is perhaps the most effective one, but it is quite useful to supply all items. Since exaggerated sections are so common, a natural scale section should be clearly labeled <u>no vertical exaggeration</u>.

There is another subtle distortion which appears when sections of regional extent are drawn as if the earth were flat. Sea level is depicted as a horizontal straight line so that the floors of sedimentary basins appear distinctly concave upward; vertical exaggeration further compounds the distortion. If true sections are drawn, it is, of course, sea level which should appear as a curve and the basin floors more nearly a straight line. This difference in basin geometry has important bearing on the mechanical properties of basin fill (Price, 1970, p. 15), and on the mechanics of basin evolution (Dallmus, 1958). Again, construct accurate, true scale sections. Then if there is a genuine reason to distort them do so with caution, and alert the reader of what you have done.

EXERCISE

Using an available geologic map, construct a true scale cross section showing both topography and structure.

20
Block Diagrams

20.1 INTRODUCTION

The block diagram is one of the best ways of presenting a wealth of geological information in a compact, three-dimensional form. Almost at a glance, the relationships between the structural data plotted on the visible surfaces of the block can be integrated into a complete spatial picture. The construction of such a diagram entails drawing a scaled block, possibly adding topography to the upper surface, and representing the geologic structures on its top and sides, and perhaps within the block.

Such a scaled block may be constructed by the methods of descriptive geometry, but the procedure is fairly involved and time consuming. With the aid of a special net, a scaled cube may be constructed directly and in any orientation, and from this unit cube a block of any proportions may be obtained. To illustrate the use of this net, we will construct a block diagram with the geologic map of Fig. 20.1 on its top surface, together with the appropriate views of the other visible sides.

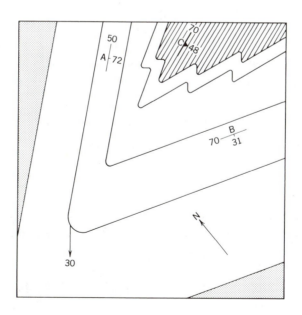

FIGURE 20.1 Geologic map for the top of a block diagram (after Lisle, 1980).

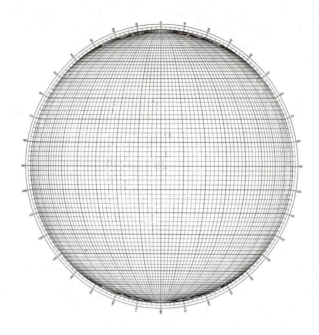

FIGURE 20.2 Orthographic net (from Wright, 1911).

20.2 ORTHOGRAPHIC NET

In both its form and use the orthographic net is closely related to the stereonet (Fig. 20.2). On its surface there is a grid composed of two families of curves.
The set of curves related to the great circles of the stereonet are semi-ellipses,
and the set related to the small circles are a series straight lines. Aside from
these differences, the method of plotting points and planes, and performing rotations is the same as on the stereonet.

With this net, a scaled unit cube in any desired orientation can be constructed
easily, as we will show in the next section. It should be noted that because of the
close packing of the grid lines near the circumference of this net, it is often
easier to count off complementary angles outward from the center than to count the
angle itself inward from the primitive.

20.3 UNIT CUBE

The construction of the unit cube may be accomplished in two equivalent ways:
by revolving the cube into any desired orientation, or by a direct plot. Because it
aids visualization, the first method will be used to introduce the use of the orthographic net.

PROBLEM

Draw a cube using the map area of Fig. 20.1 as its top surface so that the
line of sight is in a direction which plunges due north at 30°.

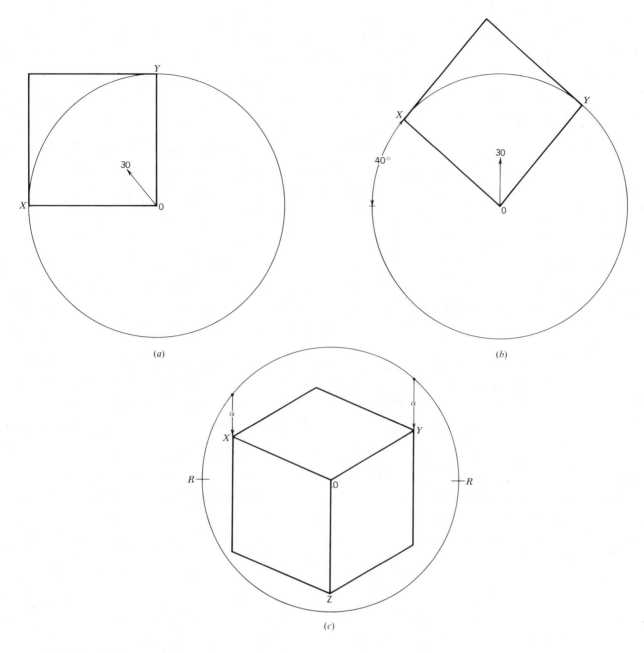

FIGURE 20.3 A unit cube on the orthographic net by rotation (after McIntyre and Weiss, 1956).

CONSTRUCTION BY ROTATION

1. On an overlay sheet with north marked, draw a square whose sides are equal to the radius of the orthographic net, located so that the front corner is at the center of the net (Fig. 20.3a).

2. First rotate this square to that the proposed line of sight trends due north. This takes a clockwise rotation of 40° (Fig. 20.3b).

3. Next rotate the block so that the plunging line of sight is at the center of the net. This manoeuver is performed in exactly the same way as it on the stereonet. First rotate the overlay 90° so the east-west rotational axis is parallel to the straight grid lines. To perform the rotation, all points move south along the straight "small circles" a distance of $\alpha = 60°$ (Fig. 20.3c).

4. The three lines x, y and z radiating from the center point 0 represent the solid angle made by the front three edges of the cube, and they appear in their correct foreshortened proportions. The cube is then completed by drawing in the other edges.

DIRECT PLOT (Fig. 20.4)

1. In its final position, the plane of the top of the block dips due north at 60°. This plane is represented by the "great circle" arc found by counting 30° outward from the center, which is then traced in.

2. To locate point X count off 50° anticlockwise from the point representing the true dip line along the arc. Point Y is similarly found by counting off 40° clockwise from this same point. As a check, the angular distance along the arc from X to Y must be 90°.

3. To locate point Z count off $\alpha = 60°$ from 0 southward along the radius of the net. A comparison with the results derived by rotation will show that they are the same. With these three points the cube can then be completed.

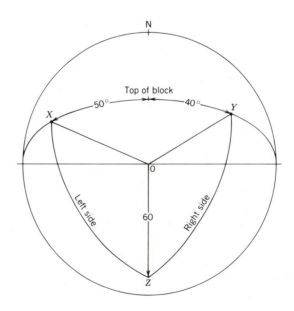

FIGURE 20.4 The unit cube by direct plot.

4. The arcs XZ and YZ representing the right and left vertical sides can be added to the diagram.

At this point, a simple proportional change in the lengths of the three lines representing the front edges of the cube can be made if the map has a rectangular shape.

20.4 GEOLOGIC STRUCTURE

By constructing a grid on the geologic map and a equivalent foreshortened grid on the parallelogram representing the top of the cube the geological boundaries are transfered from the map to the top of the cube in much the same manner used in the construction of the fold profile in Fig. 10.10, except here the spacing of both sets of grid lines must be adjusted.

The next step is to determine the orientation of the traces of the various planar structures on the top and sides of the cube, and, if desired, the orientation of lines within the block. The basic approach is to plot the structural data as points on the net and then rotate these points into the cube coordinates.

CONSTRUCTION

1. Plot the poles of the bedding at points A and B on the limbs of the fold, the pole of the axial plane cleavage at C, and the plunging hinge line F exactly in the same way as they would be plotted on the stereo-net.

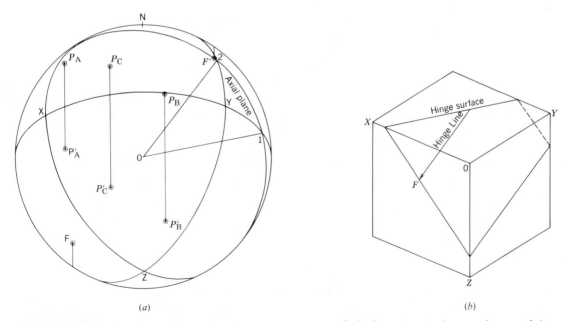

(a)   (b)

FIGURE 20.5   Traces of geologic structures: (a) found on the orthographic net; (b) transfered to the block (after Lisle, 1980).

2. Rotate these four points in the same direction and amount as the points X, Y and Z were rotated. Note that the point F moves to the primitive, reappears 180° opposite and continues its rotation.

3. With the new positions of the poles A, B and C, draw in the three corresponding great circular arcs. Only one of these planes is shown in Fig. 20.5a; it is the arc representing the axial plane cleavage at C.

4. Draw lines from the center O to the points of intersections the structural planes and the three faces of the cube. Again, only one of these is shown on the figure giving the orientation of the traces of the cleavage at C with the top (point 1) and the front right side (point 2). With these, the traces of planes parallel to C can be accurately drawn on the top and right sides. Usually the trace can be continued to the third side without further information from the net.

5. The orientation of the hinge line within the block is found by a line from O to F', and the hinge line can then drawn in from a hinge point on the map.

The completed block diagram, with the structure shown on all visible faces, as well as within the block, is shown in Fig. 20.6.

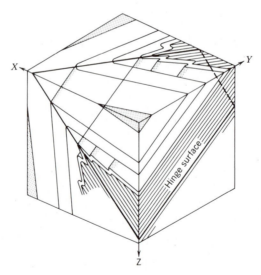

FIGURE 20.6 Completed block diagram (after Lisle, 1980).

20.5 TOPOGRAPHY

If the area has even a small amount of relief, the three-dimensional aspect of the block may be enhanced by adding topography to the diagram. A number of systems for doing this have been devised to adjust map topography systematically to the proportions and scales of the block diagram. The easiest approach method uses a rela-

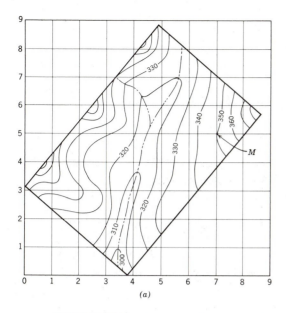

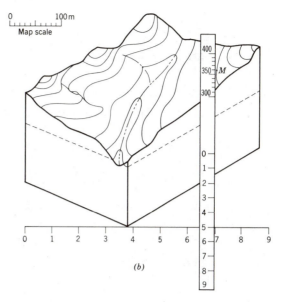

FIGURE 20.7 Topography on a block diagram: (a) topographic map with super-imposed grid; (b) method of transferring topographic detail to the block (after Goguel, 1962, p. 119).

tively simple graphic method. Given a topographic map, or any part of it, the problem is to show the surface forms on a block in any desired orientation.

CONSTRUCTION

1. Draw a square grid on the map with the ordinate in the direction of the proposed line of sight. The grid spacing should be dictated by the amount of detail to be transfered to the block (Fig. 20.7a).

2. Draw a unit cube in the require orientation. Position this cube below the map so that its front corner lies exactly along the line of sight to the corresponding front corner of the map. The cube can then be multiplied to the dimensions of the map by drawing other lines parallel to the line-of-sight line to the outside corners of the map (Fig. 20.7b).

3. The depth of the block depends on the depth of the structure to be shown. In the example, the 300 m level is placed at the top of the unit cube.

4. Along the base of the block reproduce the abcissa scale of the grid and locate it in the correct position with respect to the map grid.

5. From an oblique view, the front-to-back grid spacing is foreshortened; this contracted grid scale is related to the map grid scale by a factor

of sin p, where p is the vertical angle which the line of sight makes
with the map plane. This corrected scale is plotted along the edge of
a strip of paper.

6. The map scale, as measured vertically, is similarly reduced by a factor
   of cos β. This new scale is added to the strip.

7. The position on the block of a series of point, topographic or otherwise, are then located. For example, point M is at the corner of the
   horizontal grid number 5 and the vertical grid number 7. On the block,
   5 on the corrected grid scale is moved to 7 on the lower abscissa
   scale, keeping the measuring scale vertical. The elevation of point M
   is 350 m, and this height is located on the corrected vertical scale,
   and the point is then plotted. The procedure is continued until enough
   points have been located.

The topography on the upper surface of the block may be shown with foreshortened contours (Fig. 20.7b), or by using a simple shading technique (Fig. 20.8).

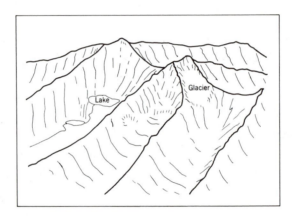

FIGURE 20.8 Topography rendered by simple shading (after Lobeck, 1958).

20.6 MODIFIED BLOCKS

In order to show certain features better, a number of modifications may be
used. The block may be cut into pieces and the pieces separated to expose internal
parts of the block. Similar cuts may be made to remove corners or variously shaped
slices to show other structural details to advantage.

Another way of emphasizing certain features is to dissect the block along a
certain structural surface. For example, a complexely folded and faulted stratigraphic horizon could be shown by artificially removing all the overlying material.
An excellent example is given by Goguel (1962, p. 134).

Especially in mountainous areas, the presence of topographic relief may hinder rather than aid the presentation, and it may be desirable to eliminate the complications of the outcrop pattern caused by it. This can be accomplished by projecting the structures to a horizontal plane. Any plane can be used, but it is often convenient to use sea level because the topographic contours also use this as datum.

PROCEDURE (Fig. 20.9)

1. On a transparent overlay sheet, rule a series of closely spaced lines parallel to the trends of the fold axes on the geologic map.

2. Select a series of points on the contact of a lithologic marker within a fold. These point should be spaced closely enough to allow the structure to be accurately sketched in.

3. Each point is projected to sea level (or other chosen level) by moving it parallel to the trend lines in the direction of the plunge through a distance equal to h/tan p, where h is the elevation of the point and p is the plunge.

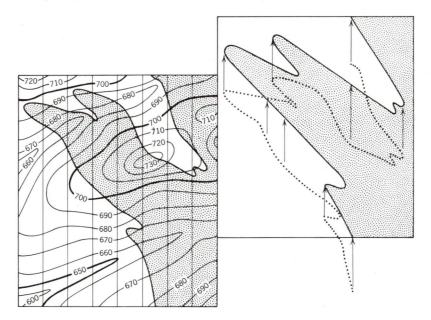

FIGURE 20.9 Projection of structural data to a horizontal plane surface (from Turner and Weiss, 1963, p. 164).

Fig. 20.10 is a famous block diagram with an artifically planar upper surface showing the plunging structures of the Pennine Nappes in the Alps. In constructing this diagram, the axial continuity of the cylindrical folds was used a guide in tracing out the structures on both the top and front of the block.

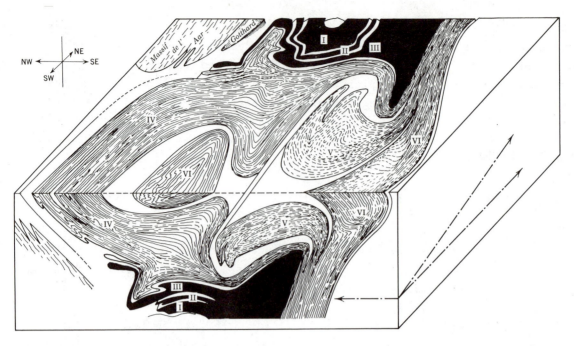

FIGURE 20.10 Block diagram showing plunging structures in the Alps (from Argand, 1911).

EXERCISE

Using an available geologic map, construct a scaled block diagram.

Appendix A
Elements of Descriptive Geometry

A.1 INTRODUCTION

Descriptive geometry is the art of accurately drawing objects, and of graphical-ly solving space problems. It is based on the idea of depicting objects in three dimensions by means of projections. Everyday examples of projections are shadows and photographs; both are the result of projecting various parts of an object to a plane by rays of light. These rays are projectors which join points on the object with the corresponding points on the image plane.

A.2 ORTHOGRAPHIC PROJECTION

The simplest type of projection, and the one used most in engineering as well as for many purposes in geology, is the orthographic projection. Orthographic means "drawn at right angles", and refers to parallel projectors that are perpendicular to an image plane. By projecting an object to an image plane, a view of that object is obtained. Alternatively, a view may be thought of as an actual picture of the object obtained along a line of sight perpendicular to the corresponding image plane. The most important property of this projection is that images of objects appear in their true shape.

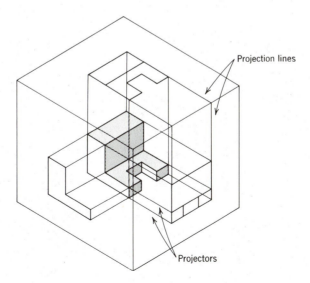

FIGURE A.1   Orthographic projection (after Warner and McNeary, 1959).

353

As in the orthographic projection of Fig. A.1, it is usually convenient to refer to three separate image planes: a horizontal plan or <u>map</u> view, and two vertical planes at right angles, often called the <u>front</u> and <u>side</u> views. Together, these constitute the principal views. Other image planes, giving <u>auxiliary</u> views, may be oriented at any angle.

In Fig. A.1 a three-dimensional object is projected orthographically to the sides of the cube, and the whole then projected to the plane of the page. In order to show the true shape of the object, it is necessary to obtain a direct or normal view of an image plane. Such a view could be obtained by rotating the cube so that our line of sight is perpendicular to a particular face of interest. Or we could obtain direct views of all image planes simultaneously by unfolding the cube, just as one unfolds a cardboard box, so that all the faces lie in a common plane (Fig. A.2). During this unfolding process, the edges which act as hinges are called <u>folding lines</u> (abbreviated FL). Any edge may act as a folding line, though it is often simpler to use lines which lie in the horizontal plane. If auxiliary planes are used it may be necessary to unfold about other lines, possibly through angles other than 90°.

In practice, of course, the three-dimensional box is never actually constructed nor is the unfolding process so literally followed. These steps are by-passed and the required orthographic views are constructed directly. Fig. A.3 shows the basic method of construction and the use of lines connecting corresponding points on different image planes, called <u>projection lines</u>. Such lines joining points on adjacent views are perpendicular to the common edges or folding lines. However, if the image planes are not adjacent, for example, the Front and Side Views of Fig. A.3, then some method of bridging the gap must be used. Circular arcs drawn from a center at the intersection of two folding lines is one convenient way of accomplishing this;

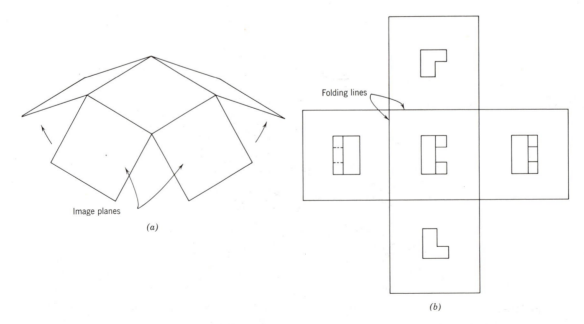

Image planes

(a)

Folding lines

(b)

FIGURE A.2   Unfolding the rectangular box of Fig. A.1.

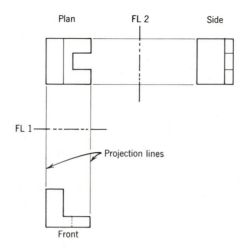

FIGURE A.3 Three principal views using folding lines and projection lines.

another is to scale off the appropriate distance along the two folding lines. As this illustration also makes clear, any additional view can be constructed if two views are given. This is the basis of the orthographic method of solving problems graphically. One simply seeks the particular view, called the normal view, which shows the required lengths or angles, or both, in their true dimensions.

A.3 GRAPHIC SOLUTIONS

Solution of problems involving points, lines and planes in a three-dimensional setting by graphical means can not be absolutely accurate. Limiting factors include the scale and accuracy of the drawing and the skill of the draftsman. With a light touch and a sharp pencil it is possible to draw a line as narrow as about 0.1 mm. The intersection of two perpendicular lines is then a small square 0.1 mm on a side. The maximum error in measuring the distance between two such points is twice this, or 0.2 mm, and the maximum error may be even greater if the intersecting lines are oblique. As the width of the line is independent of the scale, the accuracy of a given drawing depends linearly on the scale, other factors being equal. If the scale is doubled, the fractional error is reduced one half.

Theoretically, a mathematical solution is capable of absolute accuracy. But no solution can be more accurate than the original data on which it is based; therefore a graphical solution can be just as accurate if it is within the limits of the numerical observations.

The choice of method, mathematical or graphical, depends on the requirements of the problem and the nature of the observational data. In geology great accuracy is illusive. Thus graphical solutions produced under normal working conditions usually give satisfactory results for most purposes. In addition, the graphical method has one enormous advantage: the actual construction of the various orthographic views is an extremely important aid in visualizing the problem in three dimensions, and in thinking through the sequence of steps which lead to the correct answer.

If the graphical method is chosen, the accuracy of any drawing may be improved in a number of ways:

1. Enlarge the drawing; the optimum size is just slightly larger than the data requires. This insures requisite accuracy and economy of time.

2. Draw lines as narrow as possible, with a hard, sharp pencil, using light pressure.

3. Locate intersections using angles as close to 90° as possible.

4. Measure angles with a large radius protractor.

5. Avoid cumulative errors. If possible, measure the total length of a line without lifting the scale for intermediate points.

6. Quality drawing instruments help maintain a higher degree of accuracy. It is especially important for the compass to hold its setting.

7. If the drawings are to be worked on over a considerable length of time, dimensionally stable materials should be used. Often a more practical approach is to complete a construction in as short a working time as possible.

8. Mistakes may be minimized by keeping the actual construction simple and compact, and by labelling all the points on the drawing.

The solution of problems involving lengths and angles can be solved with just two fundamental constructions: the determination of the true length of a line segment; and the rotation of a plane figure into any required orientation.

A.4   TRUE LENGTH OF A LINE

To determine the true length of a line segment, a normal view of the plane containing the line must be obtained. Of the infinite number of such planes, chosing the vertical plane parallel to the trace of the line in map view is usually the most direct approach. There are two cases. In the simpler one, the line is oblique to two of the principal views and parallel to the third. This third principal view is then sought for the true length.

PROBLEM

The projection of an inclined line in Map and Side Views, each showing the line in less than true length, are given. Construct the true length of the line and its inclination in the Front View (Fig. A.4a).

APPROACH

If we had the actual cube with the given views on two of its principal faces, then by connecting the end points of the line in each of the views with projection lines to the Front View, the end points in this view would then be established (Fig. A.4b).

CONSTRUCTION (Fig. A.4c)

1. Draw folding line FL1 along the edge common to the Map and Side Views (in the particular case FL1 is perpendicular to the trace of the line in these two views).

2. Draw FL2 parallel to the edge common to the Map and Front Views (FL2 is parallel to the trace of the line in the Map View).

3. Similarly, draw projection lines from the Side to Front Views, first perpendicular to FL2, then using circular arcs with the intersection of the two folding lines as center, and finally perpendicular to FL1.

5. The corresponding intersections of the two pairs of projection lines locate the end points X and Y on the Front View.

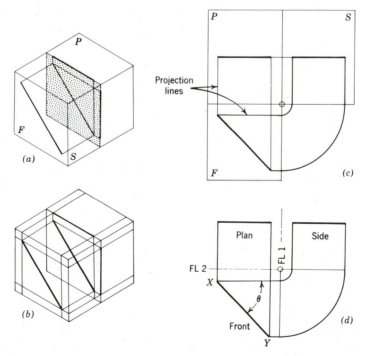

FIGURE A.4 True length of a line parallel to one principal view.

ANSWER

The segment XY is the true length of the line and the angle θ gives its inclination (Fig. A.4d).

In the second case, the line is oblique to all three principal views, and an auxiliary view is required to show the true length of the line segment.

PROBLEM

Given the projection of an inclined line in the Map and Side Views, construct the true length of the line (Fig. A.5a).

APPROACH

A vertical auxiliary view parallel to the trace of the line in Map View must be constructed. Then, as before, projection lines from the two known views to this auxiliary view establish the locations of the end points of the line segment (Fig. A.5b).

CONSTRUCTION

1. Draw FL1 parallel to the edge common to the Map and Auxiliary Views; in this case, FL1 is oblique to the trace of the line in each view.

2. Draw FL2 parallel to the edge common to the Map and Auxiliary Views (FL2 is parallel to the trace of the line in Map View).

3. Construct projection lines from the Map View to the Auxiliary View perpendicular to FL2.

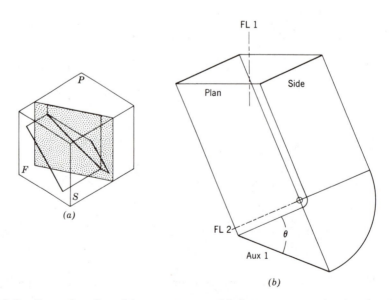

FIGURE A.5 Length of a line not parallel to any principal view.

4. Construct projection lines from the Side View to the Auxiliary View, first parallel to FL1, then using circular arcs with the intersection of FL1 and FL2 as center, then parallel to FL2.

5. The intersections of the projection lines on the Auxiliary View fix the locations of the end points A and B. The length and inclination are then measured in this view.

A.5   NORMAL VIEW OF A PLANE FIGURE

The construction of a plane figure in normal view given two principal views is a bit more involved because the orientation of the figure must first be found. While the method appears complex, in reality it is just the simpler construction of Fig. A.5 repeated several times.

PROBLEM

Determine the true shape of the triangle ABC whose projections in Front and Map Views are given.

APPROACH

Three steps are needed to determine the true shape: find the true length of one side of the triangle; then using this line as a reference direction, the figure is drawn in edge view; and finally, a third view is then constructed perpendicular to this edge view showing all sides in their true length.

CONSTRUCTION (Fig. A.6)

1. Draw three projection lines joining the points A, B and C in the Map and Front Views, and add FL1 perpendicular to these.

2. Arbitrarily chosing AB as the side to be first shown in true length, draw FL2 parallel to the trace of AB in the Map View.

3. Projection lines from the Map View perpendicular to FL2, and the corresponding projection lines from the Front View establish the location of the three points in the first auxiliary view Aux1, in which side AB is shown in true length. Note that for the center of the circular arcs the intersection of the two projection lines AA and BB is used (such centers are indicated by small circles in the figures).

4. Establish FL3 perpendicular to AB on Aux1. Projection lines from the Map View and Aux1 again locate the three corners of the triangle in the

second auxiliary view Aux2. In this view the triangle is in edge view, and the side AB appears as a point.

5. The final view is now drawn using FL4 parallel to the trace of the figure in Aux2. Projection lines from both Aux1 and Aux2 fix the location of the three corners, and the figure is now in a normal view.

ANSWER

In this normal view, all aspects of the triangle may be determined, including the lengths of all sides and their inclinations, and the three angles.

Because any plane figure can be thought of as a collection of triangles, this construction is generally applicable to the problem of determining the true shape of any such figure.

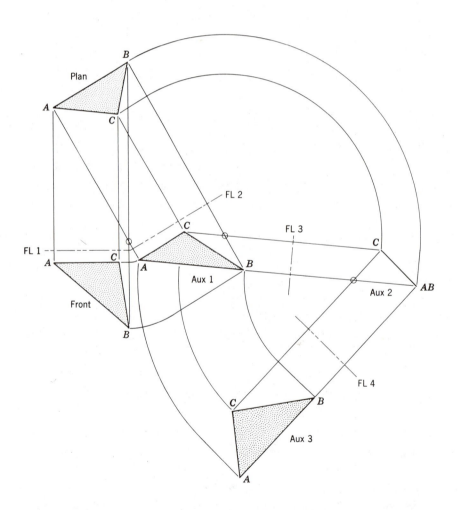

FIGURE A.6 Rotation of a plane figure into a normal view.

Appendix B
Spherical Trigonometry on the Stereonet

B.1 INTRODUCTION

A spherical triangle is a figure on the surface of a sphere bounded by the arcs
of three great circles. It has six parts: three <u>sides</u> and three <u>angles</u>. Each side
is measured by the angle it subtends at the center of the sphere, and each angle by
the dihedral angle between the two planes whose traces are the intersecting great
circular arcs.

If any three parts are known, formulas are available for determining any of the
other three parts. Plane and spherical triangles differ in two important ways: because both sides and angles have angular measures, the size of the triangle is immaterial, and the sum of interior angles is not constant.

Spherical triangles arise in many stereographic constructions, and for problems
involving parts of these triangles, the application of the methods of spherical trigonometry is an attractive way of obtaining a solution quickly and accurately, especially if the equations are programmed on a hand-held calculator.

B.2 RIGHT-SPHERICAL TRIANGLES

If one part of a spherical triangle, either an angle or a side, is 90°, the figure is known as a <u>right-spherical</u> or Napierian triangle, and the solution of any
unknown part is much simplified. Ten special cases cover all possible situations,
and the derivations of the separate formulas are given in full by Higgs and Tunell
(1959). However, long ago Napier worked out two simple rules by which all these
case may be solved. The rules are:

> Sine of middle part =
> 1. product of tangents of adjacent parts, or
> 2. product of cosines of opposite parts.

These rules may be easily remembered by noting that the letter a occurs in both
tangent and adjacent, and the letter o in both cosine and opposite. The middle,
adjacent and opposite parts are most easily identified with the aid of a simple diagram known as Napier's device.

Commonly, the right-spherical triangle contains a 90° angle. In this case, the remaining five parts of the triangle are labeled 1-5 as they are encountered going around the triangle starting with the right angle (Fig. B.1a). Each time a problem is to be solved, this device is sketched and the magnitudes of each of the known numbered parts is entered in the appropriate compartments, starting with the horizontal bar on the right; the values written to the left of the vertical line are the complementary angles of the appropriate parts (Fig. B.1b). The order may be taken in either direction so long as the pattern is consistent in both the triangle and the device; we use an anticlockwise order here.

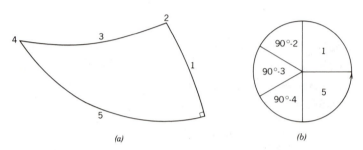

FIGURE B.1 Right spherical triangle: (a) 90° angle; (b) Napier's device.

Less commonly the 90° element may be a side. In this case, the solution of the unknown part is obtained from the corresponding <u>polar triangle</u>, which may be constructed by locating the pole of each of the three sides of the original triangle, and then connecting each pair of these poles with a great circular arc. The properties of this derived triangle are such that each side is the supplement of the corresponding original angle. In practice, the polar triangle need not actually be used. Starting the numbering scheme from the 90° side (Fig. B.2a), the five parts are entered as before, except that the supplementary angles are now used (Fig. B.2b).

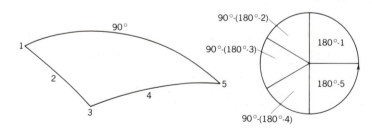

FIGURE B.2 Right spherical triangle: (a) 90° side; (b) Napier's device.

In either case, the two known parts together with any unknown third part will always be arranged in the compartments of the device so that either:

1. three parts are all adjacent, or
2. two parts are opposite a third.

In solving the equations which result from the application of Napier's rules a simplification can frequently be made by using one of the identities:

$$\sin(90°-x) = \cos x \qquad\qquad\qquad (B.1a)$$
$$\cos(90°-x) = \sin x \qquad\qquad\qquad (B.1b)$$
$$\tan(90°-x) = 1/\tan x \qquad\qquad\qquad (B.1c)$$

and

$$\sin(180°-x) = +\sin x \qquad\qquad\qquad (B.2a)$$
$$\cos(180°-x) = -\cos x \qquad\qquad\qquad (B.2b)$$
$$\tan(180°-x) = -\tan x \qquad\qquad\qquad (B.2c)$$

Several worked examples involving right-spherical triangles with 90° angles illustrate the application of this method. Fig. 5.21b shows an example of a triangle with a 90° side. Additional structural problems can be found in Higgs and Tunell (1959) and in Phillips (1971), and a number of related crystallographic problems are described by Phillips (1963).

APPARENT DIP

The attitude of a structural plane is N60°W, 20°S. What is the apparent dip in a S80°W direction?

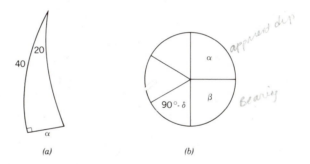

FIGURE B.3   Apparent dip problem: (a) stereogram; (b) Napier's device.

SOLUTION

1. Draw the triangle involving the elements of the problem on a stereonet, and number the parts (Fig. B.3a): Parts 1 = unknown apparent dip α, 4 = dip angle δ = 20°, 5 = bearing β = 40°.

2. Sketch Napier's device for the case of a 90° angle, and enter the values of the corresponding parts (Fig. B.3b). Note that Parts 2 and 3 are not used.

3. Applying Rule 1, $\sin \beta = \tan \alpha \tan(90°-\delta)$. With the identity of Eq. B.1c, and solving for $\tan \alpha$, we have

$$\tan \alpha = \sin \beta \tan \delta \qquad\qquad\qquad (B.3)$$

ANSWER

   With β = 40°, and δ = 20°, the apparent dip α = 13.2°. Note that Eq. B.3
   and Eq. 1.4 are identical.

PLUNGE FROM PITCH

   A plane strike due north and dips 17° to the east. What is the plunge of
   a line whose pitch is 54°S, and in what direction does it occur?

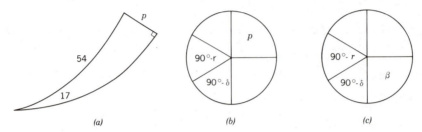

FIGURE B.4   Plunge and pitch: (a) stereogram; (b) first device; (c) second
                 device.

SOLUTION

   1. Draw and label the triangle on the stereonet (Fig. B.4a): Parts 1 =
      unknown plunge p, 3 = pitch r, 4 = dip angle δ, 5 = unknown bearing β.

   2. Sketch the first device and fill in the compartments for Parts 1, 3 and
      4 (Fig. B.4b).

   3. Applying Rule 2, sin p = cos(90°-r)cos(90°-δ) or

$$\sin p = \sin r \sin \delta \tag{B.4}$$

   4. Sketch a second device and fill in compartments for Parts 3, 4 and 5
      (Fig. B.4c).

   5. Applying Rule 1, sin(90°-δ) = tan(90°-r)tan β or

$$\tan \beta = \cos \delta \tan r \tag{B.5}$$

ANSWER

   The plunge angle p = 13.7°. The structural bearing β = 52.3° and the corresponding true bearing is S52°E.

   Triangles without a 90° element will be treated in the next section. However,
it is sometimes possible to reduce problems involving such triangles to the simpler

case of two right-spherical triangles. The problem illustrated here involves conversion of plunge and trend, expressed as azimuth, to direction cosines. To proceed it is necessary to establish a set of coordinate axes, and it is most convenient to choose the right-handed system:

+x = N, +y = E, and +z = down

DIRECTION COSINES

What are the direction cosines of a line whose attitude is 35°/070°.

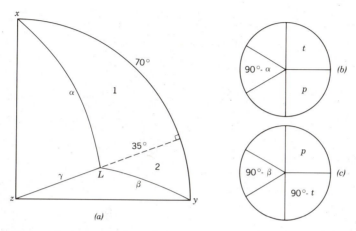

FIGURE B.5 Direction cosines: (a) stereogram; (b) first device; (c) second device.

DERIVATION

1. Plot the point L representing the plunging line, and draw in the three great circular arcs from L to each of the nearest coordinate points, in this case +x, +y and +z. The angles measured along these arcs are the direction angles α, β, and γ (Fig. B.5a).

2. The spherical triangle whose sides are α, β and the primitive does not have a 90° element. However, if the trace of the vertical plane containing the line is added to the diagram two right triangles result.

3. From Triangle 1 and the corresponding device (Fig. B.5b), we obtain

$$\sin(90°-\alpha) = \cos p \cos t$$

or

$$\ell = \cos \alpha = \cos p \cos t \qquad (B.6a)$$

4. Then from Triangle 2, and its corresponding device (Fig. B.5c),

$$\sin(90°-\beta) = \cos p \cos(90°-t)$$

or

$$m = \cos \beta = \cos p \sin t \qquad (B.6b)$$

5. Finally, from simple geometry,

$$\cos \gamma = \cos(90°-p)$$

or

$$n = \cos \gamma = \sin p \qquad (B.6c)$$

ANSWER

The three direction cosines of the given line are

$$\ell = \cos \alpha = 0.28017, \quad \alpha = 73.7°$$
$$m = \cos \beta = 0.76975, \quad \beta = 39.7°$$
$$n = \cos \gamma = 0.57358, \quad \gamma = 55.0°$$

These results may be checked by using the identity $\ell^2 + m^2 + n^2 = 1$. They may also be checked by converting back to plunge and trend using,

$$p = \text{arc } \sin(n) \qquad (B.7a)$$

$$t = \text{arc } \tan(m/\ell) \qquad (B.7b)$$

If the azimuth is used in Eqs. B.6, the signs in all quadrants will be correct, but on conversion back to plunge and trend using Eqs. B.7, most calculators will give the trend angle as a signed angle, not as an azimuth.

B.3 OBLIQUE SPHERICAL TRIANGLES

Sometimes the data are such that the resulting spherical triangle has no 90° element, and it is then termed oblique. In the usual way, angles are labeled A, B and C, and the opposite sides a, b and c. There are a large number of equations for solving for the unknown parts of such triangles, but the following list is sufficient for most purposes.

FORMULAS FOR OBLIQUE TRIANGLES

One version of the law of cosines for sides is

$$\cos a = \cos b \cos c + \sin b \sin c \cos A \qquad (B.8)$$

Solving this for cos A yields

$$\cos A = \frac{\cos a - \cos b \cos c}{\sin b \sin c} \qquad (B.9)$$

From another version of the law of cosines we obtain

$$\cos C = \frac{\cos c - \cos a \cos b}{\sin a \sin b} \tag{B.10}$$

One version of the law of cosines for angles is

$$\cos A = -\cos B \cos C + \sin B \sin C \cos a \tag{B.11}$$

and from another version,

$$\cos c = \frac{\cos C + \cos A \cos B}{\sin A \sin B} \tag{B.12}$$

The law of sines is

$$\frac{\sin A}{\sin a} = \frac{\sin B}{\sin b} = \frac{\sin C}{\sin c}$$

from which

$$\sin b = \frac{\sin B \sin a}{\sin A} \tag{B.13}$$

and

$$\sin B = \frac{\sin b \sin A}{\sin a} \tag{B.14}$$

Two of Napier's analogies are

$$\tan(c/2) = \frac{\sin[(A+B)/2] \, \tan[(a-b)/2]}{\sin[(A-B)/2]} \tag{B.15}$$

and

$$\cot(C/2) = \frac{\sin[(a+b)/2] \, \tan[(A-B)/2]}{\sin[(a-b)/2]} \tag{B.16}$$

Some derivations are given by Higgs and Tunell (1959), and more complete tabulations can be found in the CRC Standard Mathematical Tables, or in other similar references. The reason for the larger number of equations is to cover all possible labeling patterns. For example, there are three different laws of cosines with the form of Eq. B.8; the other two give expressions for the sides b and c. However, it is usually simpler to label the known and unknown elements so that a single equation will do.

In solving oblique spherical triangles, two situations arise. In the first, the distribution of the known elements is such that any unknown can be obtain unambiguously. In the second, there may be two solutions.

UNAMBIGUOUS CASES

1. All sides are known (a,b,c): solve Eq. B.9 for the unknown angle A.

2. All angles are known (A,B,C): solve Eq. B.11 for the unknown side a.

3. Two sides and included angle are known (b,A,c): solve Eq. B.8 directly for the unknown angle A, and all sides are known.

4. Two angles and included side are known (B,a,C): solve Eq. B.11 directly for A and all angles are known.

AMBIGUOUS CASES

1. Two angles and the side opposite one of them are known (a,b,A; a≠b): solve Eq. B.14 for the unknown angle B, Eq. B.15 for the unknown side c, and Eq. B.10 for the unknown angle C. If a<b there are two solutions; the second is found by replacing B with its supplementary angle arc cos(-cos B).

2. Two sides ad the angle opposite one of them are known (A,B,a; A≠B): solve Eq. B.13 for the unknown side b, Eq. B.16 for the unknown angle C, and Eq. B.12 for the unknown side c. If A<B, there are two solutions; the second is found by replacing b with its supplementary angle arc cos(-cos b).

Several examples will illustrate how to obtain solutions in both the unambiguous and ambiguous cases.

DIHEDRAL ANGLE

What is the dihedral angle between the two structural planes whose attitudes are: (1) N25°W, 30°E, (2) N35°E, 50°W?

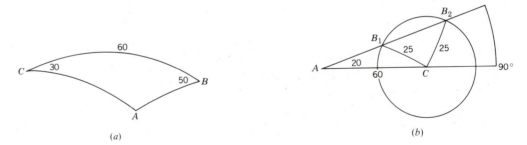

FIGURE B.6   Oblique spherical triangles: (a) dihedral angle as an example of the unambiguous case; (b) problem of the drill hole and known strike as an example of the ambiguous case.

SOLUTION (Fig. B.6a)

1. Plot the two planes as great circles. The spherical triangle containing the unknown dihedral angle is formed by arcs of these two circles and the primitive.

2. Label the corners of this triangle so that the unknown angle is A. Then $a = 60°$, $B = 50°$, and $C = 30°$, and this is the case of two angles and included side being, and we may solve for A directly.

3. Find angle A by substituting these three known values into Eq. B.11.

ANSWER

This yields $A = 111.4°$, and the acute dihedral angle is $68.6°$.

DRILL HOLE AND KNOWN STRIKE

A drill hole (30°/160°) intersects beds and the core-pole angle $\phi = 25°$. The strike of the beds is N50°E. What is the dip angle?

SOLUTION (Fig. B.6b)

1. Plot the point representing the inclined drill hole, a radius of the net in the dip direction (perpendicular to the known strike), and a small circle with ϕ as radius.

2. The two poles to bedding are located at the intersections of the radius of the net and the small circle. Labeling the parts as in Fig. B.6b, then $A = 20°$, $a = 25°$, and $b = 60°$, and this is a case of two sides and the angle opposite one of them.

3. Using Eq. B.15, find two values of B. With these results in Eq. B.16, find the two values for side c. These are the complements of the plunge angle for the two possible poles of bedding.

ANSWER

The two values of B are 44.5° and 135.5° and the corresponding dips are

$$\delta_1 = 76.8° \quad \text{and} \quad \delta_2 = 40.0°$$

References

Anderson, E.M., 1951, The dynamics of faulting: 2nd ed., Oliver and Boyd, Edinburgh, 206 p.

Argand, Emile, 1911, Les nappes de recouvrement des Alpes Pennines et leurs prolongement structuraux: Beitrage zur geologischen Karte der Schweiz, nue Folge 31, p. 1-26.

Aydin, Atilla, and A.M. Johnson, 1983, Analysis of faulting in sandstones: Journal of Structural Geology, v. 5, p. 19-31.

Aydin, Atilla, and Ze'ev Reches, 1982, Number and orientation of fault sets in the field and in experiments: Geology, v. 10, p. 107-112.

Badgley, P.C., 1959, Structural methods for the exploration geologist: Harper & Row, New York, 280 p.

Badgley, P.C., 1965, Structural and tectonic principles: Harper & Row, New York, 521 p.

Balk, Robert, 1937, Structural behavior of igneous rock: Geological Society of America Memoir 5, 177 p.

Bally, A.W., P.L. Gordy, and G.A. Stewart, 1966, Structure, seismic data, and orogenic evolution of southern Canadian Rocky Mountains: Bulletin of Canadian Petroleum Geology, v. 14, p. 337-381.

Barnes, C.W., and R.W. Houston, 1969, Basement response to the Laramide Orogeny at Coad Mountain, Wyoming: Contributions to Geology, v. 8, p. 37-41.

Barnes, J.W., 1981, Basic geologic mapping: Halsted Press, New York, 112 p.

Bayly, M.B, 1971, Similar folds, buckling and great circle patterns: Journal of Geology, v. 79, p. 110-118.

Bell, A.M., 1981, Vergence: an evaluation: Journal of Structural Geology, v. 3, p. 197-202.

Berthelsen, A., 1960, Structure contour maps applied in the analysis of double fold structures: Geologische Rundschau, v. 49, p. 459-466.

Berthelsen, A., E. Bondesen and S.B. Jensen, 1962, On the so-called wildmigmatites: Krystalinikum, v. 1, p. 31-49.

Billings, M.P., 1972, Structural geology: 3rd ed., Prentice-Hall, Englewood Cliffs, N.J., 606 p.

Borrodaile, G.J., 1976, "Structural facing" (Shackleton's Rule) and the Palaeozoic rocks of the Malaguide Complex near Velez Rubio, SE Spain: Proceedings of the Koninklijke Nederlandse Akademie van Wetenschappen, Amsterdam, series B, v. 79, p. 330-336.

Borrodaile, G.J., 1978, Transected folds: a study illustrated with examples from Canada and Scotland: Geological Society of America Bulletin, v. 89, p. 481-493.

Boyer, S.E., and David Elliott, 1982, Thrust systems: American Association of Petroleum Geologists Bulletin, v. 66, p. 1196-1230.

Bucher, W. H., 1920, The mechanical interpretation of joints, part I: Journal of Geology, v. 28, p. 707-730.

Busk, H.G., 1929, Earth flexures: Cambridge University Press, London, 106 p.

Butler, R.W.H., 1982a, Hanging wall strain: a function of duplex shape and footwall topography: Tectonophysics, v. 88, p. 235-246.

Butler, R.W.H., 1982b, The terminology in thrust zones: Journal of Structural Geology, v. 4, p. 239-245.

Buxtorf, A., 1916, Prognosen und Befunden beim Hauensteinbasis und Grenchenberg-tunnel und die Bedeutung der letzeren fur die Geologie des Juragebirges: Verhandl. Naturforsch. Ges. Basal, v. 27, p. 184-254.

Carey, S.W., 1962, Folding: Journal of the Alberta Society of Petroleum Geologists, v. 10, p. 95-144.

Casey, M., D. Dietrich and J.G. Ramsay, 1983, Methods for determining deformation history for chocolate tablet boudinage with fiberous crystals: Tectonophysics, v. 92, p. 211-239.

Charlesworth, H.A.K., and W.E. Kilby, 1981, Calculating thickness from outcrop and drill-hole data: Bulletin of Canadian Petroleum Geology, v. 29, p. 277-292.

Cloos, Ernst, 1947, Oolite deformation in South Mountain fold, Maryland: Geological Society of America Bulletin, v. 58, p. 843-917.

Cobbold, P.R., and H. Quinquis, 1980, Development of sheath folds in shear regimes: Journal of Structural Geology, v. 1/2, 9. 119-126.

Compton, R.R., 1962, Manual of Field Geology: John Wiley, New York, 278 p.

Compton, R.R., 1967, Analysis of Pliocene-Pleistocene deformation and stresses in northern Santa Lucia Range, California: Geological Society of America Bulletin, v. 77, p. 1361-1380.

Connolly, H.J.C., 1936, A contour method of revealing some ore structures: Economic Geology, v. 259-271.

Crowell, J.C., 1959, Problems of fault nomenclature: American Association of Petroleum Geologists Bulletin, v. 43, p. 2653-2674.

Cruden, D.M., 1971, Traces of a lineation on random planes: Geological Society of America Bulletin, v. 82, p. 2303-2305.

Cruden, D.M., and H.A.K. Charlesworth, 1976, Errors in strike and dip measurements: Geological Society of America Bulletin, v. 87, p. 977-980.

Cummins, W.A., and R.M. Shackleton, 1955, The Ben Lui recumbent syncline (S.W. Highlands): Geological Magazine, v. 92, p. 353-363.

Dahlstrom, C.D.A., 1969a, Balanced cross sections: Canadian Journal of Earth Science, v. 6, p. 743-757.

Dahlstrom, C.D.A., 1969b, The upper detachment in concentric folding: Bulletin of Canadian Petroleum Geology, v. 17, p. 326-346.

Dahlstrom, C.D.A., 1970, Structural geology in the eastern margin of the Canadian Rocky Mountains: Bulletin of Canadian Petroleum Geology, v. 18, p. 332-406.

Dallmus, K.F., 1958, Mechanics of basin evolution and its relation to the habitat of oil in the basin: in L.G. Weeks, ed., Habitat of oil, American Association of Petroleum Geologists, p. 883-931.

Deness, Bruce, 1970, A method of contouring polar diagram using curvilinear counting cells: Geological Magazine, v. 107, p. 55-66.

Deness, Bruce, 1972, A revised method of contouring stereograms using curvilinear cells: Geological Magazine, v. 109, p. 157-163.

Dennis, J.G., 1967, International Tectonic dictionary: American Association of Petroleum Geologists, Memoir 7, 196 p.

Dennison, J.M., and H.P. Woodward, 1963, Palinspastic maps of central Appalachians: Bulletin of the American Association of Petroleum Geology, v. 47, p. 666-680.

Den Tex, E., 1954, Stereographic distinction of linear and planar structures from apparent lineations in random exposure planes: Journal of the Geological Society of Australia, v. 1, p. 55-66.

De Paor, D.G., 1980, Some limitations of the R_f/ϕ technique of strain analysis: Tectonophysics, v. 64, p. T29-T31.

De Sitter, L.U., 1964, Structural geology: 2nd Ed., McGraw-Hill, New York, 551 p.

De Sitter, L.U., and H.J. Zwart, 1960, Tectonic development in supra- and infrastructures of a mountain chain: 21st International Geological Congress, Part 18, p. 248-256.

Dickinson, W.R., 1966, Structural relationships of San Andreas fault system, Cholame Valley and Castle Mountain Range: Geological Society of America Bulletin, v. 77, p. 707-726.

Donath, F.A., 1961, Experimental study of shear failure in anisotropic rocks: Geological Society of America Bulletin, v. 72, p. 985-990.

Donath, F.A., 1962, Analysis of Basin-Range structure, South-central Oregon: Geological Society of America Bulletin, v. 73, p. 1-16.

Donath, F.A., 1963, Fundamental problems in dynamic structural geology: in T.W. Donnelly, ed., Earth sciences--problems and progress in current research, University of Chicago Press, Chicago, p. 83-103.

Donath, F.A., 1964, Strength variation and deformation behavior in anisotropic rocks: in W.R. Judd, ed., State of stress in the earth's crust, American Elsevier, New York, p. 281-297.

Draper, Grenville, 1984, A simple device to aid plotting of pi diagrams in structural geology: Journal of Geological Education, v. 32, p. 184-185.

Dubey, A.K., 1980, Model experiments showing simultaneous development of folds and transcurrent faults: Tectonophysics, v. 65, p. 69-84.

Dunnet, D., 1969, A technique of finite strain analysis using elliptical particles: Tectonophysics, v. 7, p. 117-136.

Dunnet, D., and A.W.B. Siddans, 1971, Non-random sedimentary fabrics and their modification by strain: Tectonophysics, v. 12, p. 307-325.

Elliott, David, 1965, The quantitative mapping of directional minor structures: Journal of Geology, v. 73, p. 865-880.

Elliott, David, 1970, Determination of finite strain and initial shape from deformed elliptical objects: Geological Society of America Bulletin, v. 81, p. 2221-2236.

Elliott, David, 1976, Energy balance in thrusts and deformation mechanisms of thrust sheets: Proceedings of the Royal Society of London, v. A283, p. 289-312.

Elliott, David, 1983, The construction of balanced cross-sections: Journal of Structural Geology, v. 5, p. 101.

Elliott, David, and M.R.W. Johnson, 1980, Structural evolution in the northern part of the Moine thrust belt, NW Scotland: Transactions of the Royal Society of Edinburgh: Earth Sciences, v. 71, p. 69-96.

Faill, R.J., 1973, Kink band folding, Valley and Ridge Province, Pennsylvania: Geological Society of America Bulletin, v. 84, p. 1289-1314.

Ferguson, C.C., 1981, A strain reversal method for estimating extension from fragmented rigid inclusions: Tectonophysics, v. 79, p. T43-T52.

Ferguson, C.C., and G.E. Lloyd, 1984, Extension analysis of stretched belemnites: a comparison of methods: Tectonophysics, v. 101, p. 199-206.

Ferguson, H.G., and S.W. Muller, 1949, Structural geology of the Hawthorne and Tonopah quadrangles, Nevada: U.S. Geological Survey Professional Paper 216, 55 p.

Fleuty, M.J., 1964, The description of folds: Proceedings of the Geologists Association, v. 75, p. 461-492.

Fleuty, M.J., 1974, Slickensides and slickenlines: Geological Magazine, v. 112, p. 319-322.

Flinn, Derek, 1965, On the symmetry principle and the deformation ellipsoid: Geological Magazine, v. 102, p. 35-45.

Gage, Maxwell, 1979, A dilemma in structural mapping: New Zealand Journal of Geology and Geophysics, v. 22, p. 293.

Gill, J.E., 1971, Continued confusion in the classification of faults: Geological Society of America Bulletin, v. 82, p. 1389-1392.

Gilluly, James, A.C. Waters and A.O. Woodford, 1968, Principles of geology: 3rd ed., Freeman, San Francisco, 687 p.

Goetze, Christopher and Brian Evans, 1979, Stress and temperature in the bending lithosphere as constrained by experimental rock mechanics: Geophysical Journal of the Royal Astronomical Society, v. 59, p. 463-478.

Goguel, Jean, 1952, Traite de tectonique: Masson et Cie, Paris, 383 p.

Goguel, Jean, 1962, Tectonics: Freeman, San Francisco, 384 p.

Goodman, R.E., 1976, Methods of geological engineering in discontinuous rock: West Publishing Company, St. Paul, 472 p.

Gretener, P.E., 1981, Pore pressure: fundamentals, general ramifications and implications for structural geology (revised): American Association of Petroleum Geologists Continuing Education Course Notes Series, No. 4, 131 p.

Hafner, W., 1951, Stress distribution and faulting: Geological Society of America Bulletin, v. 62, p. 373-398.

Hamblin, li5W.K., 1965, Origin of "reverse drag" on the downthrown side of normal faults: Geological Society of America Bulletin, v. 76, p. 1145-1164.

Handin, John, 1969, On the Colomb-Mohr failure criterion: Journal of Geophysical Research, v. 74, p. 5343-5349.

Hansen, Edward, 1971, Strain facies: Springer Verlag, New York, 207 p.

Hansen, W.R., 1960, Improved Jacob staff for measuring inclined stratigraphic intervals: Bulletin of the American Association of Petroleum Geologists, v. 44, p. 252-255.

Harrison, J.M., 1963, Nature and significance of geologic maps: in C.C. Albritton, Jr., ed., The fabric of geology, Addison-Wesley, Reading, p. 225-232.

Hewett, D.F., 1920, Measurement of folded beds: Economic Geology, v. 15, p. 367-385.

Higgins, C.G., 1962, Reconstruction of flexure fold by concentric arc method: Bulletin of the American Association of Petroleum Geologists, v. 46, p. 1737-1739.

Higgs, D.V., and George Tunell, 1959, Angular relations of lines and planes with applications to geologic problems: W.C. Brown Co., Dubuque, 43 p.

Hill, M.L., 1959, Dual classification of faults: American Association of Petroleum Geologist Bulletin, v. 43, p. 217-221.

Hill, M.L., 1963, Role of classification in geologuy inC.C. Albritton, ed., The fabric of geology, Addison-Wesley, Reading, Mass., p. 164-174.

Hobbs, B.E., W.D. Means, and P.F. Williams, 1976, An outline of structural geology: John Wiley, New York, 571 p.

Hossack, J.R., 1979, The use of balanced cross-sections in the calculation of orogenic contraction: a review: Journal of the Geological Society of London, v. 136, p. 705-711.

Hossain, K.M., 1979, Determination of strain from stretched belemnites: Tectonophysics, v. 88, p. 279-288.

Hudleston, P.J., 1973, Fold morphology and some implications of theories of fold development: Tectonophysics, v. 16, p. 1-46.

Hsu, K.J., 1968, Principles of melanges and their bearing on the Franciscan-Knoxville paradox: Geological Society of America Bulletin, v. 79, p. 1063-1074.

Hubbert, M.K., 1951, Mechanical basis for certain familiar geologic structures: Geological Society of America Bulletin, v. 62, p. 365-372.

Ickes, E.L., 1923, Similar, parallel and neutral surface types of folding: Economic Geology, v. 18, p. 575-591.

Jaeger, J.C., 1969, Elasticity, fracture and flow: 3rd ed., Methuen, London, 268 p.

Jaeger, J.C., and N.G.W. Cook, 1979, Fundamentals of rock mechanics: 3rd ed., Methuen, London, 593 p.

Kalsbeek, F., 1963, A hexagonal net for the counting out and testing of fabric diagrams: Neues Jahrbuch fur Mineralogie, Monatshefte, v. 7, p. 173-1776.

Kilby, W.E., and H.A.K. Charlesworth, 1980, Computerized downplunge projection and the analysis of low-angle thrust-faults in the Rocky Mountain Foothills of Alberta, Canada: Tectonophysics, v. 66, p. 287-299.

Kitson, H.W., 1929, Graphic solution of strike and dip from two dip components: American Association of Petroleum Geologists Bulletin, v. 13, p. 1211-1213.

Kottlowski, F.E. 1965, Measuring stratigraphic sections: Holt, Rinehart & Winston, New York, 253 p.

Kranck, E.H., 1953, Interpretation of gneiss structures with special reference to Baffin Island: Proceedings of the Geological Association of Canada, v. 6, p. 59-68.

Kupfer, D.H., 1966, Accuracy in geologic maps: Geotimes, v. 10, no. 7, p. 11-14.

Laing, W.P., 1977, Structural interpretation of drill core from folded and cleaved rocks: Economic Geology, v. 72, p. 671-685.

Langstaff, C.S., and David Morrill, 1981, Geologic cross sections: International Human Resources Development Corporation, Boston, 108 p.

Laubscher, H.P., 1977a, An intriguing example of a folded thrust in the Jura: Eclogae geologicae Helvetiae, v. 70, p. 97-104.

Laubscher, H.P., 1977b, Fold development in the Jura: Tectonophysics, v. 37, p. 337-362.

LeRoy, L.W., and J.W. Low, 1954, Graphic problems in petroleum geology: Harper & Row, New York, 238 p.

Leverson, I.A., 1960, Paleogeologic maps: Freeman, San Francisco, 174 p.

Lisle, R.J., 1977, Clastic grain shape and orientation in relation to cleavage from the Aberystwyth grits, Wales: Tectonophysics, v. 39, p. 381-395.

Lisle, R.J., 1980, A simplified work scheme for using block diagrams with the orthographic net: Journal of Geological Education, v. 29, p. 81-83.

Lobeck, A.K., 1958, Block diagrams: Emerson-Trussell, Amherst, 212 p.

Low, J.W., 1957, Geologic field methods: Harper & Row, New York, 489 p.

Lowe, K.E., 1946, A graphic solution for certain problems of linear structures: American Mineralogist, v. 31, p. 425-434.

Macelwane, J.B., and F.W. Sohon, 1932, Introduction to theoretical seismology, part II—seismometry: St. Louis University, St. Louis, xxx p.

Mackin, J.H., 1950, The down-structure method of viewing geologic maps: Journal of Geology, v. 58, p. 55-72.

Mackin, J.H., 1962, Structure of the Glenarm series in Chester County, Pennsylvania: Geological Society of America Bulletin, v. 73, p. 403-410.

Marjoribanks, R.W., 1974, An instrument for measuring dip isogons and fold layer shape parameters: Journal of Geological Education, v. 22, p. 62-64.

Marsh, O.T., 1960, A rapid and accurate contour interpolator: Economic Geology, v. 55, p. 1555-1560.

Matthews, P.E., R.A.B. Bond, and J.J. Van den Berg, 1974, Analysis and structural implications of a kinematic model of similar folding: Tectonophysics, v. 12, p. 129-154.

McIntyre, D.B., 1963, Rotation of spherical projections: Technical Report No. 7, Department of Geology, Pomona College, Claremont, California, 7 p.

McIntyre, D.B., and L.E. Weiss, 1956, Construction of block diagrams to scale in orthographic projection: Proceedings of the Geologists Association, v. 67, p. 145-155.

McKinstry, H., 1961, Structure of the Glenarm Series in Chester County, Pennsylvania: Geological Society of America Bulletin, v. 72, p. 557-578.

Mertie, J.B., Jr., 1922, Graphic and mechanical computation of thickness of stratum and distance to a stratum: U.S. Geological Survey Professional Paper 129-C, 52 p.

Mertie, J.B., Jr., 1940, Stratigraphic measurement in parallel folds: Geological Society of America Bulletin, v. 51, p. 1107-1133.

Mertie, J.B., Jr., 1947, Delineation of parallel folds and measurement of stratigraphic dimensions: Geological Society of America Bulletin, v. 58, p. 779-802.

Mogi, Kiyoo, 1971, Fracture and flow of rocks under high triaxial compression: Journal of Geophysical Research, v. 76, p. 1255-1265.

Muller, Leopold, 1933, Untersuchungen uber statistische Kluftmessung: Geologie und Bauwesen, v. 5, no. 4, p. 185-255.

O'Driscoll, E.S., 1962, Experimental patterns in superposed similar folding: Journal of the Alberta Society of Petroleum Geologists, v. 10, p. 145-167.

O'Driscoll, E.S., 1964, Cross fold deformation by simple shear: Economic Geology, v. 59, p. 1061-1093.

Oertel, Gerhard, 1962, Extrapolation in geologic fabrics: Geological Society of America Bulletin, v. 73, p. 325-342.

Peach, C.J., and R.J. Lisle, 1979, A FORTRAN IV program for the analysis of tectonic strain using deformed elliptical markers: Computers & Geoscience, v. 5, p. 325-334.

Pfiffner, O.A., 1981, Fold-and-thrust tectonics in the Helvetic Nappes (E Switzerland): in K.R. McClay and N.J. Price, eds, Thrust and nappe tectonics, Special Publication No. 9, Geological Society of London, p. 319-327.

Phillips, F.C., 1963, An introduction to crystallography: 3rd ed., John Wiley, New York, 340 p.

Phillips, F.C., 1971, The use of stereographic projection in structural geology: 3rd ed., Edward Arnold, London, 90 p.

Powell, C. McA., 1974, Timing of slaty cleavage during folding of Precambrian rocks, northwest Tasmania: Geological Society of America Bulletin, v. 85, p. 1043-1060.

Price, N.J., 1966, Fault and joint development in brittle and semi-brittle rocks: Pergamon, Oxford, 176 p.

Price, N.J., 1970, Laws of rock behavior in the earth's crust: in W.H. Somerton, ed., Rock mechanics--theory and practice, Procedings of the 11th Symposium on Rock Mechanics, American Institute Mining, Metallurgical and Petroleum Engineering, New York, p. 3-23.

Pumpelly, Rafael, J.E. Wolfe and T.N. Dale, 1894, Geology of the Green Mountains in Massachusetts: U.S. Geological Survey Monograph 23, 206 p.

Ragan, D.M., 1969a, Introduction to concepts of two-dimensional strain and their application with the use of card-deck models: Journal of Geological Education, v. 17, p. 135-141.

Ragan, D.M., 1969b, Structures at the base of an ice fall: Journal of Geology, v. 77, p. 647-667.

Ragan, D.M., and M.F. Sheridan, 1972, Compaction of the Bishop Tuff, California: Geological Society of America Bulletin, v. 83, p. 95-106.

Raleigh, Barry, and Jack Evernden, 1981, Case for low deviatoric stress in the lithosphere: in N.L. Carter, M. Friedman, J.M. Logan and D.W. Stearns, eds., Mechanical behavior of crustal rocks, Geophysical Monograph 24, American Geophysical Union, p. 173-186.

Ramsay, J.G., 1961, The effects of folding upon the orientation of sedimentary structures: Journal f Geoloy, v. 69, p. 84-100.

Ramsay, J.G., 1962, Interference patterns produced by the superposition of folds of similar type: Journal of Geology, v. 70, p. 466-481.

Ramsay, J.B., 1963, Structure and metamorphism of the Moine and Lewisian Rocks of the Northwest Caledonides: in M.R.W. Johnson and F.H. Stewart, eds., The British Caledonides, Oliver and Boyd, Edinburgh, p. 143-175.

Ramsay, J.G., 1964, The uses and limitations of beta-diagrams and pi-diagrams in the geometrical analysis of folds: Quarterly Journal of the Geological Society, v. 120, p. 435-454.

Ramsay, J.G., 1967, Folding and fracturing of rocks: McGraw-Hill, New York, 568 p.

Ramsay, J.G., 1969, The measurement of strain and displacement in orogenic belts: in P.E. Kent, G.E. Satterthwaite, and A.M. Spencer, eds., Time and place in Orogeny, Special Publication No. 3, Geological Society of London, p. 43-79.

Ramsay, J.G., 1980, Shear zone geometry: a review: Journal of Structural Geology, v. 2, p. 83-99.

Ramsay, J.G., and M.I. Huber, 1983, The techniques of modern structural geology--volume 1: strain analysis: Academic Press, London, 307 p.

Reches, Ze'ev, 1978, Analysis of faulting in three-dimensional strain field: Tectonophysics, v. 47, p. 109-129.

Reches, Ze'ev, D.F. Hoexter, and Francis Hirsch, 1980, The structure of a monocline in the Syrian Arc System, Middle East--surface and subsurface analysis: Journal of Petroleum Geology, v. 3-4, p. 413-425.

Rickard, M.J., 1971, A classification diagram for fold orientation: Geological Magazine, v. 108, p. 23-26.

Rickard, M.J., 1972, Fault classification--discussion: Geological Society of America Bulletin, v. 83, p. 2545-2546.

Robinson, G.D., 1959, Measuring dipping beds: Geotimes, v. 4, p. 8-9, 24-27.

Schryver, K., 1966, On the measurement of the orientation of axial plane of minor folds: Journal of Geology, v. 74, p. 83-84.

Schweinfurth, S.P., 1969, Contour finder--inexpensive device for rapid, objective contouring: U.S. Geoological Survey Professional Paper, 650-B, p. B147-B148.

Schwerdtner, W.M., P.J. Bennett, and T.W. Janes, 1977, Application of L-S fabric scheme to structural mapping and paleostrain analysis: Canadian Journal of Earth Science, v. 14, p. 1021-1032.

Screven, R.W., 1963, A simple rule of V's of outcrop patterns: Journal of Geological Education, v. 11, p. 98-100.

Seymour, D.B., and C.A. Boulter, 1979, Tests of computerised strain analysis by the analysis of simulated deformation of natural unstrained sedimentary fabrics: Tectonophysics, v. 58, p. 221-235.

Shackleton, R.M., 1958, Downward-facing structures of the Highland Border: Quarterly Journal of the Geological Society of London, v. 113, p. 361-392.

Sibson, R.H., 1974, Frictional constraints on thrust, wrench and normal faults: Nature (Physical Sciences), v. 249, p. 542-543.

Simpson, Carol, and S.M. Schmid, 1983, An evaluation of criteria to deduce the sense of movement in sheared rocks: Geological Society of America Bulletin, v. 94, p. 1281-1288.

Stauffer, M.R., 1966, An empirical-statistical study of three dimensional fabric diagrams as used in structural analysis: Canadian Journal of Earth Science, v. 3, p. 473-498.

Stauffer, M.R., 1968, The tracing of hinge-line ore bodies in areas of repeated folding: Canadian Journal of Earth Science, v. 5, p. 69-79.

Stauffer, M.R., 1973, New method for mapping fold axial surfaces: Geological Society of America Bulletin, v. 84, p. 2307-2318.

Suter, H.H., 1947, Exaggeration of vertical scale of geologic sections: Bulletin of the American Association of Petroleum Geologists, v. 31, p. 318-339.

Suppe, John, 1983, Geometry and kinematics of fault-bend folding: American Journal of Science, v. 283, p. 684-721.

Terzaghi, R.D., 1965, Sources of error in joint surveys: Geotechnique, v. 15, p. 287-304.

Thiessen, R.L., and W.D. Means, 1980, Classification of fold interference patterns: a reexamination: Journal of Structural Geology, v. 2, p. 311-316.

Thompson, G.A., 1960, Problem of late Cenozoic structure of the Basin Ranges: 21st International Geological Congress, Copenhagen, v. 18, p. 62-68.

Thomson, William, and P.G. Tait, 1867, Treatise on Natural Philosophy, vol. 1, Oxford, England.

Thomson, William, and P.G. Tait, 1962, Principles of mechanics and dynamics, Part I: Dover Publications, New York, 508 p.

Threet, R.L., 1973a, Classification of translational fault slip: Geological Society of America Bulletin, v. 84, p. 1825-1828.

Threet, R.L., 1973b, The down-dip method of viewing geologic maps to obtain sense of fault separation: Geological Society of America Bulletin, v. 84, p. 4001-4004.

Treagus, S.H., 1981, A simple-shear construction from Thomsom & Tait (1867): Journal of Structural Geology, v. 3, p. 291-293.

Treagus, S.H., 1982, A new isogon-cleavage classification and its application to natural and model fold studies: Geological Journal, v. 17, p. 49-64.

Turner, F.J. and L.E. Weiss, 1963, Structural analysis of metamorphic tectonites: McGraw-Hill, New York, 545 p.

Verhoogen, John, F.J. Turner, L.E. Weiss, Clyde Wahrhaftig and W.S. Fyfe, 1970, The earth: Holt, Rinehart & Winston, New York, 748 p.

Vistelius, A.B., 1966, Structural diagrams: Pergamon, Oxford, 178 p.

Warner, F.M., and M. McNeary, 1959, Applied descriptive geometry: 5th Ed., McGraw-Hill, New York, 243 p.

Weiss, L.E., 1969, Flexural-slip folding of foliated model materials: in A.J. Baer and D.K. Norris, eds., Research in Tectonics, Geological Survey of Canada, Paper 68-52, p. 294-357.

Wegmann, C.E., 1929, Beispiele tektonischer Analysen des Grundgebirges in Finnland: Bulletin de la Commission geologique de Finlande, no. 87, p. 98-127.

Wellman, H.G., 1962, A graphic method for analysing fossil distortion caused by tectonic deformation: Geological Magazine, v. 99, p. 348-352.

Wentworth, C.K., 1930, The plotting and measurement of exaggerated cross-sections: Economic Geology, v. 25, p. 827-831.

Weijermars, Ruud, 1982, Definition of vergence: Journal of Structural Geology, v. 4, p. 505.

Wernicke, Brian and B.C. Burchfiel, 1982, Modes of extensional tectonics: Journal of Structural Geology, v. 4, p. 105-115.

Williams, G.D., 1984, The calculation of horizontal thrust transport using excess area in cross-sections: Tectonophysics, v. 104, p. 177-182.

Williams, G.D., and T.J. Chapman, 1979, The geometrical classification of noncylindrical folds: Journal of Structural Geology, v. 1, p. 181-185.

Williams, G.D., and T.J. Chapman, 1983, Strains developed in the hanging walls of thrusts due to their slip/propagation rate: a dislocation model: Journal of Structural Geology, v. 5, p. 563-571.

Wilson, Gilbert, 1967, The geometry of cylindrical and conical folds: Proceedings of the Geologists Association, v. 78, p. 179-210.

Wilson, Gilbert, 1961, Tectonic significance of small scale structures and their important to the geologist in the field: Annales de la Societe Geologique de Belgique, v. 84, p. 423-548.

Wilson, Gilbert, 1982, Introduction to small-scale geological structures: George Allen & Unwin, London, 128 p. [revision of Wilson, 1961]

Woodcock, N.H., 1976, The accuracy of structural field measurements: Journal of Geology, v. 84, p. 350-355.

Wright, F.E., 1911, The methods of petrographic microscope research: Publication No. 158, Carnegie Institution of Washington, 204 p.

Zimmer, P.W., 1963, Orientation of small diameter drill core: Economic Geology, v. 58, p. 1313-1325.

Author Index

Subject Index

Structure sections, 332
Superimposed deformations, 156
Superimposed folds, 248, 253
Symbols, see map symbols
Syncline, 209
Synform, 209
Symmetrical fold, 204
Synoptic diagram, 282
Tectonite, 237
Tensile strength, 134
Terzaghi's relationship, 126
Thickness, 19
    axial plane, 208
    orthogonal, 208
    stratigraphic, 19
Three-point problem, 39
Thrust, 135
Tight fold

Tilting, 75
Torque, 111
Traction, 112
Trajectories, 165
Translation, 149
Trend, 1
Trilobite, 195
Trough, 202
Unconformity, 331
Vertical exaggeration, 336
Vertical fold, 210
Wave length, 204
Wild fold, 252
Worm's eye map, 331
Wrench fault, 135
Wulff net, 63
Zenithal point, 60

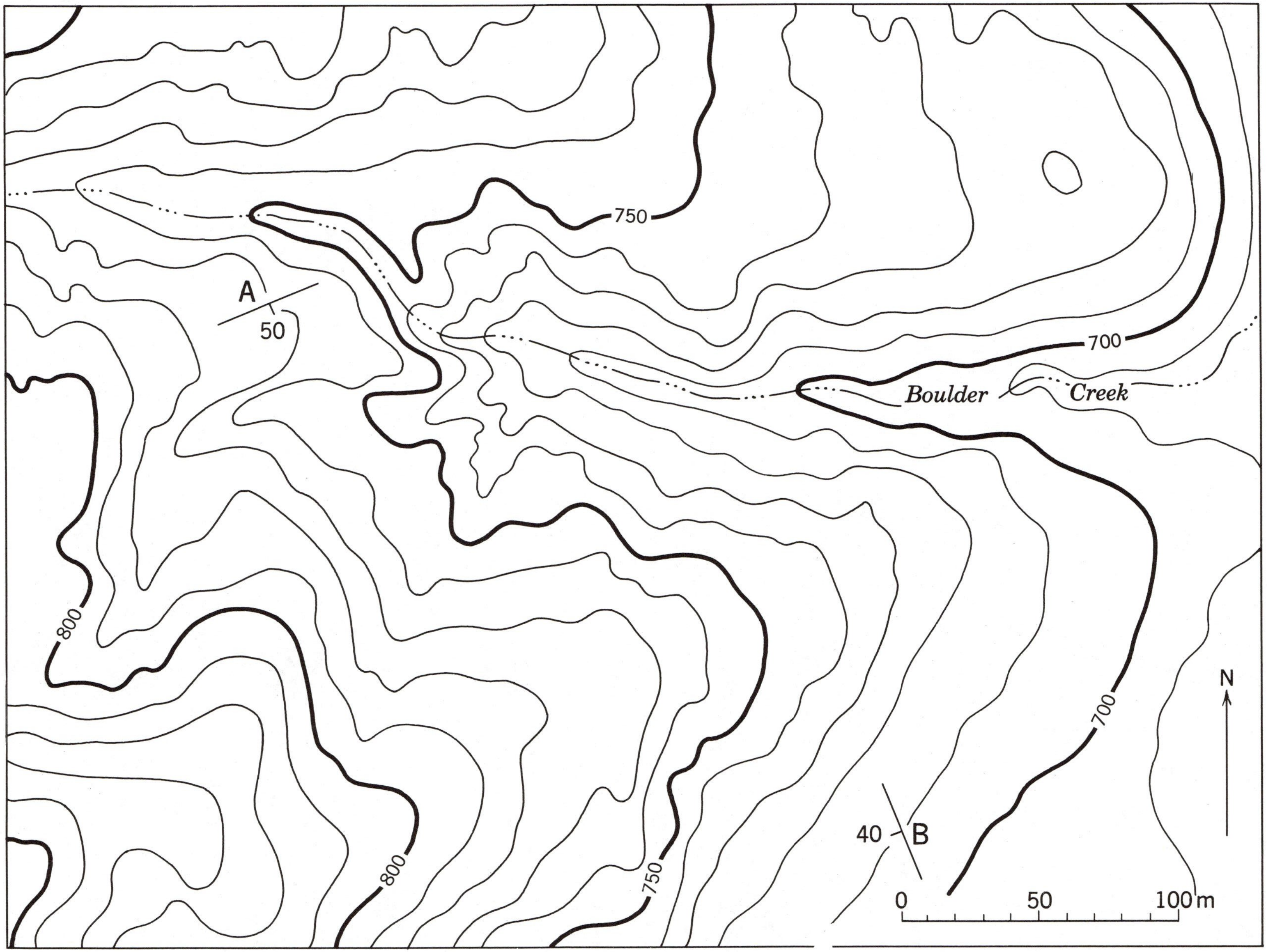

FIGURE X 4.1

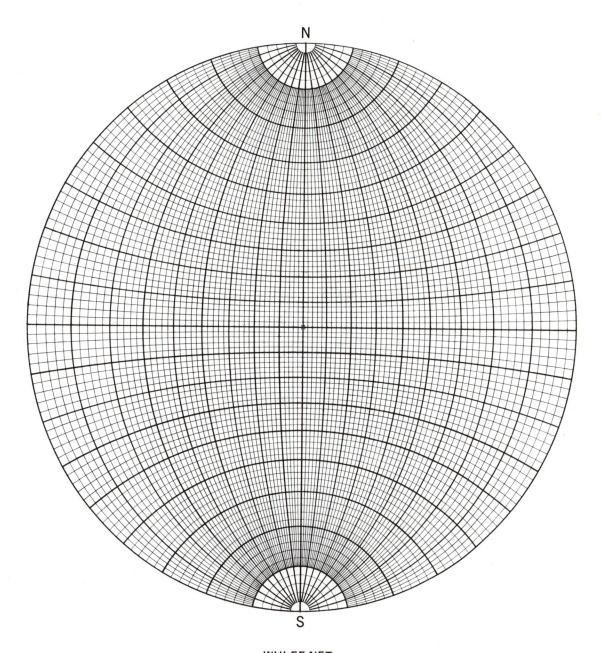

N

S

WULFF NET

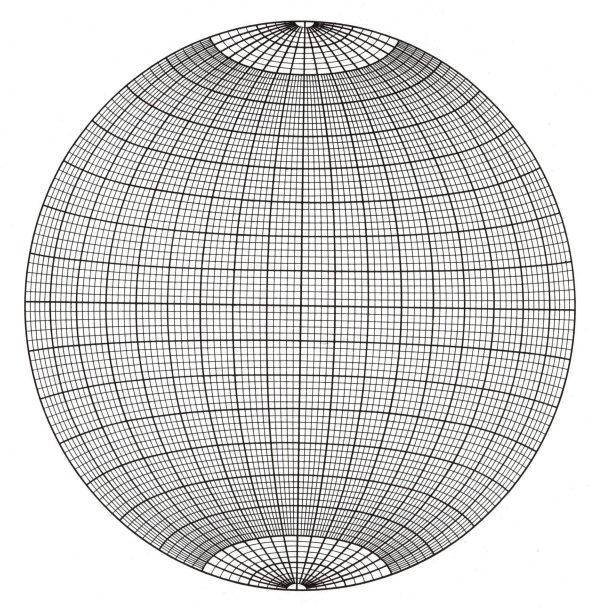

SCHMIDT NET

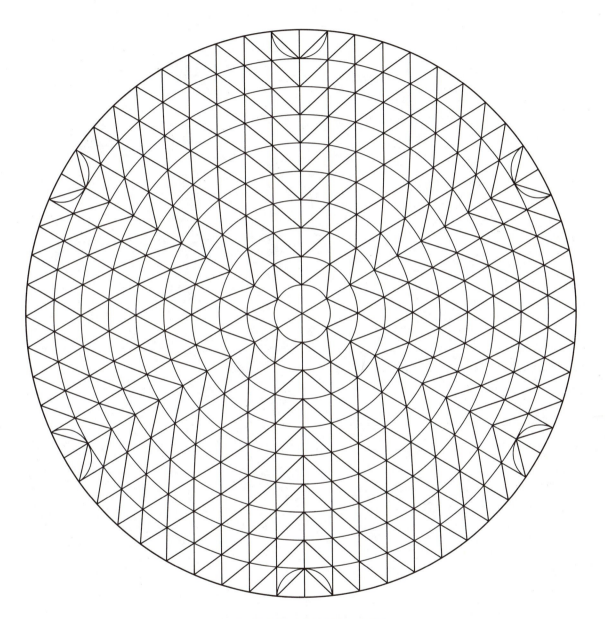

KALSBEEK COUNTING NET

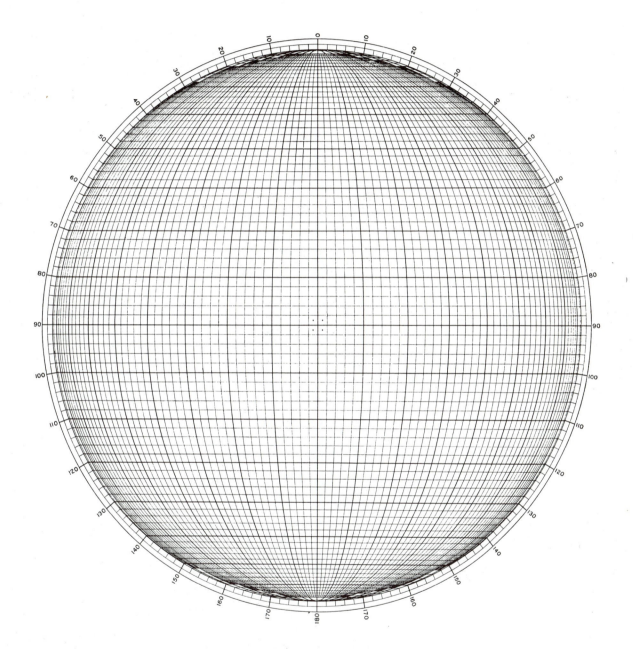

ORTHOGRAPHIC NET